CHEMISTRY RESEARCH AND APPLICATIONS

SPECTROSCOPIC TECHNIQUES FOR SECURITY, FORENSIC AND ENVIRONMENTAL APPLICATIONS

CHEMISTRY RESEARCH AND APPLICATIONS

Additional books in this series can be found on Nova's website
under the Series tab.

Additional e-books in this series can be found on Nova's website
under the e-book tab.

CHEMISTRY RESEARCH AND APPLICATIONS

SPECTROSCOPIC TECHNIQUES FOR SECURITY, FORENSIC AND ENVIRONMENTAL APPLICATIONS

YASHASHCHANDRA DWIVEDI
SHYAM BAHADUR RAI
JAGDISH PRASAD SINGH
EDITORS

nova publishers
New York

For permission to use material from this book please contact us:
Telephone 631-231-7269; Fax 631-231-8175
Web Site: http://www.novapublishers.com

Library of Congress Cataloging-in-Publication Data

ISBN: 978-1-63117-404-9

Published by Nova Science Publishers, Inc. † New York

CONTENTS

PREFACE

Spectroscopy covers a wide range of applications those related directly to the civil Society, has become more prominent over the past two decades. In recent years, our society is under growing threat of issues related to civil crime and defence. To date accurate warning, tracking, and predictive network of chemical and biological agent sensors have become important in current world and these can be used for the protection of military installations, population centres and transportation hubs. The available field-kits for military personnel and laboratory-type instruments do not offer the necessary combination of sensitivity, resistance to false alarm, stand-alone operation, and economical cost necessary for implementation of a widespread network of sensors, which evoke much enthusiasm to invent newer technologies. In addition to challenges in security front and precise forensic investigations, much wider spreading and global problem is the environmental diagnosis. Under these circumstances, a variety of new spectroscopic techniques significantly contributing in such diverse areas of sensors, communications, forensic science, defence with unequalled precision have come up in recent years using lasers. Spectroscopy provides analytical tools for analysis of chemicals and materials in any state and permit use of these techniques to detect the presence of any chemical harmful to the species even in trace amount. Laser fluorescence, Raman and infrared spectroscopic techniques are regularly used in the studies on semiconductors, microelectronics, catalysis and nano-materials. Photoacoustic spectroscopy makes it possible to detect trace amount of explosive powders and residues on surfaces from a distance. Newer sophisticated Laser Induced Breakdown Spectroscopy technique help for real time, precise identification and detection of trace amounts of all elements and virtually all hazards, both molecular and biological at a standoff distance.

To discuss various aspects of aforesaid issues we had organised an international conference during January 18-20, 2013, in Physics Department at Banaras Hindu University, Varanasi, India, in collaboration with Mississippi State University, USA which was much appreciated by leading scientific community. This book has emerged in part from fruitful discussion made during the conference with objectives to explore the applications of spectroscopic techniques to matters related to national security with in-depth analysis of various spectroscopic techniques and their relative merits. Though, it is practically impossible to discuss all the spectroscopic techniques related to these issues in one book, we have discussed some of the major spectroscopic techniques like Raman scattering, Microwave, LIBS, Photoacoustic, Lanthanide spectroscopy etc. These techniques are of great importance in detection of explosives, chemical and biological agents; and in meeting the defence threats, environmental and forensic applications.

The idea of international workshop was conceived and planned by great visionary late Professor D. K. Rai. Unfortunately, he died on July 12, 2012, during the preparation for the workshop. His contribution in spectroscopy teaching and research at B.H.U., Varanasi for five decades had been a great source of inspiration for many of us. We have also lost our great friend and guide Prof. Amar Nath Singh (on June 29, 2013) while this book was finalized. We would like to dedicate this book to these two spectroscopists.

We gratefully acknowledge the help rendered by our teacher honourable Professor Surya Narayan Thakur, Banaras Hindu University for his valuable support and guidance throughout the workshop and book.

In: Spectroscopic Techniques for Security, Forensic ... ISBN: 978-1-63117-404-9
Editors: Y. Dwivedi, S. B. Rai and J. P. Singh

Chapter 1

SPECTROSCOPIC TECHNIQUES FOR HIGH SENSITIVITY HEAVY METAL DETECTION IN THE LIQUID PHASE

Robert Fedosejevs*, Yogesh Godwal, Zhijiang Chen, Shyama Banerjee and Ying Tsui
Department of Electrical and Computer Engineering, University of Alberta, Edmonton, AB, Canada

ABSTRACT

Heavy metal toxins and pollutants are a serious environmental and security concern affecting safe drinking water and public health in general. LIBS provides a powerful tool for detection of such threats but generally requires enhanced techniques in order to achieve the required ppb sensitivity levels for detection of such threats. A number of enhanced LIBS techniques have been developed recently and will be discussed in this chapter. These techniques include a dual laser pulse technique, Laser-Ablation Laser-Induced Fluorescence (LA-LIF), and sample pre-concentration techniques employing either absorbent sampling sticks or electrolytic deposition of species on metal electrodes. These techniques can be deployed using low energy laser pulses of the order of 4 μJ to 4 mJ which would be suitable for the development of portable systems using high efficiency diode pumped, high repetition rate lasers. In particular the detection of lead in water is discussed in detail and current resultant limits of detection (LOD) presented. The limits of detection of all three techniques were determined to be on the order of 7 to 30 ppb (3 sigma) allowing for the detection of lead contamination at the 15 ppb North American drinking water standard.

* Email: rfed@ualberta.ca.

1. Introduction

Heavy metals represent an important set of elements required for many processes and functions in modern society. However, at the same time they represent a significant risk to human health and environment [1-4]. In some cases they could also pose a security threat to our water supplies and systems. Thus it is important that simple, robust and preferably real time techniques be developed to both properly manage but also properly monitor these elements in our environment. The required levels of detection must be at a sensitivity level sufficient to measure these elements at the maximum safe threshold levels which are allowable in the environment, which typically are in the parts per billion (ppb) level [5-6].

Emission spectroscopy is a very powerful and sensitive technique for detecting any element in the periodic table. However, in general the species must be prepared in some manner in order to carry out the spectroscopic measurement. Laser ablation is one general method of vaporizing and atomizing samples from any starting state whether it be liquid, solid or gaseous. It requires no sample preparation and in principle allows any element to be easily measured by recording the emission spectrum from the excited atoms or ions in the vapour plume.

In general, the sensitivity of standard ablation spectroscopy techniques such as Laser Induced Breakdown Spectroscopy (LIBS) [7-11] is of the order of several parts per million (ppm) which is well above the parts per billion (ppb) sensitivity required for detecting hazardous heavy metals in the environment. Thus extra techniques must be employed in order to achieve the required sensitivities. These could include excitation at ultraviolet wavelengths, the use of resonant excitation of the atomic species or sample concentration techniques as outlined in this chapter. At the same time, ablation of liquid samples generally is more difficult than ablation of solid surfaces due to the high transmission of most liquid samples requiring high intensities to achieve breakdown. Additional complications can arise due to mechanical disruption of the fluid surface which can lead to splashing, difficulty in maintaining the focus at the surface of the sample and contamination of optical surfaces. Various means to mitigate such effects include the use of liquid jets, liquid droplets or microfluidic platforms.

With the aim of developing techniques which can be employed in field measurements it is also desirable to employ low energy microjoule to a few millijoule laser pulses which can be readily generated in compact, efficient, diode pumped laser systems. In order to improve the overall signal to noise one can take advantage of the potential repetition rates of such systems to achieve low limits of detection by averaging over thousands of shots. Such systems are currently available either in the form of microchip or fibre-optic lasers.

While detection of environmental contaminants at the part per billion level is still challenging and generally requires special techniques to enhance the spectroscopic signals, recent results have indicated that such detection is readily achievable. We have investigated three such techniques in detail: 1) using laser ablation followed by laser induced fluorescence (LA-LIF) [12]; 2) using absorbent wooden sampling sticks to concentrate the species on the surface after evaporation [13, 14] and; 3) using electrolytic sample concentration techniques [62, 64].

This chapter will review and outline the above approaches to measuring heavy metals in liquid samples and the limits of detection achieved to date with such techniques. It will conclude with a comment on future directions and the possible practical implementation of these techniques to serve the needs of protecting our health and the environment.

2. PREVIOUS WORK

Optical spectroscopic techniques probing atomic transitions in atoms can readily be used to detect and identify individual atomic elements due to the unique energy transitions and spectral signatures of each atomic element. LIBS has been applied to the analysis of a number of contaminants in water as reported in several of studies in the past and outlined below. One particular species of interest is monitoring of lead in drinking water because of its prevalence in a number of city water systems with old metal distribution pipes. On-line real-time monitoring of hazardous contaminants in water is possible with LIBS but, in general, analysis of liquid samples is not as well studied as solid target samples because of increased difficulty in breaking down liquid samples. According to the United State Environmental Protection Agency, the action level (AL) for lead in drinking water is 15 ppb by weight [15]. Excessive ingestion of lead can result in physical or mental development delays for children and high blood pressure and kidney problems for adults [16]. Common techniques for measuring the lead content is by inductively coupled plasma mass spectrometry (ICPMS) or graphite furnace atomic absorption spectrometry (GPAAS). These techniques can achieve sensitivities ranging from 0.01 to 0.1 ppb [17,18]. However, the testing can only be conducted in a laboratory, which requires that samples be taken and sent to laboratories for testing, typically with inherent delays of several days before results are know. Fiddler *et al.* [19] has also described the different laser spectroscopy techniques to detect elements in solid, liquid, gases and aerosols. In contrast, LIBS is potentially a real-time portable technique which can be applied directly in the field.

There are a number of advantages of using water samples over solids such as, aqueous solutions do not suffer from spatial inhomogeneities caused by micron scale size precipitates and there is no permanent damage of the sample by the LIBS process. Thus, water samples provide good test cases for the investigation of the fundamental properties of LIBS. One drawback of using liquid as a target sample is that often the samples are somewhat transparent with little absorption at the entrance interface and thus higher fluences are normally required to achieve breakdown in liquids compared to solid surfaces. Detection of contaminants in water has been reported using LIBS in a variety of sampling geometries including water jets [20-25] and breakdown within a bulk water sample [26-27]. One of the earliest studies dealing with analysis of liquid solutions using LIBS was reported by Cremers *et al.* [28]. Using Nd:YAG laser pulses and detection of spectrally resolved light by a photomultiplier tube (PMT) they achieved a limit of detection for many trace elements at levels better than 1 μg/mL. The best LOD of 0.006 μg/mL was achieved for Li. In this case the laser was focused inside the liquid samples. Arca *et al.* focused 1064 nm laser pulses on the surface of water and measured LOD's of few ppm for Ca, Mg and ~100 ppb for Cr [26]. P. Fichet *et al.* used a pulsed laser and focused with a tilted angle on the surface of the liquid to determine the concentration for twelve elements (Pb, Si, Ca, Na, Zn, Sn, Al, Cu, Ni, Fe, Mg and Cr) [29]. Two different liquid matrices were used such as water and oil. The detection limits achieved were in the range of 0.3–120 μg/mL with Ca having the best LOD of 0.3 μg/mL.

Typical limits of detection (LODs) for trace elements in water samples are in the ppm range well above the required sensitivities for monitoring drinking water and pollution in lakes and rivers. Also if one wants to employ lower energy pulses the LOD climbs higher. For example, a pulse energy of 260 μJ [30] gives an LOD of 75 ppm by weight whereas a pulse energy of 22 mJ gives 12.5 ppm [27]. In the latter report, using an excimer (λ=308 nm) pumped dye-laser system producing 28ns, 22 mJ pulses, Knopp *et al.* has achieved the best limit of detection (LOD) for lead in water using conventional LIBS to date of 12.5 ppm [27]. They also obtained LOD's for Li, Na, Ca, Ba and Cd in aqueous solutions of 0.0075 to 500 ppm. Another study using UV laser pulses by Lo *et al.* achieved an LOD as low as 300 ppb [25]. In this study, an ArF laser at 193 nm was used, and the LIBS signal was accumulated over 500 shots with 14 mJ laser energy per pulse. In their earlier study [31], it was found that the 193 nm laser pulses can enhance the atomic emission and reduce the continuum background in aqueous sample, compared with 532 nm pulses due to the very short absorption length of the order of 20 microns in the solution of this far UV wavelength [25]. Charfi *et al.* have reported LOD's of 1 and 2 ppm by weight for Mg and Na solutions respectively using 200 mJ, 7 ns laser pulses [32]. Loudyi *et al.* have obtained LOD values of 19 ppm for Pb and 37 ppm for Fe in their LIBS measurement by averaging over 100 laser shots [33]. Yuan Feng *et al.* demonstrated detection of Pb in a $Pb(NO_3)_2$ aqueous solutions with an LOD of 60 ppm when observing the Pb I emission line at 405.78 nm by using 532 nm, 10 ns laser pulses of 57 mJ energy [34].

Further enhancement of signals can be obtained using dual pulse LIBS. Rifai *et al.* have used dual pulse LIBS to detect Pb, Fe and Au where they showed that LOD's of 6 ppm could be obtained for Pb in water using 32 mJ energy for the first pulse and 200 mJ for the second pulse whereas, conventional single shot LIBS gave only 74 ppm using laser pulses of 32 mJ energy [35]. In their experiment the plasma was generated by using a UV (266 nm) frequency-quadrupled 7 ns Nd:YAG laser pulse and then reheated by a 7 ns 1064 Nd:YAG laser in a quasi-collinear geometrical configuration. In general, it is found that direct LIBS or dual pulse LIBS has a sensitivity level in the ppm range which is sufficient to monitor heavy metal contaminants in industrial processing situations or in the monitoring of direct effluent streams from such processing before dilution in the external environment. However, higher sensitivities are required for assessment of drinking water and environmental pollution levels.

One technique to obtain higher sensitivity is the direct excitation of a particular transition in the element of interest. In order to do this requires that the element be in an atomic state where excitation to an upper state from the ground state can be achieved. Fluorescent emission directly back to the ground state or to another lower excited state then can be detected as a signature of the transition. This process has been called Laser induced Fluorescence (LIF) [36] or Laser Excited Atomic Fluorescence Spectroscopy (LEAFS) [37-39]. One technique for preparation of the element in the atomic state is by means of laser ablation of a small sample spot and waiting until the ablation plume has cooled upon expansion but has not started to recombine into molecular species or clusters. At this optimum point in time the atoms can be excited by a narrow linewidth laser pulse tuned precisely to a resonant transition and the fluorescent emission observed promptly thereafter. This process which is closely related to LIBS has been called Laser Ablation Laser Excited Atomic Fluorescence Spectroscopy (LA-LEAFS) [12,40] or Laser Ablation Laser Induced

Fluorescence (LA-LIF) [13,41]. A limited number of previous studies have been reported to date using these techniques.

LA-LIF has proven to be an effective technique for measurement of trace elements in both solid and liquid samples. In one of the earliest reports, Kwong and Measures used the LA–LIF technique in the year 1979 to determine the concentration of chromium in metal alloys and organic substances at the ppm level [42]. After this initial work there have been a number of reports in literature that use the technique to improve the sensitivity over that of single pulse LIBS. Sdorra *et al.* carried out LA–LIF on steel samples [43] and Bondybey and English used the LA–LIF technique to study the spectroscopic and structural properties of metallic diatomic molecules [44]. Gornushkin et al. used this technique to determine concentrations of cobalt in soil, steel, and graphite samples in open air [45]. Detection of uranium in glass sample at a concentration of 462 ppm was demonstrated by Shen *et al.* [46]. In another publication they measured concentrations of lead in the range 0.15–750 ppm by weight in copper, brass, steel, and zinc by laser excited atomic fluorescence combined with UV laser ablation in a low-pressure (130 mbar) argon atmosphere [40]. The mass of the material ablated was determined by repetitive weighing of the target prior to and after approximately 10,000 ablating shots. A relative limit of detection of 72 ng/g was obtained for the determination of lead in copper, brass, steel, and zinc. The technique was also used to determine the concentration of lead in ultrafine aerosols by Neuhauser *et al.* [47].

For water samples, Koch *et al.* used the technique to measure the concentration of indium in bulk water samples [48] at a level of 10 ppm. Ho and Cheung demonstrated a sensitivity of 0.25 ppb for lead in colloid form, which propagated slower in the plume thus resulting in a more concentrated region in the expansion plume [20]. Atomic fluorescence from this region was emitted when excited by a 193 nm ArF laser. Loudyi *et al.* obtained LOD's of 39 ppb and 65 ppb for Pb and Fe respectively [33] using over 100 laser shots and laser energies of the order of 40 mJ for the ablation pulse and few ten's of μJ for the excitation pulse. In a different experiment determination of Pb concentration in water with an LOD of 4.3 ppm was achieved using 100 shots [30] and 35 ppb when 10,000 shots were employed [12]. LeBihan *et al.* have employed laser fluorescence with an optical parametric oscillator (OPO) to develop an analytical method for the direct determination of lead content in seawater by electrothermal atomization–laser-excited atomic fluorescence (ETA–LEAF) [49]. The best LOD for lead achieved by this technique was 0.3 ng/L by averaging 20 shots, which is the best reported laser induced fluorescence sensitivity for detection of Pb in water that we are aware of. It can be seen that the LA-LIF technique can achieve higher sensitivities into the ppb range as compared to LIBS itself and thus would be one approach to achieving sensitivities sufficient to assess the safety of drinking water directly.

In a related technique, Atsushi Kitano *et al.* demonstrated highly sensitive liquid electrode plasma atomic emission spectrometry (LEP-AES) by using a combination of a quartz glass microfluidic chip and sample flow system to investigate the LOD's for Cd and Pb [50]. In this process a sample solution is put into a microfluidic channel with a very narrow central space (~ 100 μm width) between two reservoirs. Two electrodes are in the two reservoirs at the ends of this channel. A high voltage, few millisecond pulse of ~1500V was applied across the two electrodes heating the solution and creating a vapor bubble in the narrow channel region which ionizes and produces a discharge plasma. By analyzing the spectrum of that plasma they achieved LOD's of 0.52 μg/L for Cd and 19.0 μg/L for Pb under optimized conditions.

Another approach to achieving higher sensitivities with LIBS is to concentrate the analyte to be measured prior to carrying out the LIBS probing. In the past the results from a few such sample concentration techniques have been reported. These include the pre-concentration of analytes on a variety of porous substrates, such as paper [51-55], membranes [56-58], wood sampling sticks [14, 59], carbon disks [60] and bamboo charcoal [61]. In addition, electrolytic methods can be used to concentrate the metal cations from the aqueous sample to the surface of a metal [62-66]. Although the pre-concentration approach requires extra extraction steps before the LIBS analysis, it converts the liquid sample to a solid leading to an increase of the effective analyte density and avoids the water splashing and quenching of the plasma plume which can occur in the liquid sample.

Table 1. List of limits of detection (LODs) reported using different absorbent substrates (*Reported in mg/L, which is converted to ppm by weight in the aqueous sample)

Substrate	Analyzed material and LOD (ppm)	Ref	Comments
Wood	Cr (0.034), Cu(0.029), Cd(0.59), Pb(0.076)	59*	
Wood	Pb (0.113)	14	Different laser wavelengths were used
Paper	Ag(1), Al(4), Cd(4), Cr(12), Cu(1), Fe(3), Mn(3), Mo(5), Ni(20), Ti(5), V(2), Zn(2), Pb(3)	51	Oil samples. Double laser pulses LIBS
Ion Exchange Membranes	Ba(0.13), Cd(0.21), Cr(0.13), Cu(0.0042), Pb(1.1), Hg(2), Ni(3.1), Ag(0.43), Zn(0.85)	56*	Filtration through a membrane
Membrane-based filter paper	Sm(1.3), Eu(1.9), Gd(2.3)	52	
Filter paper	Pb(0.075), Cr(0.018)	53	Totally evaporated a glass of water. Argon flow assisted
Paper	Ca(1.9), Mg(3.2)	54*	
Filter paper	Na(0.9), Fe(0.5), Cu(0.4)	55	Oven assisted pre-concentration. Ethanol samples
PVC membrane	Cu(0.015)	58*	pH 6.0 sodium acetate buffer solution
Amorphous graphite	Mg(0.01), Al(0.01), Si(0.01), Ca(0.01), Ti(0.01), Cr(0.1), Fe(0.01), Co(>0.1), Zn(1), As(5), Cd(0.1), Hg(10), Pb(2)	60	75 Torr argon buffer gas
Bamboo charcoal	Pb(8.5)	61*	

Vander Wal *et al.* deposited the lead solution sample onto a carbon planchet and then evaporated the water. The residue left was probed by LIBS, and a sensitivity of 2 ppm for lead was obtained [67]. In a different approach, X. Hou *et al.* [68] used preconcentration of Cd, Ni, Cu, and Pb metals on a chelating solid-phase extraction disk and then determined the LOD by X-ray fluorescence spectrometry. The limits of detection for Cd, Ni, Cu, and Pb were 3.8, 0.6, 0.4, and 0.3 ng/mL, respectively. The concentration of samples by absorption in a porous medium such as paper or wood has been studied by a few groups and the reported LODs are in the ppm and sub-ppm. A summary of such studies is given in the Table 1. Among these studies, the best LOD of lead was found to be 0.075 ppm using filter paper by Y. Lee *et al.* [69]. In this work, a filter paper was placed at the bottom of a glass of sample solution, and the solution was totally evaporated to concentrate all the analyte on a filter paper. In the other works, the absorbent substrates were placed in the aqueous sample to absorb some of the analyte, or in a few cases a few drops of sample were dropped on the substrate.

A second major technique for sample concentration is through electrolytic techniques which is particularly suitable for metal ions in solution. In this technique, two metal plates with a potential difference are placed in the solution; the metal ions can then be attracted and pre-concentrated on the surface of the cathode. In general, the electrolytic method has sensitivity in the ppb level, which is better than using absorbent substrate, according to the reported LODs as listed below. This is due to the fact that 1) the applied electric field can guide the metal ions actively to the surface and 2) the pre-concentrated ions stays mostly on the surface of the cathode, where the LIBS technique is suitable for surface analysis. It is also noteworthy that a sub-ppb LOD was achieved from the work of Z. Chen *et al.* [62] and F. Zhao *et al.* [65]. In their studies, the laser spot was slightly defocused to excite a larger amount of analyte for each shot. A summary of such studies is given in the Table 2.

Table 2. List of detection limits reported using electrolytic pre-concentration
(*Reported in ppb, which is converted to ppb by weight in aqueous sample)

Analyzed material and LOD (ppb)	Ref	Comments
Cr (0.57), Mn(0.37), Cu(0.083), Zn(5.6), Cd(0.53), Pb(0.52)	62*	Laser pulse was slightly defocused to increase the excited area
Cu(16)	63	Compared ion replacement reaction with electric field guiding pre-concentration
Pb(6.4)	64	Microchip laser of 4 μJ and 8 kHz repetition rate was used
Cr(0.32), Mn(0.18), Cu(0.16), Zn(1.4), Cd(0.79), Pb(0.57)	65*	Sample was prepared with tap water. Laser pulse was slightly defocused
Hg(11)	66*	Circulation of liquid was appled

Other attempts to convert the liquid sample to solid phase have also been reported. Pace *et al.* have converted aqueous solutions of Cr, Pb, Cd and Zn into solid pellets of calcium hydroxide by mixing with CaO. Therefore, liquid sample analysis was replaced by solid matrix analysis [70]. They used 160 mJ Nd:YAG laser pulses to obtain detection limits for Cr, Pb, Cd and Zn in the range of 1–120 ppm. Caceres *et al.* have carried out analysis of trace elements in frozen water samples using a quick freeze drying procedure [71], which shows the possibility of detection of Na and Al in water sample with the sensitivity limits of the order of a few ppm. According to Sorbal *et al.* [72], after converting the water sample to ice, the LIBS sensitivity can be improved by 3 to 9 times, and the LODs of Cd, Fe, Mg, Cr, Cu, Hg and Pb reached the 0.3-3.7 ppm level.

Overall past results have indicated that the detection of contaminants in water is possible at the ppb level. However, optimization of techniques and system parameters is required in order to achieve such results. In the following sections we present recent studies on the LA-LIF and sample concentration techniques with low energy laser pulses that would be feasible to carry into the field for real time, in-situ measurements.

3. LA-LIF TECHNIQUE

LA-LIF has been carried out with a variety of laser systems observing different emission lines in the past [12, 14, 40-48]. In the interest of developing field portable systems for measurement of water contamination in the future we have focused on an initial investigation of lead contamination in water with the use of low energy, 100 μJ to 300 μJ laser pulses, a regime referred to as microLIBS [73]. Such a technique should be amenable to the development of portable systems based on diode pumped high repetition rate lasers (either fiber or microchip based) together with precisely tuned excitation wavelength lasers based on current fiber optic wavelength division multiplexing techniques [74]. An initial goal is the reliable assessment of Pb contamination at the15 ppb level which is a common standard in North America for allowed levels in drinking water.

To achieve such sensitivity accumulations of the order of 1,000 - 10,000 laser shots may be required. In the technique of Laser-Ablation Laser-Induced Fluorescence (LA-LIF) an ablation plume is created from the edge of a water jet sample containing a low concentration of Pb. After the plume has expanded in air for several hundred nanoseconds a second pulse tuned to a resonant absorption line of atomic lead at 283.3 nm is used to probe the plume with resultant fluorescence detected at 405.8 nm.

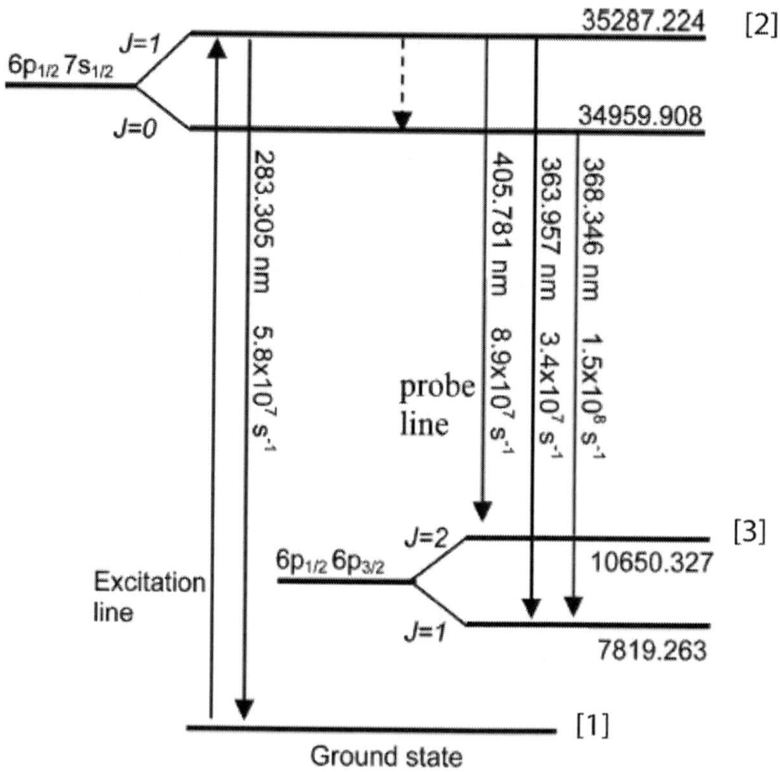

Figure 1. Energy levels of a Pb atom indicating the excitation at a wavelength of 283.305 nm and observation of emission at 405.781 nm [13].

The excitation scheme used for lead is shown in Figure 1 and is typical for excitation of species using the LA-LIF technique. In this case the excitation is at a wavelength of 283.305nm and emission is observed at the fluorescent transition at 405.871 nm. The excitation pulse was generated in a XeCl excimer laser pumped dye laser. The absorption cross sections are very strong for transitions from the ground state and thus it is relatively easy to excite the free atoms to the upper excited state with a few microjoules of energy and one can start to saturate the free atoms in the ablation plume with only 10 microjoules of excitation laser energy [13].

The experimental layout used for the LA-LIF experiments probing lead in water is shown in Figure 2. The laser ablation pulse was generated by a frequency-quadrupled flash lamp pumped Q-switched Nd:YAG laser (Big Sky Ultra CFR) at 266 nm with a pulse width of 10 ns. Typical pulse energies of 100 to 300 microjoules were used. The pulses were focused by a microscope objective lens (Optics For Research, 10x, 15 mm working distance, NA 0.25) to approximately a 10 micron spot on the water jet surface. The objective was protected by a microscope cover slip slide. It was found that this cover slip required cleaning on the order of every several 10,000's of shots. For longer experimental shot sequences it is expected that either a movable cover slip or a thin disposable plastic film cover could be employed instead. The liquid sample was flowed through a 1 mm nozzle and the laser focused onto the side of the liquid stream. The Pb was introduced into the water in the form of a lead salt, typically lead nitrate, The concentrations of solutions are given in ppm or ppb by weight and at low concentrations were checked by measuring the samples using Inductively Coupled Mass Spectroscopy (ICPMS) analysis.

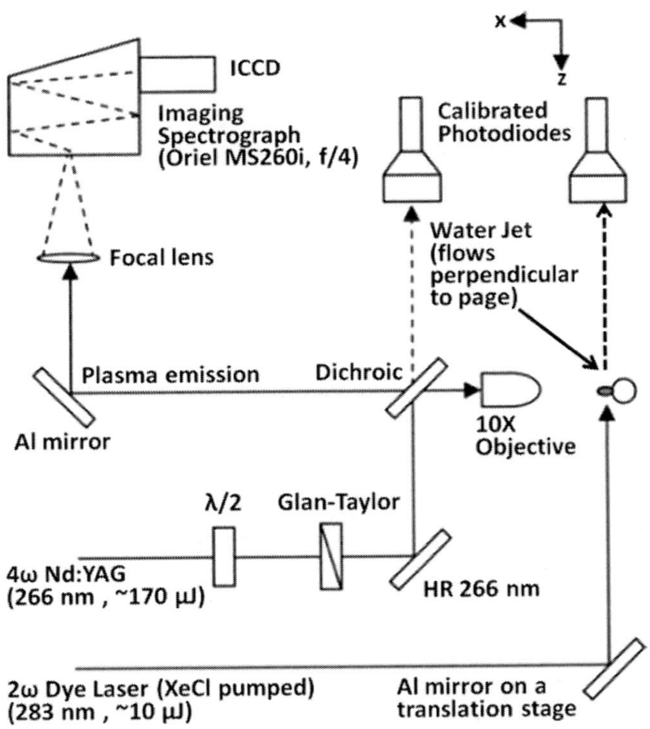

Figure 2. The experimental layout for the LA-LIF measurement [13].

An excitation probe pulse of the order of 10 μJ in energy was produced by frequency doubling the output of a XeCl pumped tunable dye laser (Lumonics HD-500, rhodamine 575 dye). This 9 ns probe pulse was focussed with a long focal length lens to a spot size of the order of 240 μm diameter at right angles through the ablation plume. The emission light was collected through the 10x objective and focused with a 20 cm focal length lens onto the entrance slit of a 0.25 m f/3.9 imaging spectrometer (Oriel MS260i) equipped with an intensified CCD (Andor iStar DH720). A grating of 2400 line mm^{-1} was used and the slit width was typically set at 300 μm, yielding a spectral resolution of the order of 0.45 nm. Precise tuning of the excitation pulse was obtained by observing the LIBS emission from the water sample at higher Pb concentrations where the three lines at 280.19 nm, 282.31 nm and 283.305 nm are observed as shown in Figure 3(a). The probe wavelength was then tuned to precisely overlap the third emission line. An example of the probe spectrum is shown in Figure 3(b). Since the ablation plume is still hot at the time of probing, the linewidth is expected to still be significantly broadened which reduces the requirement for a very narrow linewidth source and very precise frequency alignment. The temporal separation ΔT between the laser pulses was controlled by a delay generator (Stanford DG535) and the ICCD was also gated at slightly earlier time before the probe pulse with a 400 ns gate width to collect the emission.

Initial studies optimized the gate delay and width for the standard single pulse LIBS signal alone indicating a 3 sigma limit of detection (3σ-LOD) for a 100 shot accumulation of the order of 70 ppm by weight for a 260 μJ excitation pulse. The 3σ-LOD is defined as the sample concentration which would give an emission signal which is 3 times the standard deviation noise in the measured signal [13].

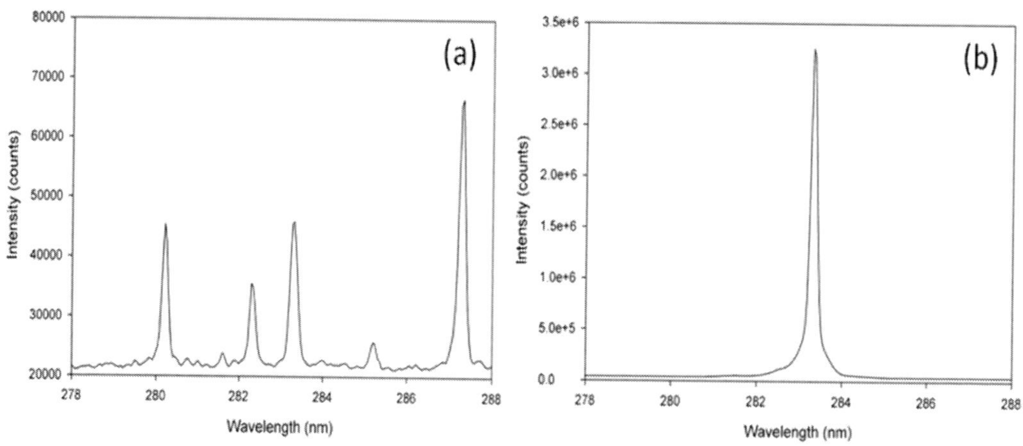

Figure 3. (a) The LIBS emission spectrum of Pb is shown. Three emission lines at 280.19 nm, 282.31 nm and 283.305 nm are observed from Pb. (b) The probe pulse spectrum from the dye laser at 283.305 nm is shown. A 2400 l/mm grating was used in both cases [13].

A number of measurements were taken to determine the 3σ-LOD for Pb in water. These include measuring the SNR for various concentrations of Pb in water over the range of 1 to 1000 ppm and taking 100 or 1000 shot accumulations. Also the SNR for a 5ppm sample of Pb was measured with accumulations of 100 to 10000 shots. The latter measurement demonstrated the $N^{1/2}$ scaling of SNR expected for statistical sampling with an increasing number of shots, N, as shown in Figure 4. This scaling with number of shots is a critical result which can be exploited by using high repetition rate, low energy excitation lasers in the future. The LOD determined from three different experiments are given in Table 3 where it can be seen that a consistent value of the order of 35 ppb was obtained when scaled to 10,000 shot accumulations.

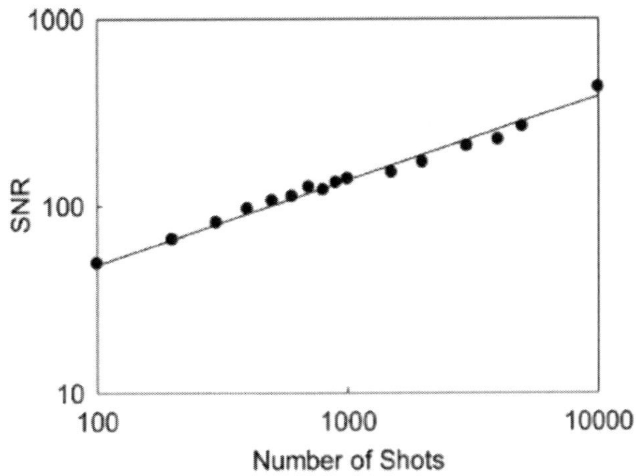

Figure 4. Relationship between number of accumulated shots and the S/N. Data were taken for a Pb concentration of 5 ppm and ΔT=300 ns. The line is a least squares fit to an $N^{1/2}$ slope scaling law.

Table 3. Summary of the LODs from the measurements scaled to 10000 shot accumulations using a $N^{1/2}$ SNR scaling law with number of shots [12]

Measurement Series	LOD scaled to N = 10,000 (ppb)
100 shot data	35 ± 7
1000 shot data	36 ± 7
Scaling with $N_{shot}^{1/2}$	35 ± 8
Average	35 ± 7

A summary of typical sensitivities, both LIBS and our LA-LIF result, for measuring Pb in water by previous groups and ourselves is given in Table 4 when scaled to 100 shot accumulations. It can be seen that the use of LA-LIF has increased the sensitivity by the order of 200 times as compared to straight LIBS using our 260 μJ excitation energies. The use of a double pulse LIBS by Pu *et al.* also shows fairly high sensitivity and again is on the order of 3 times more sensitive than the single pulse LIBS results of Lo *et al.* at similar excitation pulse energies of 16 to 18 mJ. This double pulse sensitivity is slightly better than our LA-LIF result but with combined laser pulse energy of 74mJ which is 274 times our combined laser pulse energy of 270 μJ. Thus the LA-LIF technique allows ppb level sensitivity using 100 microjoule level laser pulses which otherwise would require 10 - 100 mJ level laser pulses which are not very compatible with portable kHz repetition rate systems.

Table 4. Summary of the LODs reported for Pb from literature including our results scaled to 100 shot accumulation

Group	LOD	Laser Pulse Energy	Technique	Reference
Present work	75 ppm	260 μJ	μLIBS	30
Knopp	13 ppm	27 mJ	LIBS	27
Taschuk	6 ppm	45 mJ	LIBS	73
Arca	100 ppm	500 mJ	LIBS	75
Lo	0.894 ppm	16 mJ	LIBS	25
Pu	0.304 ppm	18 mJ +56 mJ	Double Pulse LIBS	76
Present work	0.353 ppm	260 μJ + 10 μJ	LA-LIF	12

Further investigations were carried out using filtered photomultiplier and filtered avalanche photodiode array detectors. Since the LA-LIF technique is looking for one element at a time the use of a filtered single channel detector gives almost as much information as a spectrometer and is much simpler and less expensive to implement. In addition they allow the collection of light over a much larger solid angle and thus in principle can lead to higher signal to noise ratio with near single photon detectors.

The first system investigated was based on a standard (Hamamatsu R7518) photomultiplier tube (PMT) operated at a bias of 900V and an electron gain of ~10^6. A large collection angle, f/0.8, aspheric lens was used to collect the emitted light and image onto the PMT. A line filter with a center wavelength of 405.8 nm and a bandwidth of 2 nm was used to filter the incident light on the detector. A suitable delay time must be chosen so that both the self plasma emission has decayed away and the detector itself has recovered from the initial plasma emission. In the present case a delay time of 300 ns was employed for a 260 μJ excitation pulse. A 1 ppm solution of Pb was measured yielding a 100 shot accumulation SNR of 43 leading to a 3σ-LOD of 70 ppb. A typical signal is shown in Figure 5(a). In this case there is no limit to the potential repetition rate used since a fast gated integrator can be used to save the signal on every shot and summed in a second integrator system. Thus for a 10,000 shot accumulation a 3σ-LOD of 7 ppb could be achieved. In principle, a higher gain photomultiplier could be employed with a high voltage gating circuit to reduce sensitivity to the initial plasma emission, potentially yielding a more sensitive LOD.

The second detector investigated was a multi-pixel photon counter (MPPC) which basically is an avalanche photodiode (APD) array where single photon sensitivity could be obtained on each single photodiode element. The detector used had an array of 10 by 10 elements each 100 μm square (Hamamatsu model S10362-100U). The same light collection system was employed for the MPPC as for the PMT detector but the MPPC has a much smaller active detection area of 1 mm^2 whereas the beam size of the imaged emission region was ~6 mm^2. Thus only a fraction of the photons imaged were detected in this case. One advantage with the MPPC is the high quantum efficiency (45%) which leads to a higher signal compared to the ICCD and the PMT. Another advantage in using the MPPC over the PMT is the lower power requirements for the MPPC for the same gain. Thus the MPPC would be ideal in a portable water monitoring system for Pb. A typical signal pulse measured by the MPPC detector is shown in Figure 5 (b). The measured 100 shot accumulation 3σ-LOD for a 1 ppm Pb solution was 103 ppb. If we had collected all the light onto the APD surface then it would be expected that the measured 3σ-LOD would improve to 42 ppb for 100 shots making it potentially the most sensitive detector combination available. This would lead to a potential 10,000 shot 3σ-LOD of 4 ppb. Again, the detector would have no trouble measuring at high repetition rate as the measured recovery time for the array was on the order of 130 ns. This quick recovery time also allowed the recovery from the initial continuum emission from the plasma before the probe time which was set at a delay time of 300 ns after the initial plasma formation.

Overall, the results indicate that LA-LIF can meet the requirements for detection of heavy metals such as Pb in water at sensitivity levels below those for safe drinking water if high repetition rate lasers are employed.

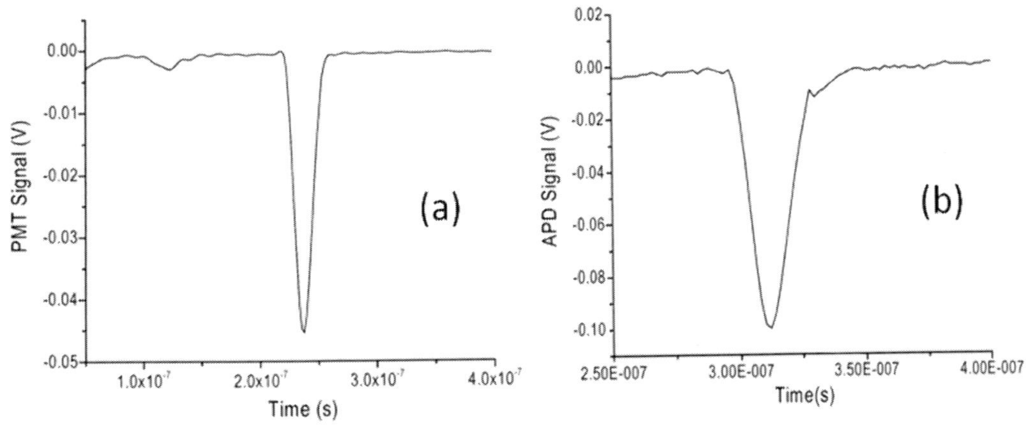

Figure 5. Measured laser induced fluorescence signals. (a) Pb emission signal measured using a PMT for an accumulation of 100 shots. The ablation laser energy was 260 μJ the probe laser energy was 8 μJ and (b) Pb emission signal measured using a MPPC for an accumulation of 100 shots. The ablation laser energy was 260 μJ the probe laser energy was 8 μJ. The interpulse separation was 300 ns in both cases.

4. Sample Concentration on Porous Surfaces

A simple technique for concentration of aqueous samples is by dipping a porous absorbing sampling plate into the liquid sample and letting the sample plate dry. As it dries, the absorbed solution is wicked back to the surface concentrating any dissolved or suspended constituents back to the surface. This then leads to an enhanced concentration of such constituents on the surface with the advantage of carrying out measurements on solid surfaces, which is usually easier than on liquid surfaces. As discussed above this technique has already been investigated by a few groups with encouraging results. To date, wood slices, carbon substrates and paper absorbing filters have been explored. These are all relatively easy to handle and are capable of achieving a low LOD. For example, it has been reported that, by use of wood substrates, different types of heavy metal, such as lead, chromium, copper, and cadmium in water can be detected to an LOD at the sub-ppm level using single laser pulses of 1064nm with an energy of 100mJ [59]. In more recent work we have focused on investigating this technique with lower laser pulse energies of ~2-5 mJ, which is more favorable for portable LIBS systems for in situ analysis

Measurements were carried out using small, 7 mm by 50 mm, wooden sampling sticks (coffee stirring sticks) which were immersed in the water solution for a period of 15 minutes and then dried for a period 10 to 20 minutes on a hotplate until the absorbed water evaporated, leading to enhannced concentration of any contaminants on the surface of the sticks due to the wicking of the water to the surface during drying. LIBS measurements were then performed on the surface of the sample using 2-4 mJ energy laser pulses of 10 ns duration with wavelengths of 266, 532 and 1064 nm. The whole procedure could be carried out in a period of approximately 30 minutes and could be sequenced for a continuous sequential stream of samples if working in the field.

The experimental layout for measuring the samples is illustrated in Figure 6. The emission from the plasma plume was imaged onto an ICCD spectrograph. Millijoule energies were required here to obtain consistent breakdown on the porous surface of the sampling sticks. Samples were prepared by dissolving lead nitrate in de-ionized water, and the solutions were homogenized by a magnetic stirrer before use. We prepared solutions with concentrations from 0.2 to 50 ppm for the measurements.

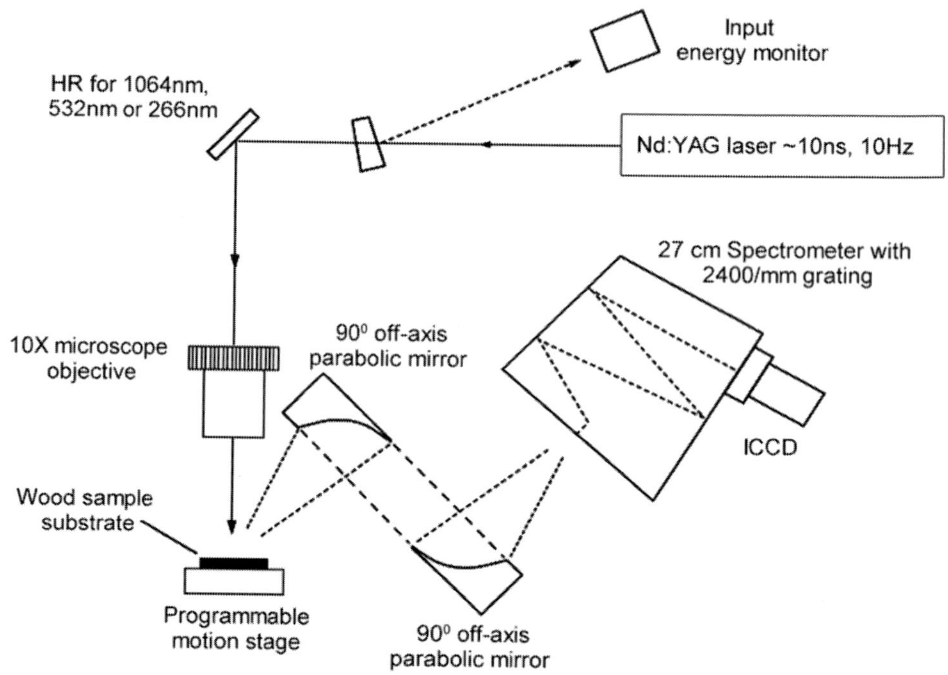

Figure 6. Illustration of the experimental setup for scanning the porous samples and measuring the LIBS spectra.

The laser pulses were focused with a 10 x microscope objective onto the sample surface with a focal spot of the order of 10 μm in size on the surface with ablation spot size varying from 15 to 50 μm depending on the energy and wavelength of laser pulses [14]. The samples were mounted on a motorized stage which was translated at 1.2 mm/s leading to a separation of the order of 120 μm between two consecutive spots to guarantee no overlap of spots. The

405.8 nm line of lead was measured with integration of the emission in a 0.75nm bandwidth line in the ICCD spectrometer detector.

For the three different wavelengths, 266, 532, and 1064 nm, the breakdown thresholds were determined to be ~13 μJ and 800 μJ and 2 mJ respectively. For the 1064 nm wavelength there was approximately a 60% probability of breakdown at 2.2mJ pulse energies since the initiation of breakdown is more stochastical at longer wavelengths. After optimization of gate widths and sample times the LOD was found for the case of 100 shot accumulations as shown in Table 5. It can be seen that the best LODs obtained were for the case of 532 nm illumination with 3.5 mJ laser pulses where an LOD of 113 ppb was achieved which could lead to an LOD of 11.3 ppb if a 10,000 shot accumulation was used.

Table 5. Limits of detection measured for different wavelength excitation pulses with sample concentration on wooden sample sticks using 100 shot accumulations with sample concentrations ranging from 200 ppb to 50 ppm. Also, the LOD determined from a single sample of 20 ppm Pb taken with 1000 shots is given

Wavelength (nm)	266	532	1064
100 shot LOD (ppb)	147	113	131
Expected 1000 shot LOD (ppb)	45	36	41
1000 shot LOD of 20 ppm samples (ppb)	33	35	34

In order to confirm this sensitivity a sample of 200 ppb Pb, in water, was measured using the 532 nm excitation wavelength as shown in Figure 7.

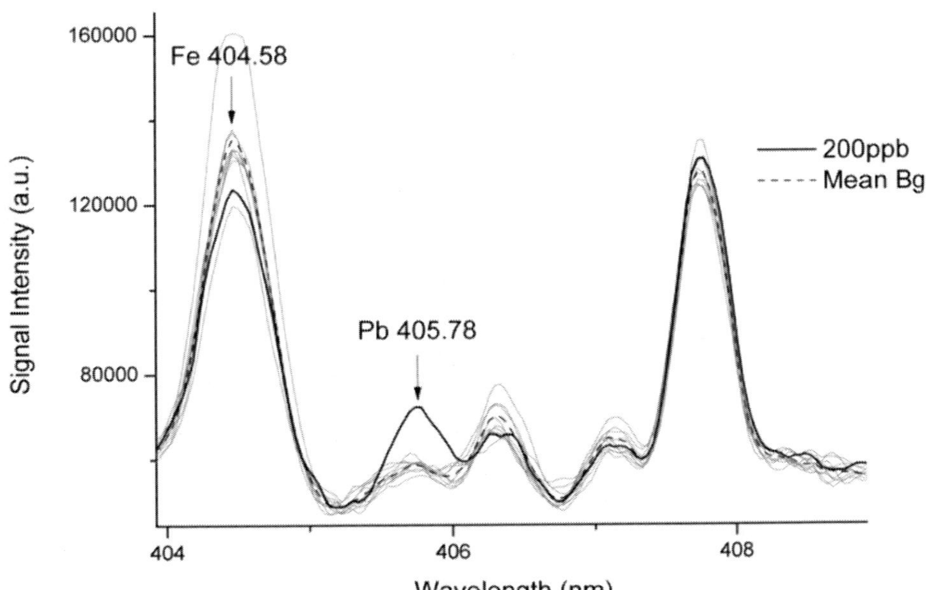

Figure 7. Measurement of the background emission of 10 different runs (light grey lines) and the emission from the wood substrate that has been dipped in 200 ppb lead solution (solid back line) by 100 shot accumulations with 532 nm pulses at an average energy of 3.5 mJ. The average of background signal is given by the black dashed line. An SNR of 6 is obtained.

The range of fluctuations in the background signal can be observed form the multiple samples of background taken. One can observe both the continuum emission (baseline) and line emission of other elements which lead to variations in the region of the 405.78 nm Pb line. However the Pb line is clearly much higher than the fluctuation level in this wavelength region. The background emission around the lead emission line (405.7 nm) also contains weak contributions possibly from the manganese lines at 405.55 and 405.89 nm. By subtracting the mean value of background emission from the lead line emission signal, the residual signal yields a SNR of 6 and a 100 shot 3σ-LOD of 100 ppb. If the wood chips would have no background line emission around 405.8 nm, the background fluctuations could be further minimized, and the SNR could be further improved.

The concentration by evaporation could lead to different response to species in dissolved form versus those in particle or nanoparticle form. It is conceivable that the latter species could be concentrated more on the surface than even the dissolved species. In addition, it is known that the use of femtosecond laser pulse LIBS leads to enhanced sensitivity to a thin surface layer [77] with little or no damage on the substrate [78], and thus might yield even higher sensitivity for the concentrated species on the surface. The results might also be improved further by use of artificial manmade porous surfaces that are finer in grain size, more uniform from batch to batch and with little background signal compared to wooden sampling sticks. Thus, it is expected that the sensitivity of this technique can be improved further. Additional studies in the future will be required in order to investigate these possibilities.

5. Sample Concentration by Electrolytic Techniques

The second technique for concentrating the elemental species dissolved in water is by means of electrolytic deposition on pure metal electrodes. This technique coupled with LIBS was demonstrated by Chen *et al.* [59] in 2008. They demonstrated sub-ppb LOD's for elements such as Cr, Cu, Mn, Zn, Cr and Pb. The trace metal atoms were electrically deposited on the surface of a high purity aluminium plate for a few minutes and quantitatively analyzed by LIBS using using multi-mJ laser pulses. For ultrahigh purity testing this technique can be very effective as the preconcentration is very sensitive to the dissolved ions in the water.

In our recent investigation of this technique we have again focused on extending these results to the use of microjoule laser pulses at high repletion rate. The samples were collected by using two aluminum electrodes in water and passing current through the electrodes for a period of 15 minutes. A voltage of 6 V was applied across the electrodes in order to drive the current. It should be noted that the operating voltage and current depends on the conductivity of the solution and thus requires some adjustment and control if higher concentration or more conductive samples are measured.

After accumulation, the cathode electrode was dried and mounted on a spinning stage for LIBS analysis using an 8 kHz repetition rate microchip laser as shown in Figure 8. In this case the spinning stage was taken from a hard disk drive from a personal computer system and rotated at 5,400 rpm. The passively Q-switched Nd:YAG microchip laser delivered 4 μJ, 1.3 ns duration pulses at 1064 nm at a pulse repetition rate of 8 kHz. The signal was integrated for 20 ms at a time and 1000 such signals added together giving an effective accumulation time of 20 seconds with no time gating using the ICCD spectrometer system. Thus the signal corresponded to an accumulation of 160,000 laser shots.

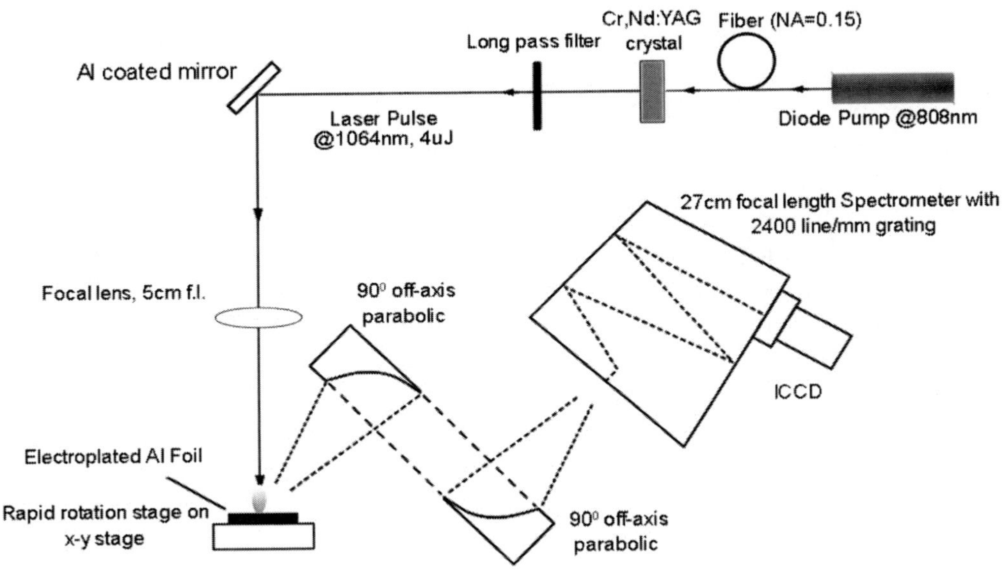

Figure 8. Experimental setup for μLIBS using electrolytic pre-concentration. A microchip laser operating at 8 kHz repetition rate, 1064 nm and with 4 μJ pulse energy is also shown in the diagram.

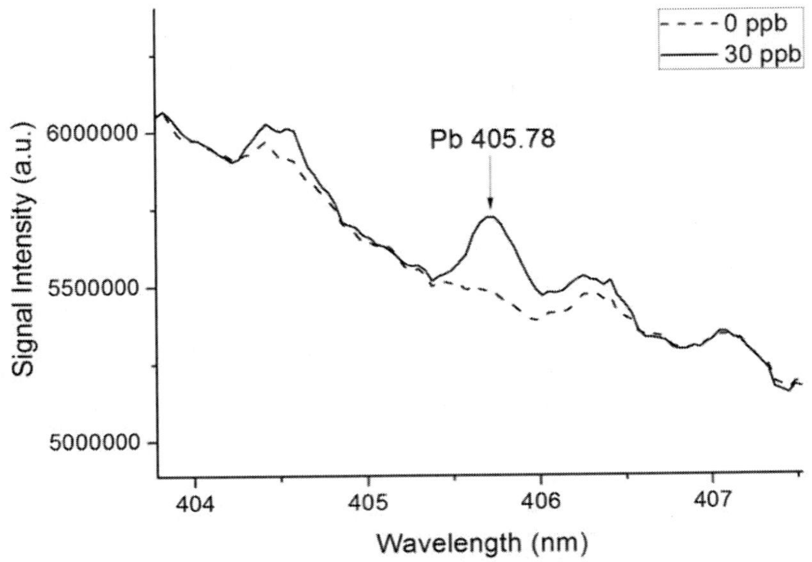

Figure 9. Measured signal and background using electrolytic concentration from a 30 ppb sample of Pb in water.

Figure 9 shows the measured signal and the background signal obtained using this setup for a sample containing 30 ppb Pb dissolved in water. The concentration of the sample was verified using ICP-MS. The SNR of this spectrum was 14 which correspond to a 3σ–LOD of 6.4 ppb. This initial result indicates this technique has great potential to increase the sensitivity of μLIBS and make it a more widely acceptable technique. However during these measurements we observed that if the solution contained high concentrations of the heavy metals, more gas bubbles would be generated. These gas bubbles attached onto the surface of the aluminum plate and thus interfered with the further deposition of the metals. A similar observation was reported by [59]. In the study by Zhao *et al.* [65], tap water was used as background solution to simulate the actual condition of drinking water, and a rotational cathode was used to remove the bubbles during the pre-concentration efficiently. Their study also indicated that the hardness of the water can affect the sensitivity of this technique, and the calibration requires the knowledge of hardness of the aqueous sample. Further work is required in order to address these issues when implemented in real applications.

6. Future Directions

While the results discussed above already indicate the growing potential for enhanced LIBS techniques in the realtime analysis of heavy metals in water there is still considerable scope for further improvements in sensitivities obtained and to expand the scope of techniques employed.

One major area of investigation is the extension of the above measurements to other important metal samples such as Cr, Cu, Zn, Mo, Mn, Cd, Va and Hg and also to non-metallic contaminants such as As. The later is of particular importance for countries where arsenic is a significant contaminant in ground water. Thus suitable excitation and emission wavelengths must be chosen and studied for each element for the LA-LIF technique. For many species the upper state lifetimes are on the order of nanoseconds and thus it may be possible to re-excite the same atom several times in one ablation plume with longer excitation laser pulses of the order of 100 ns, leading to an increased number of detected photons and improved LOD. An optimization of the collection of light onto APD array detectors should also lead to improved signal to noise ratios and more sensitive detection of contaminants. In the experiments of Zhao *et al.*, trace heavy metal ions in tap water were measured by electrolytic deposition on the surface of a high purity aluminum rod and then analyzed with LIBS. The LOD's for Cr, Mn, Cu, Zn, Cd and Pb in tap water samples yielded sub ppm LODS [65].

Alternative techniques for delivering liquid samples for the LA-LIF measurement techniques are also under development. Droplet liquid sources have been used in many different laser ablation applications including laser driven high harmonic [79] and xuv radiation sources for next generation short wavelength EUV lithography [80]. Such sources can be timed to be synchronous with high repetition laser sources. Also the size of microdroplets can be as small as 20 microns allowing over 100,000 spectra to be accumulated per microliter of liquid sample if every droplet is irradiated. Our studies using a piezoelectircally driven microdroplet source [13] have indicated that the LA-LIF signal increases up to a droplet size of around 20 microns and then levels off for the conditions used for the LA-LIF experiments reported above.

Another delivery system of interest for micro samples is the use of microfluidic systems. While LIBS or laser ablation is difficult to carry out directly inside the channels it is possible to extrude droplets out of the channel under command similar to an inkjet printer nozzle. This can be done either using a piezoelectric pulser or a thermal heater. Initial experiments have been carried out to investigate this technique using a custom designed thermal heater to extrude droplets on demand out of an outlet hole on the top of the microfluidic chip and LIBS spectra were demonstrated on these extruded droplets [81]. It was determined that the droplet extrusion time was on the order of 25 μs and thus sampling rates of up 10 kHz should be possible for systems. Just as for the piezo droplet source the actual amount of liquid sample required could be in the range of several microliters for data accumulations of 100,000 shots.

While microchip lasers and fiber lasers already are available as ablation pulses for the LA-LIF technique, the generation of suitable narrow linewidth 10 μJ class excitation pulses at the precise wavelengths required will require further development work. One potential architecture would be to employ diode seed laser sources which can be amplified and frequency converted to the required wavelengths. An initial demonstration of such a diode pumped fiber laser system has been developed with a frequency output at 386 nm, the 4th harmonic of 1544 nm, as illustrated in Figure 10. In this case, an ablation pulse energy of 50 μJ at 1550 nm and an excitation pulse energy of 1 μJ at 386 nm was obtained in 100 ns width pulses at a repetition rate of 10 kHz. The system was fabricated mainly from telecommunication type components which are relatively inexpensive and very robust. In future improvements the energies can be increased by a factor of 4 times by the addition of another large mode area fiber amplifier stage and the pulse width shortened by 10 to 100 times by the use of a faster pulsed DFB diode laser. The system illustrates that such systems are feasible and could easily be packaged in a portable battery powered system. The most difficult challenge is obtaining the laser source wavelengths such that the 4th harmonic will match the required excitation wavelength. For example to meet the requirement for the excitation of lead atoms at 283.3 nm requires a source laser at 1133.2 nm. This is just at the end of the possible tuning range for Yb:fiber lasers and might require mixing of one photon from a 1550 nm laser and 3 photons from an 1040 nm Yb: fiber laser (frequency summing and then frequency summing again). While this would require two narrow linewidth source lasers and associated amplifiers it is still readily achievable using current fiber laser technology. With current advances in laser technology we expect that such techniques could be employed in the near future for real time detection of heavy metal contamination in the environment and drinking water.

In the case of concentration of contaminants on porous sample collectors, a number of additional improvements can be envisaged. More uniform manmade porous substrates such as porous carbon. Ion exchange membranes [52, 56] may have better uniformity than natural substrates and perhaps could lead to lower background signals and more reproducible signals from shot to shot which could improve the measured signal to noise ratio. In addition the use of a dark surface for such collectors, which would have a lower breakdown threshold, would allow lower energy laser pulses to be used in such systems. Femtosecond lasers have demonstrated enhanced detection of surface contaminants due to their limited interaction depth and thus could prove valuable for both the porous sample and electrolytic sample concentration techniques since they both concentrate the analyte onto a solid surface.

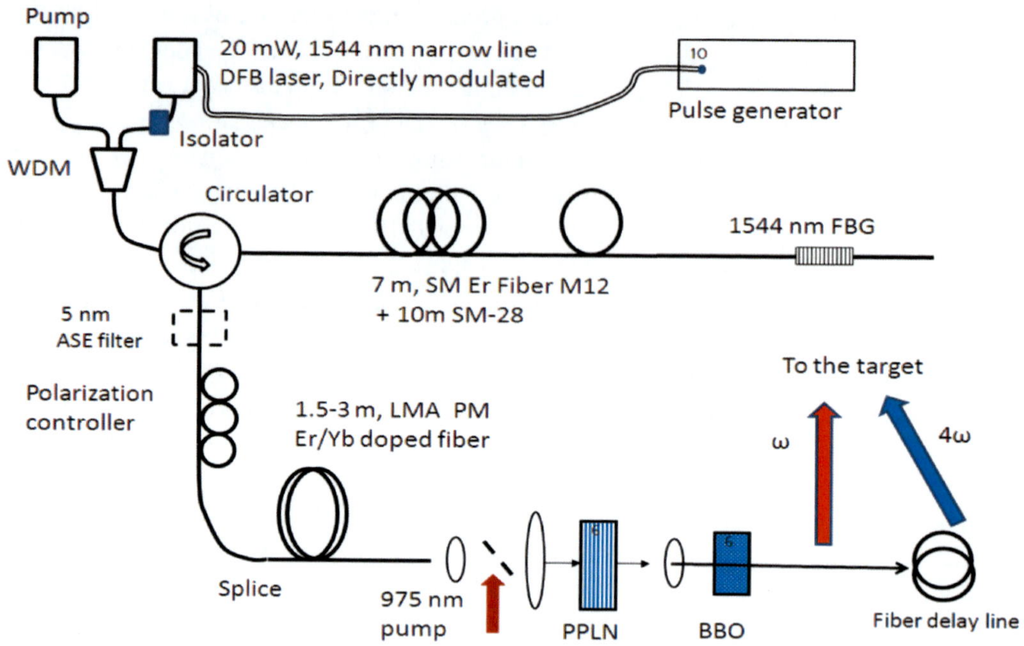

Figure 10. High repetition rate fiber laser based system as a prototype system for LA-LIF detection of heavy metals [74].

CONCLUSION

Studies of detection of heavy metals in water have been carried out using microjoule to few millijoule laser pulses suitable for future portable detector systems. Using the LA-LIF technique, measurements have been carried out at different concentrations of lead and it has been found that 3 sigma limits of detection of 35±5 ppb 7±2 ppb and 10±3 ppb could be obtained for the cases of ICCD, PMT and APD detectors respectively when scaled to 10,000 shots. Using sample concentration on the wooden sampling sticks, accumulated measurements of 100 shots were carried out leading to a 3 sigma LOD of 113 ppb which would scale to 11 ppb for accumulation of 10,000 shots. Finally, when using electrolytic concentration technique a signal was acquired at a repetition rate of 8 kHz for a period of 20 s with 3σ-LOD of 6 ppb. These results are consistent with previous work reported in the literature, mainly using much higher energy pulses.

It appears that enhanced LIBS and Laser induced emission spectroscopy techniques can be applied to the measurement of heavy metal contaminants in water at the ppb level. The development of portable measurement systems can now be envisaged in the near future for realtime detection of such heavy metal contamination in the environment and rapid analysis of liquid samples of all types from drinking water and environmental samples to clinical samples in medicine.

REFERENCES

[1] S. K. Gupta, M. K. Vollmer, R. Krebs, *Sci. Total Environ. 178* (1996) 11-20.

[2] L. Järup, *Brit. Med. Bull.* 68 (2003) 167–182.

[3] R. L. Hough N. Breward, S. D. Young, N. M. J. Crout, A. M. Tye, A. M. Moir, I. Thornton, *Environ. Health Perspect.* 112 (2) (2004) 215–221.

[4] J. R. P. –Videaa, M. L. Lopeza, M. Narayana, G. Saupea, J. G.-Torresdey, *Int. J. Biochem. Cell Biol.* 41 (2009) 1665–1677.

[5] T. C. Benner, *Environ. Sci. Technol.* 38(13) (2004) 3457–3464.

[6] D. A. Basketter, G. Angelini, A. Inqber, P. S. Kern, T. Menne, *Contact Dermatitis* 49 (2003) 1–7.

[7] A.P.M. Michel, *Spectrochimica Acta Part* B 65 (2010) 185–191

[8] A. K. Pathak, R. Kumar, V. K. Singh, R. Agrawal, S. Rai, A. K. Rai, *Appl. Spectrosc. Rev.,* 47 (2012) 14-40.

[9] D. W. Hahn, N. Omenetto, Appl. Spectrosc. 66(4) (2012) 347-419.

[10] J. P. Singh, S. N. Thakur, "*Laser Induced Breakdown Spectroscopy,*" Elsevier Science, First Edition 2007.

[11] R. Noll, "*Laser Induced Breakdown Spectroscopy: Fundamentals and Applications,*" Springer, 2012.

[12] S.L. Lui, Y. Godwal, M. T. Taschuk, Y. Y. Tsui, and R. Fedosejevs*al.*, *Anal. Chem.* 80 (2008) 1995–2000.

[13] Y. Godwal, *Laser Ablation Laser Induced Fluorescence for Sensitive Detection of Heavy Metals in Water*, PhD thesis, University of Alberta (2010)

[14] Z. Chen, Y. Godwal, Y. Y. Tsui, R. Fedosejevs, *Appl. Opt.* 49 (2010) 87-94.

[15] DC 2006. *United States Environmental Protection Agency,* U.S. Government Printing Office: Washington. Lead and Copper Rule State File Review: National Report. EPA, Publication No. EPA816R06001, 2006.

[16] *National Primary Drinking Water Standards.* Publication No. EPA816F03016 Office of Water, United States Environmental Protection Agency, U.S. Government Printing Office, Washington DC, 2003.

[17] S. Bakirdere, F. Aydin, E. G. Bakirdere, S. Titretir, I. Akdeniz, I. Aydin, E. Yildirim, Y. Arslan, *Appl. Spectrosc. Rev.,* 46 (2011) 38-66.

[18] A. Leykin, S. Nelms. *Inductively Coupled Plasma Mass Spectrometry in Environmental Analysis* In Book *Comprehensive Environmental Mass Spectrometry*; Ed. by Albert T. Lebedev, ILM Publications, 2011

[19] M. N. Fiddler, I. Begashaw, M. A. Mickens, M. S. Collingwood, Z. Assefa and S. Bililign, *Sensors* 2009, 9, 10447-10512.

[20] Nai-Ho Cheung, E.S. Yeung, *Appl. Spectrosc.* 47(7) (1993) 882-886.

[21] Nai-Ho Cheung, E.S. Yeung, *Anal. Chem.* 66 (7) (1994) 929-936.

[22] S. K. Ho, N. H. Cheung, *Anal. Chem.* 77 (2005) 193-199.

[23] N. H. Cheung, C. W. Ng, W. F. Ho, E. S. Yeung, *Appl. Surf. Sci. 127*–129 (1998) 274-277.

[24] O. Samek, D.C. S. Beddows, J. Kaiser, S. V. Kukhlevsky, M. Liska, H. H. Telle, J. Young, *Opt. Eng.* 39(8) (2000) 2248-2262.

[25] K. M. Lo, N. H. Cheung, *Appl. Spectrosc.* 56(6) (2002) 682-688.

[26] G. Arca, A. Ciucci, V. Palleschi, S. Rastelli, E. Tognoni, *Appl. Spectrosc.* 51(8) (1997) 1102-1105.
[27] R. Knopp, F. J. Scherbaum, J. I. Kim, *Fresen. J. Anal. Chem.* 355 (1996) 16-20.
[28] D. A. Cremers, L. J. Radziemski, T. R. Loree, *Appl. Spectrosc.* 38(5) (1984) 721-729.
[29] P. Fichet, P. Mauchien, J-F Wagner, C. Moulin, *Anal. Chim. Acta.* 429 (2001) 269-278.
[30] Y. Godwal, S. L. Lui, M. T. Taschuk, Y. Y. Tsui, R. Fedosejevs, *Spectrochim. Acta* B 62 (2007) 1443-1447.
[31] W. F. Ho, C. W. Ng, N. H. Cheung, *Appl. Spectrosc.* 51(1) (1997) 87-91
[32] B. Charfia, M. A. Harith, *Spectrochim. Acta B* 57 (2002) 1141-1153.
[33] H. Loudyi, K. Rifai, S. Laville, F. Vidal, M. Chakera and M. Sabsabib, *J. Anal. At. Spectrom.* 24 (2009) 1421-1428.
[34] Y. Feng, J. Yang, J. Fan, G. Yao, X. Ji, X. Zhang, X. Zheng, Z. Cui, *Appl. Optics* 49(13) (2010) C70-C74.
[35] K. Rifai *et al., J. Anal. At. Spectrom.* 27 (2012) 276-283.
[36] P. A. Demchenko, *Introduction to Fluorescence Sensing.* Springer Science and Business Media: New York, NY, USA, 2009.
[37] X. Hou *et al., Laser-Excited Atomic Fluorescence Spectrometry: Principles, Instrumentation, and Applications*; Wiley-VCH: New York, NY, USA, 1987
[38] P. Stchur *et al., Spectrochim. Acta B* 56 (2001) 1565-1592.
[39] D. J. Butcher, *Appl. Spectrosc. Rev.* 40 (2005) 147-164.
[40] I. B. Gornushkin, S. A. Baker, B. W. Smith, J. D. Winedordner, *Spectrochim. Acta B* 52 (1997) 1653-1662.
[41] S. Laville *et al., Spectrochim. Acta B* 64 (2009) 347-353.
[42] H. S. Kwong, R. M. Measures, *Anal. Chem.* 51(3) (1979) 428-432.
[43] W. Sdorra, A. Quentmier, K. Niemax, *Mikrochim. Acta* 98 (1989) 201–218.
[44] V. E. Bondybey, J. H. English, *J. Chem. Phys.* 74 (12) (1981) 6978-6979.
[45] I. B. Gornushkin *et al., Appl. Spectrosc.* 51(7) (1997) 1055-1059.
[46] X. K. Shen, Y. F. Lu *Appl. Optics* 47(11) (2008) 1810-1815.
[47] R. E. Neuhauser *et al., Anal. Chim. Acta* 346 (1997) 37-48.
[48] S. Koch, W. Garen, W. Neu, R. Reuter, *Anal. Bioanal. Chem.* 385 (2006) 312-315.
[49] Le Bihan *et al. Anal. Bioanal Chem.* 375 (2003) 815-819.
[50] A. Kitano *et al., Anal. Chem.* 83 (2011) 9424-9430.
[51] P. Yaroshchyk, R. J. S. Morrison, D. Body, B. L. Chadwick, *Spectrochim. Acta B* 60 (2005) 1482-1485.
[52] D. Alamelu, A. Sarkar, S. K. Aggarwal, *Talanta* 77 (2008) 256-261.
[53] Y. Lee, S-W Oh, S-H Han, *Appl. Spectrosc.* 66 (2012) 1385-1396.
[54] D. Zhu *et al., Appl. Opt.* 50 (2011) 5695-5699.
[55] J. Cortez, C. Pasquini, *Anal. Chem.* 85 (2013) 1547-1554.
[56] N. E. Schmidt, S. R. Goode, *Appl. Spectrosc.* 56 (2002) 370-374.
[57] T. Kim, M. L. Ricchia, C-T Lin, *J. of the Chin. Chem. Soc.* 57, (2010) 829-835.
[58] K. M. Santos *et al., Math. Journal* 110 (2013) 435-438.
[59] Z. Chen, H. Li, F. Zhao, R. Li, *Spectrochim. Acta* B 63 (2008) 64-68.
[60] R. L. Vander Wal, T. M. Ticich, J. R. West, Jr., NASA/CR 2004–212980 (National Center for Microgravity Research, 2004).
[61] D. Zhu, J. Chen, J. Lu, X. Ni, *Anal. Meth.* 4 (2012) 819-823.

[62] Z. Chen, H. Li, F. Zhao, R. Li, *J. Anal. At. Spectrom.* 23 (2008) 871-875.
[63] Y. Lu, Y. Li, J. Wu, S. Zhong, R. Zheng, *Appl. Opt.* 49(13) (2010) C75-C79.
[64] Z. Chen *et al.*, *Conference on Lasers and Electro-Optics (CLEO) and Quantum Electronics and Laser Science Conference* (QELS), May 16-21, 2010, San Jose, CA, USA, (Optical Society of America, 2010), paper AWA3.
[65] F. Zhao *et al.*, *Anal. Meth.* 2 (2010) 408-414.
[66] X. Fang, S. R. Ahmad, *Appl. Phys.* B 106 (2012) 453-456.
[67] R. L. Vander Wal, T. M. Ticich, J. R. West Jr., P. A. Householder, *Appl. Spectrosc.* Vol. 53(10) (1999) 1226-1236.
[68] X. Hou, H. L. Peters, *Appl. Spectrosc.* 57(3) (2003) 338-342.
[69] Y. Lee, S-W Oh, S-H Han, *Appl. Spectrosc.* 66 (2012) 1385-1396.
[70] D. M. Díaz Pace, C. A. D'Angelo, D. Bertuccelli, G. Bertuccelli, *Spectrochim. Acta* B 61 (2006) 929-933.
[71] J. O. Cáceres, J. T. López, H. H. Telle, A. G. Ureña, *Spectrochim. Acta* B 56 (2001) 831-838.
[72] H. Sobral, R. Sanginés, A. Trujillo-Vázquez, *Spectrochim. Acta* B 78 (2012) 62-66.
[73] M. T. Taschuk, *Quantification of Laser-Induced Breakdown Spectroscopy at Low Energies,* PhD thesis, University of Alberta (2007).
[74] S. Sindhu, *Erbium Fiber Laser Development for Applications in Sensing*, MSc thesis, University of Alberta (2012).
[75] G.Arca *et al. IEEE Geoscience and Remote Sensing Symposium* (1996) 520-522.
[76] X. Y. Pu, N. H. Cheung, *Appl. Spectrosc.* 57 (2003) 588-590.
[77] M. T. Taschuk, Y. Y. Tsui, R. Fedosejevs, *Appl. Spectrosc.* 60(11) (2006) 1322-1327.
[78] S. P. Banerjee, Z. Chen, I. Utkin, R. Fedosejevs, *Appl. Phys.* B 111 (2013) 791-798.
[79] A. Flettner, T. Pfeifer, D. Walter, C. Winterfeldt, C. Spielmann, G. Gerber, *Appl. Phys.* B 77 (2003) 747-751.
[80] C. Keyser, G. Schriever, M. Richardson, E. Turcu, *Appl. Phys.* A 77 (2003) 217-221.
[81] Y. Godwal, G. Kaigala, V. Hoang, S. L. Lui, C. Backhouse, Y. Tsui, R. Fedosejevs, *Opt. Express* 16(17) (2008) 12435-12445.

In: Spectroscopic Techniques for Security, Forensic ...
Editors: Y. Dwivedi, S. B. Rai and J. P. Singh
ISBN: 978-1-63117-404-9

Chapter 2

APPLICATION OF LASER INDUCED BREAKDOWN SPECTROSCOPY (LIBS) TO CARBON SEQUESTRATION RESEARCH AND DEVELOPMENT

Dustin L. McIntyre[1,*], Jinesh C. Jain[2], Christian L. Goueguel[3] and Jagdish Prasad Singh[4,5]

[1]U.S. Department of Energy, National Energy Technology Laboratory, Morgantown, WV, US

[2]URS-Washington Division, National Energy Technology Laboratory, Pittsburgh, PA, US

[3]Department of Civil and Environmental Engineering, Carnegie Mellon University, Pittsburgh, PA , US

[4]Institute for Clean Energy Technology, Mississippi State University, Starkville, MS, US

[5]Department of Physics, King Saud University, Riyadh, KSA

ABSTRACT

This chapter presents the application of laser induced breakdown spectroscopy (LIBS) to different sensing scenarios that highlight the applicability and need for such a diverse and multi-modal sensing technique for carbon capture and sequestration applications. The use of LIBS to sense, qualify, and quantify atomic concentrations in solids, liquids, and gases at the parts per million (ppm) level without any significant difference in the experimental setup is a testament to the versatility and robustness of this technique when applied to such a complex range of measurement environments. The LIBS technique was calibrated and applied to the measurement of CO_2 concentrations in ambient air as well as over a wide range of concentrations. The relative standard deviation (RSD) of the measurements was found to range from 1-10% and significantly depended on the number of data points collected. The sampling time ranged from 8 seconds (single measurement) with a RSD of analysis approximately 10% to a sampling time of 6 minutes (40 measurements) with a RSD of approximately 1.24%.

* dustin.mcintyre@netl.doe.gov.

The LIBS technique was also calibrated and applied to the measurement of total carbon in soil samples. This technique required minimal sample preparation and provided measurements which had a relative accuracy that was comparable to a commercial carbon analyzer and shown to be 5.44% or less. This technique also explored multiple linear regression models to improve the detection capabilities of the system.

The LIBS technique was lastly applied to measure the concentration of calcium in various saline solutions to determine the effect, if any, that the amount of sodium in solution has on the measurement of the calcium concentration. The limit of detection for the calcium concentration was measured to be 17 +/- 0.6 mg/L (ppm). It should also be noted that the measurement of the calcium at low concentration became possible as the increased concentration of sodium in the solution induced a significant increase in the intensity of calcium emission line. The liquid measurement technique was designed to produce a LIBS spark within the bulk liquid and not at the liquid air interface.

1. INTRODUCTION

The levels of greenhouse gases such as CO_2 in the atmosphere have been increasing steadily since the beginning of the industrial revolution and are causing noticeable climate change. Due to the potential harm that the continued release of CO_2 may cause to the Earth, many nations are currently looking into ways of separating CO_2 from industrial/electrical power generation flue gas streams where the concentrated gas can then be liquefied and pumped underground for permanent storage [1, 2]. A significant portion of the carbon sequestration research being performed in the United States involves the risk assessment of injecting large quantities of CO_2 into different underground formations, and the consequences involved if the injected CO_2 moves into different formations or if it leaks into aquifers or back into the atmosphere.[3-7]

During carbon capture and storage (CCS) operations the CO_2 will be pumped into the storage formations as a supercritical fluid. If a leakage pathway develops the CO_2 will most likely escape the containment formation as a combination of the supercritical fluid and the native fluid, with dissolved CO_2. Once either the supercritical fluid and/or the CO_2 saturated brine reach a portion of the subsurface with appropriate temperature and pressure conditions the CO_2 will become a gas and continue its leakage as a gas. CO_2 can be detected in either the gas or liquid phase and at a number of locations in the subsurface as well as in the atmosphere once it has left the containment formation. The gas can be measured in water wells, once it has degassed, it can be measured in the air of mines, it can be measured in gas and oil production wells, and in the atmosphere above the leakage point.

During CCS injection the leakage of CO_2 into upper formations and possibly into the vadose zone and into the plant bearing soil area will need monitoring due to possible negative effects on soil and/or vegetation. Significant natural leakages of CO_2 have been known to devastate large areas of forest [8, 9]. Expected leakage from an industrial scale injection site is thought to be minimal to negligible, however to mitigate possible widespread damage it would be prudent to understand the condition of the soils in overlying areas before and during the course of CCS injection.

When sensing potential leakage sites for carbon sequestration sites, there is need for a wide range of modality in sensing such as atmospheric measurements, subsurface air measurements, surface and subsurface water measurements, and soil measurements all at varying temperature and pressure conditions. There are a number of technologies which are being employed however they consist of either single point field deployable systems or samples taken back to a lab based piece of equipment [10-13]. The currently available methods typically do not have the ability to sample and measure the particular constituents at varying pressure and temperature so there can be a great deal of error introduced into the measurement when the sample is prepared at laboratory conditions. Optical systems such as open path Fourier-Transform Infrared (FTIR) and Light Detection and Ranging (LIDAR) have been developed to scan and observe gases in open areas above suspected leakage sites. These systems have an advantage of wide area monitoring however they are very expensive, complex and not necessarily suitable for the diverse environments where sequestration may take place. Single point measurements can be made to very high precision with Gas Chromatographs (GC), Mass Spectrometers (MS), and Cavity-Ring-Down Spectrometers (CRDS) but due to their size, cost, and need for constant maintenance it is usually best to retrieve samples from the field and relay them back to these types of analysis devices. In addition, the quantification of CO_2 remains a challenge and there is a need for a real-time and long-term monitoring technique that can also address the leak detection threshold.

The current technology gap warrants development of new monitoring techniques that are affordable, robust, and easily distributable and deployable in many different locations to study a variety of different aspects of the environment. LIBS is a promising tool for detection of solid, liquid and gaseous samples [14-16]. A LIBS set up is essentially comprised of a high intensity laser as an excitation source and a time-gated spectrometer to collect the signal. Generally, a laser beam is focused onto a solid sample or within a gaseous or liquid sample to create high temperatures, dissociating the sample to form plasma. Radiation from the plasma is then collected by the spectrometer. Subsequent analysis of the radiation gives qualitative and quantitative information of chemical species present in the sample.

2. Application of LIBS to Gaseous CO_2 Measurement

The release of CO_2 into the atmosphere has been found to significantly contribute to global climate change. In the case of injection well failures or leakage of abandoned wells, a sudden and rapid release of CO_2 could result in an elevation of CO_2 concentration in the air near injection sites. If the leakage is not detected in a timely manner it could be hazardous to human health and the biota in the vicinity of the release. Currently, CO_2 in the gas phase is measured by various methods [17]; the two most commonly used methods are GC-MS and infrared spectroscopy. The working principle of GC-MS is based on selective adsorption and elution of the gas molecules of interest in the GC column and the subsequent analysis by a mass-spectrometer. Within, the sample gas is bombarded with a high-energy electron beam, causing fragmentation of the molecules in the sample. These fragmented molecules are accelerated by a magnetic field and separated by virtue of their different mass to charge ratios. Infrared (IR) absorption spectroscopy is an optical technique, which is most commonly used technique for CO_2 detection. Carbon dioxide has a non-zero dipole moment which causes it to absorb specific wavelengths of light in the infrared region. Thus analysis of absorption spectral bands gives the CO_2 concentration in the sample. This technique has a distinct disadvantage in the form of interference from water vapor and carbon monoxide.

Various authors have used LIBS for detection of gaseous samples [18-23]. While Hahn *et al.* [18] have used LIBS for hydrogen leak detection, Winefordner *et al.* [19] and McNaghten [20] have utilized it for detection of gaseous/particulate fluoride and helium, and argon in binary and ternary gas mixtures with nitrogen, respectively. Recently, Eseller *et al.* [21] used the technique for monitoring Ar, He and Oxygen impurities in hydrogen. F. Ferioli *et al.* [22] have applied LIBS to measure the equivalence ratio of a spark-ignited engine, where they showed that the ratio of either C(711.3nm) or CN(707-734nm) peaks with any of the N(746.3 and 743.8nm), O(776.6nm) spectral lines can be used to estimate the equivalence ratio. In fact, the LIBS calibration curve of gaseous samples shows a better linearity and reproducibility than solids and liquids; this may be attributed to weak interactions relevant to the matrix effect in low density gas sample at atmospheric pressure and to the homogeneity of the gaseous sample.

Recent experiments in our laboratory utilized LIBS for determination of CO_2 in air [24]. The measurement was made by creating a laser spark in the air and subsequently measuring the carbon signal at C-247.86nm; a line used by other researchers [25, 26] for carbon measurement in soil. A schematic of the experimental set up is shown in Figure 1. It was assumed that the main source of the carbon signal in air is from carbon dioxide (the concentrations of CO_2, methane, and carbon monoxide in air are approximately 390 ppm, 1.79 ppb, and 0.1 ppm, respectively).

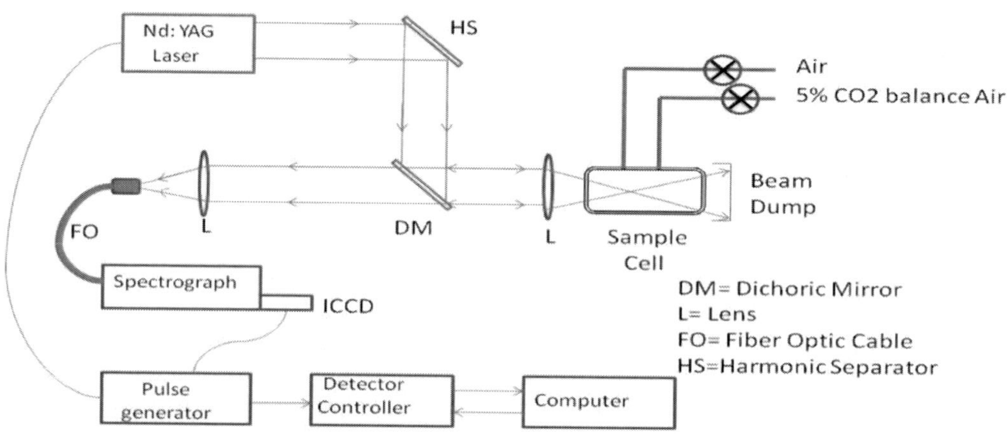

Figure 1. Schematic of experimental set up for CO_2 measurement [24].

A two-way cylindrical cell was used to hold the sample during the measurement. The cell was fitted with a quartz window at both ends to allow the laser beam to enter and exit the cell. The sample cell contained ancillaries to allow inflow and venting of gases, and measurement of the cell pressure. A rotary vane pump and a Baratron absolute pressure gauge were also connected to the cell. The Nd:YAG laser with a wavelength of 532 nm was used in this investigation. It was operated in Q-switched mode with a pulse repetition rate of 10 Hz. The pulse width (FWHM) and the maximum pulse energy was 8 ns and 180 mJ respectively. The beam diameter was 6.5 mm and had a Gaussian beam profile. A spherical plano-convex fused silica lens with a 10 cm focal length was used for focusing the laser beam into the center of the sample cell to create a plasma spark within the gas mixture. The emission from the laser-induced plasma was focused into a fiber optic cable by a fused silica lens with a 10 cm focal length. The other end of the optical fiber was coupled to a UV–visible Echelle optical spectrometer. The spectrometer had a spectral range of 200–780 nm (though it had some spectral gaps). It had a linear dispersion of approximately 5–19 pm/pixel. A 1024 X 1024 element intensified charge-coupled device with a pixel width of 24 microns which was cooled by Peltier elements and attached to the exit of the spectrometer which was used to detect the light from the laser spark. The detector was operated in gated mode using a dedicated high-voltage fast-pulse generator that was synchronized with the laser pulse. The data acquisition and analysis were performed using a personal computer with *ESAWIN* software. Two certified gas mixtures supplied by Nexair were used in the experiments. These comprised a cylinder of compressed air and another of CO_2 (5% CO_2 balanced with air). The certified gas mixtures were used to prepare the gas sample for the system calibration.

The initial efforts were concentrated on investigating the effect of experimental parameters (laser energy, gate delay, and choice of focusing lens) on the intensity of the carbon 247.85 nm line and are illustrated in Figure 2.

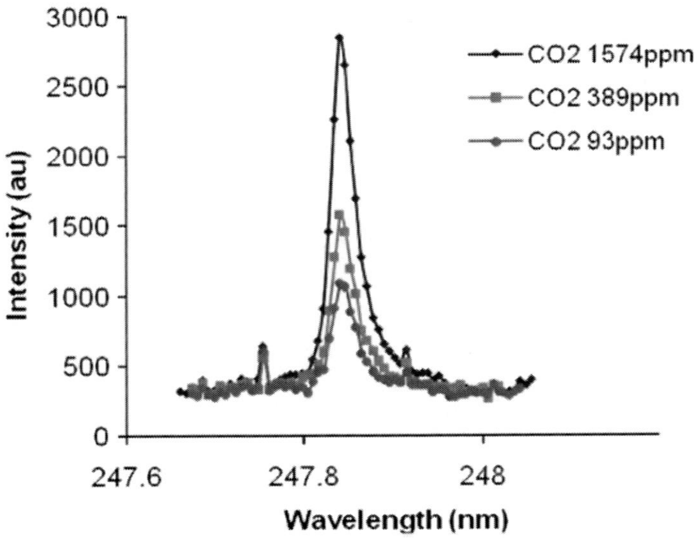

Figure 2. Spectral *line of carbon at 247.84 nm at various CO_2 concentrations in the cell.* Smooth lines connecting the points are to guide the eye [24].

The most persistent carbon line that could be read with the spectrometer was found to be C-247.85 nm. The intensity of the selected line was read directly from the accompanying software, which calculated the intensity by subtracting the background from the peak maximum. First, the effect of two different focusing lenses on the signal strength was investigated. The laser beam was focused with both a 10 cm and 20 cm focal length lens and a spark was created in air. The best signal to noise ratio, without saturation, was found with an on-chip accumulation of 40. The signal intensity was recorded with different pulse energies and gate delays and is shown in Figure 3 & 4. The gate delay and gate width were fixed at 0.5 μs and 5 μs respectively for the data illustrated in Figure 4. Each data point is an average of 10 measurements; on-chip accumulation was 40. The use of a 10 cm lens gave a stronger signal than that of the 20 cm lens (Figures 3 & 4), which is to be expected due to the smaller focal spot size and associated higher focal intensity.

The laser pulse energy and the gate width were respectively fixed at 145 mJ and 5 μs. Each data point is an average of 10 measurements and on-chip accumulation was 40. A linear least square fit is shown for data.

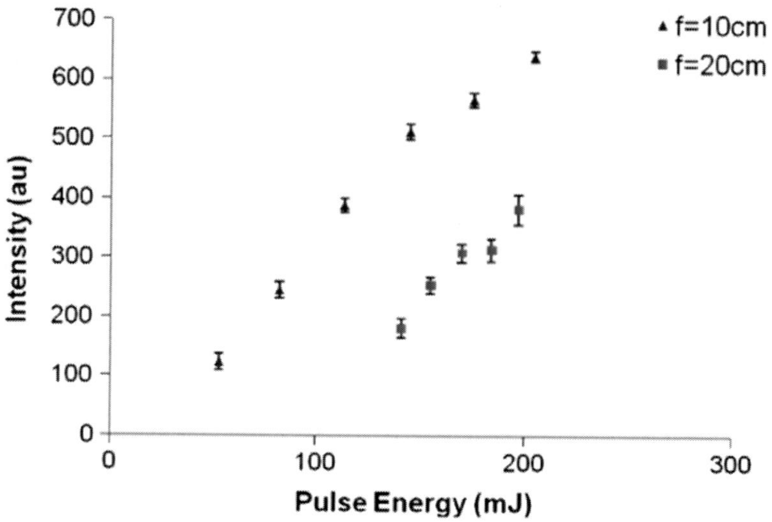

Figure 3. The scatter plot shows the effect of using focusing lenses of different focal lengths where discrete values of carbon spectral line intensity (C-247.85 nm) at different laser pulse energies are shown [24].

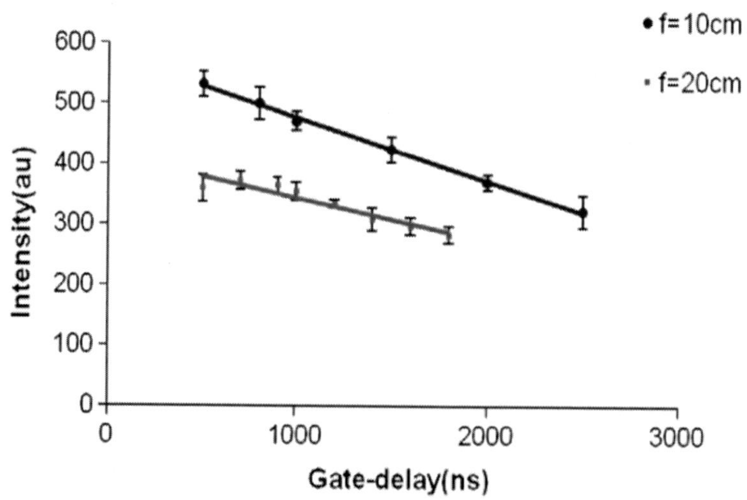

Figure 4. Variation of C247.85 nm line intensity with gate-delay using two different focusing lenses (focal 10cm and 20cm) is shown [24].

The plasma spark size observed in the experiment was smaller in the case of the 10 cm lens. The tight focusing of the laser beam in a small volume by a 10 cm lens had resulted in greater irradiance and thus a better signal. The maximum signal-to-noise ratio (S/N) was found to be approximately 60 for a 10 cm lens and 27 for a 20 cm lens. Figure 4 also shows that maximum signal strength can be obtained by setting the gate delay to 500 ns.

For studying the nature of plasma in air and to select the appropriate pulse energy, the intensity ratio of the C 247.86 nm line to various nitrogen lines against the pulse energy were plotted (Figure 5). The gate delay and gate width was set at 0.5 μs and 5μs. Each data point is an average of 10 measurements and on-chip accumulation was 40.

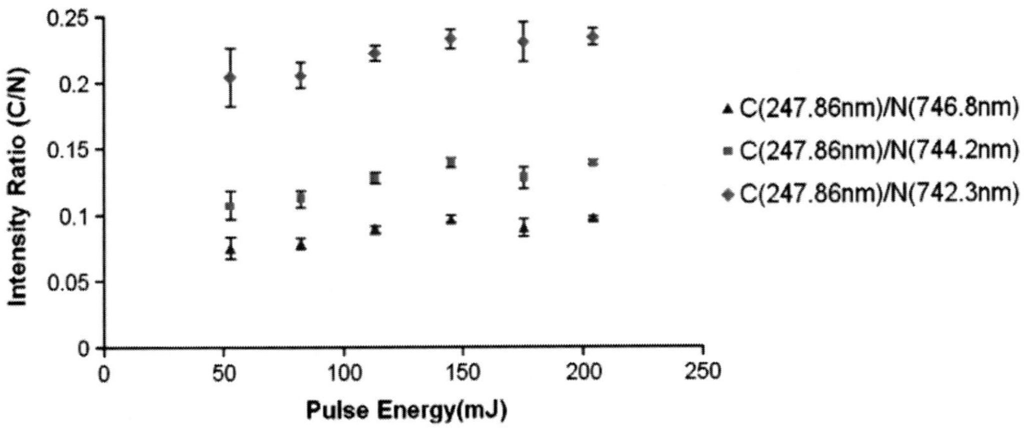

Figure 5. Variation of intensity ratio of different spectral lines with laser pulse energy is shown [24].

The ratio plots using three different nitrogen lines show a similar trend; this suggests that the plasma is being formed uniformly. The ratio of C (247.86) line intensity to the nitrogen lines increase with increasing laser pulse energy. When the laser energy reaches above ~145 mJ, the ratio of C (247.86) line intensity to the nitrogen lines becomes constant and serves as a secondary indication of saturation both for carbon and nitrogen lines. Based on this study, laser pulse energy of 145 mJ was selected for the instrument calibration.

The schematic of the experimental set up used for calibration is shown in Figure 1. The pressure inside the sample cell was set at atmospheric pressure. The experimental parameters used for the instrument calibration are as follows: laser pulse energy of 145 mJ, gate delay of 0.5 μs, and gate width of 5 μs. ICCD on-chip accumulation was set at 40, thus 40 shots were needed for a single measurement, which took 8 seconds. Ten such measurements were averaged to remove shot-to-shot variation. The C 247.86 nm line was selected as the analyte line for CO_2 measurement. The observed analyte line at three different CO_2 concentrations is shown in Figure 2. The desired gas samples for the instrument calibration were prepared by diluting the 5% (50000 ppm) CO_2 mixture with air. When we plotted the calibration curve for entire concentrations ranging from 50 ppm to 5000 ppm the calibration curve was found to be non-linear. However, when the whole concentration region was divided into three sub-regions the calibration curve showed linearity in each separate concentration region (data not shown).

The calibration curve utilizing the CO_2 concentrations up to 1500 ppm is shown in Figure 6. To ensure the repeatability of the measurement, the calibration data were collected on five different days and calibration curves were made. To demonstrate the fact that the calibration curves overlap well we divided these data sets into two groups: training data and test data. The training data set was comprised of data from the first four days and the test dataset was comprised of data obtained on the fifth day. The training data set was used to plot the calibration curve shown in Figure 6a, which was further used to predict the concentrations of the test data set. 'Actual' versus 'predicted' concentrations from the test data were plotted in Figure 6b. The data presented in the both Figures show that the measurements are repeatable.

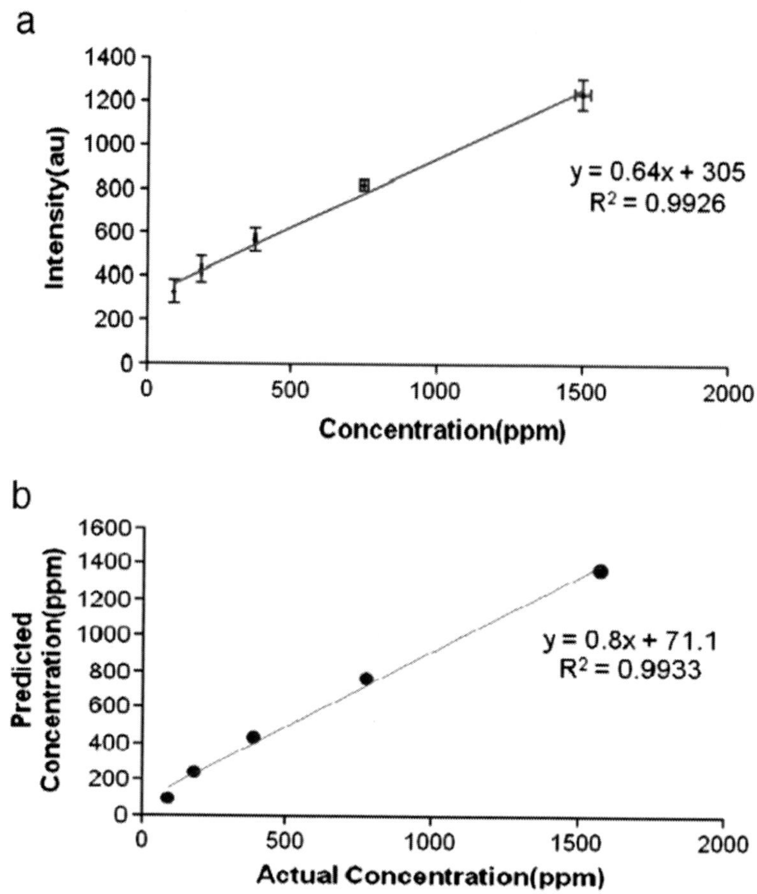

Figure 6. The experiment was repeated five times; first four datasets were used to make the calibration plot shown in (a), this calibration curve was used to predict concentrations of the fifth dataset. Actual concentrations and predicted concentrations are plotted in (b) [24].

A high R^2 value ensures that CO_2 concentration is directly proportional to the intensity of C 247.85nm spectral line. The concentration can be estimated from the intensity if the slope and intercept of the calibration line are known. The limit of detection (LOD) of CO_2 was evaluated using the standard method shown in Equation 1

$$LOD = \frac{3\sigma}{m} \qquad (1)$$

Where, σ is the standard deviation of a blank measurement and m is the slope of the calibration line. In this experiment mean standard deviation was 10 and the mean slope was 0.83 (produced from the calibration data). The estimated LOD found was ~36 ppm.

The precision and rapid real-time detection has been demonstrated by CO_2 measurement of the laboratory ambient air. The sample cell was filled with ambient air and measurements were made continuously for 30 minutes. Figure 7 shows CO_2 concentration measurements made at various time instances. *The laser pulse energy, the gate delay and the gate width were set at 145 mJ, 0.5 μs, and 5μs respectively.*

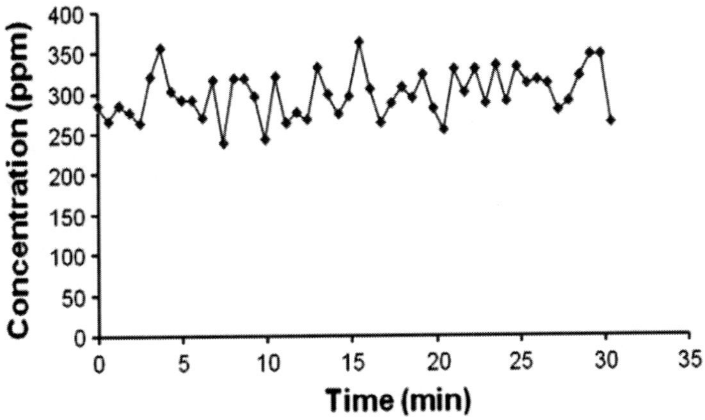

Figure 7. Variation *of CO_2 concentration of ambient air with time is shown. Each point is an average of five spectra. Lines connecting the points are to guide the eye [24].*

The relative standard deviation (RSD) found was 3.6%. The average CO_2 concentration over a period of 30 minutes was 298.82 ppm. An important pattern seen in the data was that more precision can be achieved by averaging more data points shown in Table 1.

Table 1: Shot-to-shot intensity variation, which is inherent in LIBS, can be reduced by averaging an increased number of measurements, which is reflected by reduced standard deviation

No. of measurements averaged	Relative Standard Deviation(RSD)	Time taken (sec)*
1	9.2%	8
5	3.6%	40
10	2.5%	80
20	1.6%	160
40	1.2%	360

While the RSD corresponding to a single measurement was 9.2% and the time required for analysis was 8 seconds; averaging 40 such measurements reduced the RSD to 1.24% and the time required rose to 360 seconds.

Analytical results presented in this section support that LIBS can be used to measure CO_2 in air. Precision and repeatability of the analysis is ensured by a low RSD, which is comparable to commercially available IR CO_2 sensors (RSD 1% to 3%). The accuracy and the LOD can be further improved by employing more accurate sample preparation methods for calibration and also by modifying the experimental set up. Major errors in IR based technique for atmospheric CO_2 measurement are generally introduced during sample collection, drying of the sample, and due to minute leakage [27]. The LIBS based method requires no sample collection and uses a simple experimental setup, which eliminates such errors. Therefore this technique can be used as a complementary data collection technique with other techniques for atmospheric CO_2 measurement to provide reliable and rapid results.

3. Application of LIBS to Total Carbon Measurement in Soil

CO_2 saturation of soils is known to increase the soil carbon content primarily through the acidification of moisture held within the soil. Therefore it is becoming increasingly important to measure total carbon content of soils to monitor and verify the effectiveness of carbon abatement efforts [28]. Measurement of total carbon content in soils before, during, and after CCS activities can evaluate if any CO_2 leakage is occurring. Use of accurate and reliable measurement techniques is important to know how much carbon is typically present in a particular scenario [5, 8]. Quantification of total carbon content in soil can be carried out using many conventional analytical techniques [12]. However, most of those techniques are laborious, time consuming and require a great deal of sample preparation depending on the complexity of the soil matrix [13]. LIBS is a widely used technique for environmental chemical analysis including analysis of soil samples [8, 26, 29-35]. Since little or no sample preparation is required LIBS provides a rapid and cost effective measurement technique that can also be used for in situ analysis.

To demonstrate that LIBS is a viable tool for measuring total carbon concentration in soil samples we analyzed samples containing a wide range of total carbon content [36] and compared the results with commercially available carbon analyzer [34]. The study used carbon powder to simulate higher carbon concentrations within the soil samples. The soil samples were finely ground, filtered, mixed with carbon powder in 6 different known concentrations and made into pellets. LIBS spectra of the pellet samples were collected and the calibration curves were developed using modified simple linear regression (SLR) and multiple linear regression (MLR) calibration models. Both the SLR and MLR models successfully predicted the concentration of the unknown sample with reasonable correlation coefficient values and acceptable uncertainty values. The details of instrumentation and sample analysis are provided in the following paragraphs. The experimental setup used to record the LIBS signal from the pellet samples is shown in Figure 8.

A frequency doubled, Q-switched Nd:YAG laser was used as an excitation source. The laser light at 532 nm (26.4 mJ energy per pulse) was focused onto the sample surface by a fused silica plano convex lens of 30 cm focal length, and the emitted light from the laser-induced plasma was collected with a pick-up lens aligned at approximately 45° with respect to the laser beam and directed into a 100 μm optical fiber. The distance of the pick-up lens was 6 cm from the focal point of the laser beam. The collected signal was sent to a broadband spectrograph, which was integrated with the ICCD. The detector head used a built-in digital delay generator (DDG) that controlled the gate delay (0.5 μs) and gate width (5 μs). Andor iStar software was used for the data acquisition. To obtain data with a better S/N ratio, ten spectra were collected from each sample and each spectrum recorded was an average of five shots (50 shots total). The sample pellets were mounted on a rotating platform to ensure data collection from various parts of the prepared sample and LIBS data were collected for various rotation speeds from 1 to 3 rotation/min.

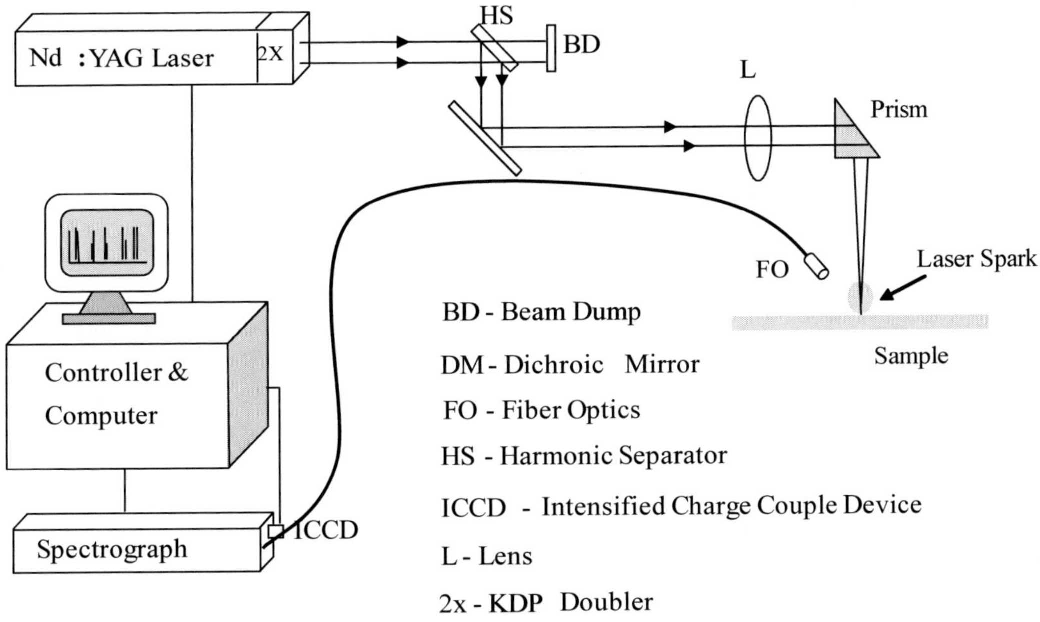

Figure 8. Schematic diagram of LIBS Experimental Setup [36].

A soil sample from Starkville, MS, USA was used in this study. The soil sample was finely ground and filtered with a No. 60 sieve. The filtered soil sample was mixed with varying amounts of carbon powder to produce 5 calibration samples that contained carbon powder concentrations from 3.74% to 9.74%. A sample was prepared without mixing any carbon powder and termed as blank, and finally one unknown sample was prepared with carbon powder. The pellets were made using the same binder and pressure reported by Lal *et al.* [37]. The soil sample (5 g) was mixed with 0.4 ml PVA (2 wt% in distilled water) binder. The soil sample was then hand mixed again to make sure that the binder was evenly dispersed into the mixture. This mixture was placed in a 25-mm-diameter die and pressed into a pellet under 3000 psi. The prepared pellets were approximately 5 g weight, 3 mm thickness and 25 mm in diameter and were kept on a heating plate at 60^0 C for 90 sec to remove the moisture from the binder. Finally, the pellets were kept in sealed plastic bags to avoid contamination from the moisture in the ambient air. The sample pellets were analyzed in triplicate by a commercially available carbon analyzer and it was found that the total carbon concentration of the blank soil sample, including binder, was approximately 2.43%. The triplicate concentrations measured from the carbon analyzer were averaged and considered as a reference while developing the calibration models. The concentration values of the different samples from the carbon analyzer and the contributions of weight from the soil and samples are shown in Table 2.

Table 2: Total weight percent of the carbon in the prepared pellets

Sample	Carbon analyzer Value (wt. %)	Weight	
		Soil (g)	Carbon powder (g)
Blank	2.43±0.05	5.0	0.0
1	3.74±0.04	4.9	0.1
2	5.22±0.14	4.8	0.2
3	6.88±0.18	4.7	0.3
4	8.31±0.07	4.6	0.4
5	9.74±0.23	4.5	0.5
Unknown	5.75±0.11	4.7	0.2

Thirty spectra from each calibration sample were used to develop the calibration model and ten additional spectra of each calibration sample were used for validation of the calibration model. The two calibration models evaluated in this study were the SLR and MLR analytical models [38, 39]. The SLR assumes a linear relationship between the peak intensity (area) of the analyte line and the elemental concentration. However instead of using the individual analyte lines as a function of the elemental concentration to calibrate the SLR model, ratios of the analyte lines to other present elements as a function of the concentrations were used to negate interference effects. The MLR model was also evaluated for LIBS calibration to account for various interference effects caused by other components of the sample matrix or random experimental errors. The MLR uses more analyte lines to produce the calibration data and was found by previous researchers to perform much better than the SLR [40]. The MLR calibration curve was obtained by fitting the linear equation below to a set of experimental data consisting of the ratio of the measured peak areas of C and Fe, and known concentrations of carbon. The MLR procedure estimates a linear equation of the form shown in Equation 2:

$$C_p = b_0 + \sum_{i}^{n} (b_i A_i) \tag{2}$$

where C_p is the concentration of the carbon in the pellet, b_0 is the intercept, b_i is the coefficient corresponding to A_i and A_i is the ratio of peak areas of the C and Fe lines. The multiple lines can be used in the regression to improve the measurement accuracy by correcting for the matrix effect.

The predictive quality of the calibration models were evaluated by calculating the correlation coefficient (R^2) and relative accuracy. The relative accuracy (RA %) is calculated as shown in Equation 3

$$RA(\%) = \frac{|C_{libs} - C_{true}|}{C_{true}} \bullet 100\% \tag{3}$$

where C_{libs} is the value from the LIBS measurement and C_{true} is the known value.

The LIBS spectra of pelletized soil samples were recorded using various experimental conditions such as lens to sample distance, laser energy per pulse, gate delay, gate width and sample rotation speed which were all determined to be optimal for this study. These parameters were optimized to achieve the best S/N ratio and lowest possible relative standard deviation. A representative spectrum is shown in Figure 9 indicating the emission lines selected for LIBS analysis.

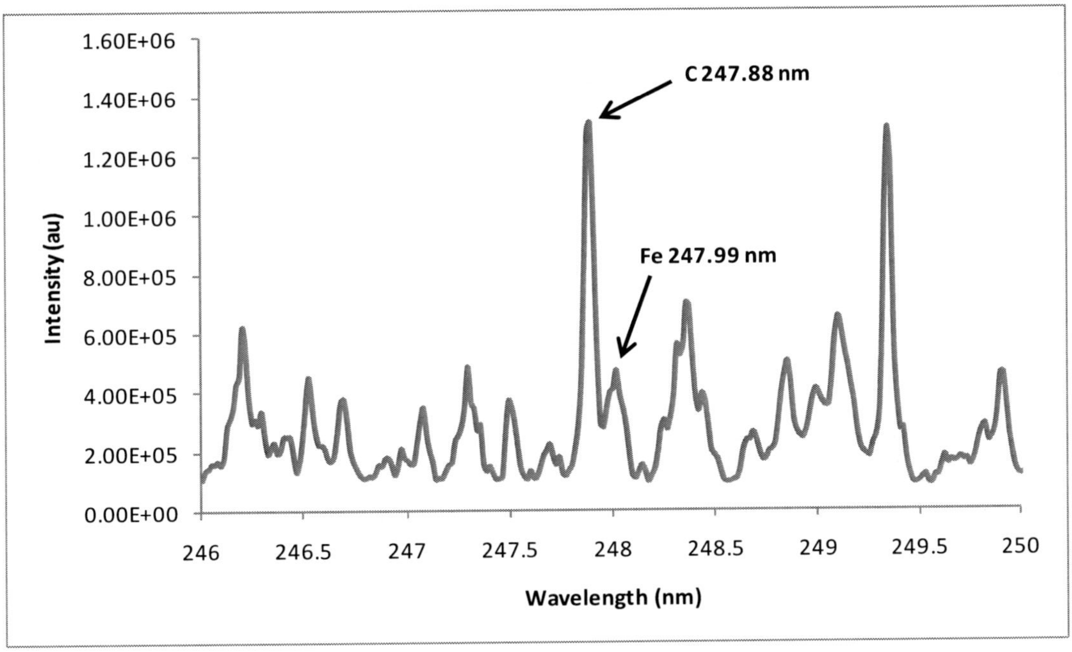

Figure 9. Spectrum of emission lines selected for LIBS analysis [36].

Da Silva *et al.* [28] reported that the carbon line at 247.88 nm shows strong interference with Fe emission lines at 247.86 nm and 247.95 nm. If the selected analyte line has spectral interference due to overlapping or self-absorption, the correlation between the line intensity and analyte content will be overly complicated. For this reason they have used the carbon line at 193.03 nm for calibration purposes. Figure 9 shows the carbon pellet spectra of present study and confirms the same. In our case, the spectrometer could only provide a wavelength range of 200 – 975 nm. Due to this limitation of the spectrometer, the carbon line at 193.03 nm could not be used; therefore the carbon line at 247.88 nm was used instead. To account for interference with Fe emission lines a different approach to the SLR method was applied. The total peak area under C 247.88 nm and Fe 247.99 nm were considered as a single peak area for C 247.88 nm and used in the analysis. The ratio of total peak area of C 247.88 nm with that of Fe 246.51 nm and Fe 247.48 nm (as a function of carbon concentration measured from the carbon analyzer) were used to develop the calibration curves. The peak area of each selected line was extracted from the spectral data of each pellet with the aid of functions provided in the Andor iStar software.

The LIBS data were collected on 6 samples which varied in carbon concentration from 2.43% to 9.74% to develop the calibration curves. The data from the 9.74% sample shows a lower intensity ratio value as well as strong saturation effects due to the high concentration of carbon. Zhang *et al.* [41] investigated the carbon concentration effects and reported that the intensity of the analyte line with higher carbon concentration emitted a much higher intensity than those of the lower concentrations. The calibration curve with the 9.74% sample data produced poor results due to the strong saturation and self-absorption effects. To improve the calibration results, the 9.74% sample was omitted from the data and the calibration curves were developed from the remaining samples (2.43% to 8.31%). Figure 10 shows the calibration curves for carbon using the SLR method, the error bars represent a single standard deviation.

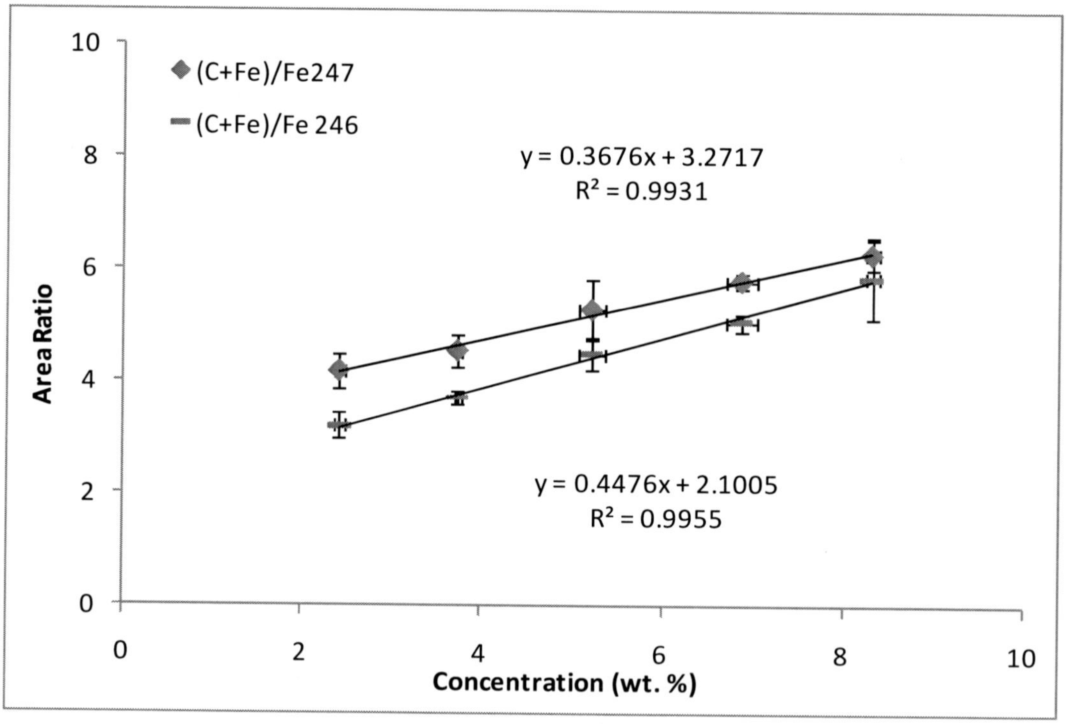

Figure 10. Calibration plots based on line-intensity ratios of Carbon using modified SLR method [36].

The correlation coefficient of the ratio of C 247.88 nm line with Fe 246.51 nm and Fe 247.48 nm line were 0.996 and 0.993 respectively. The calibration data from both the C247/Fe246 and C247/Fe247 line ratios were used to predict the concentrations of the validation data. The uncertainties and RA% were calculated and reported in Table 3 for comparison.

Even though the R^2 values of the SLR method with the ratio of both the Fe lines are acceptable, the ratio of C 247.88 nm line with Fe 246.51 nm has the uncertainty within the limit, i.e. less than the 20%, than the ratio of C 247.88 nm line with Fe 247.48 nm for the validated data. Therefore the calibration data for the C 247.88 nm / Fe 246.51 nm ratio was used for the analysis of an unknown sample.

Table 3: Comparison between LIBS analysis (SLR model) and carbon analyzer data

Sample	C analyzer value (wt. %)	LIBS value (wt. %)			
		C 247/Fe 246	RA%	C 247/Fe 247	RA%
blank	2.43±0.05	2.49±0.44	2.41	2.56±1.24	5.44
1	3.74±0.04	3.76±0.42	0.58	3.70±1.91	1.19
2	5.22±0.14	5.40±0.69	3.31	5.26±1.85	0.76
3	6.88±0.18	6.53±0.49	4.98	7.02±0.95	2.12
4	8.31±0.07	8.50±0.84	2.33	8.74±3.68	5.22

To better account for effects such as saturation, interference, and random experimental error, the MLR technique was employed. The MLR can be a good approach to predict the unknown concentration if the elemental concentration can be well correlated with the intensity of a few spectral lines. However, this method can be inefficient or inappropriate if many spectral lines are needed or if there is significant collinear relation between the spectral lines. Therefore, the selection of analyte lines when employing the MLR is very important [40]. The ratio of C 247.88 nm line with Fe lines (246.51 and 247.48 nm) were selected for the linear regression. The correlation coefficient obtained was 0.996 and uncertainty and RA% values for validation data using the MLR model is given in Table 4 for comparison.

Table 4: Comparison between LIBS analysis (MLR model) and carbon analyzer data

Sample	C analyzer value (wt. %)	LIBS value (wt. %)	RA%
Blank	2.43±0.05	2.54±0.43	4.47
1	3.74±0.04	3.55±0.62	5.14
2	5.22±0.14	5.13±0.55	1.72
3	6.88±0.18	6.96±0.31	1.28
4	8.31±0.07	8.38±0.70	0.86

The uncertainty and RA% values from MLR model for most of the samples improved when compared to the SLR model output. The ability of LIBS to produce accurate quantitative data with the calibration models was tested by examining the LIBS spectra of an unknown sample under the same experimental conditions. The uncertainty and RA% value for the unknown sample when using the SLR with a ratio of Fe 246.51 nm line and MLR calibration methods were acceptable and within their limit. The calibration results for the unknown sample for both the SLR and MLR models are reported in Table 5 for comparison.

Table 5: Comparison between LIBS analysis of SLR and MLR models and carbon analyzer analysis for unknown sample

	SLR		MLR	
C analyzer value (wt. %)	Value (wt. %)	RA%	Value (wt. %)	RA%
5.75±0.11	5.50±0.71	4.26	5.34±0.51	7.10

The MLR model provided better uncertainty and RA% results for most of the samples used for the validation data as compared to the SLR model. In the case of the unknown sample, the MLR model has given better uncertainty values than the SLR model. The accuracy of the MLR model can depend on the analyte lines and combination of lines chosen; the best results for the available lines were reported here.

These results are part of our research endeavor towards initial development of a miniaturized portable LIBS analyzer for CO_2 leak detection in the field. Because the actual saturation of soil samples with CO_2 was not possible with our current laboratory configuration a straight forward approach of adding carbon powder to the samples to simulate increased carbon content was chosen. The use of carbon powder to simulate higher carbon concentrations within the soil samples appears to be an appropriate and acceptable simulation technique. The results show that LIBS can be successfully applied to quantitative measurement of carbon in soil with a relative accuracy better than 5%. Both SLR and MLR models successfully predict the concentration of the unknown sample with the reasonable correlation coefficient values and acceptable uncertainty values. Further experiments will incorporate the development of a general calibration model that can be used for rapid *in situ* monitoring of the carbon in soil samples for qualitative analysis.

4. Application of LIBS to Subsurface Brine Measurement

The ability of subsurface fluid reservoirs to retain stored CO_2 is an important issue in CCS technologies [42]. The CCS approach is critical in the portfolio of solutions available to combat global temperature increases, because it allows significant reductions in CO_2 emissions from anthropogenic sources. It is believed that the deep saline formations (brine-filled) are one of the most promising options for geologic CO_2 sequestration (GCS) because of their large storage capacity [43]. To ensure the success of GCS, it is important that the CO_2 that is injected underground remains there (99% permanence over 1000 years) [44]. However, risks of CO_2 leakage are latent and could be due to degradation/alteration of seal and/or reservoir integrity. In order to maintain safe and efficient GCS operation, it is necessary to monitor leaks (CO_2 or brine leakage) from the storage sites. LIBS appears to be a suitable technique for monitoring potential leaks from the storage sites. Simplicity of the technique and its capability of performing measurements in various environments make it an ideal tool for monitoring brine composition in storage sites.

In different deep saline formations brine cation compositions vary greatly by location [45, 46]. Based on field site investigations sodium is typically the most abundant element in deep saline waters constituting from 70% to 90% of total cation mass. While calcium is generally the second most abundant cation [45], in addition to magnesium and potassium and some other trace elements like sulfur, bromine, iodine and carbon. It is widely known that one of the main challenges for an efficient quantitative analysis by LIBS is due to the matrix effects. Matrix compounds can have considerable effects on the way LIBS analysis is conducted and on the precision and accuracy of the results obtained. These effects are caused by numerous factors and are extensively investigated in the LIBS literature. They have been observed for the analysis of a wide range of analytes, over many diverse matrices [14, 17, 47-54]. Analytical procedures for correction have been proposed, and in most cases, normalization procedures are the most common method used [47].

The origins of matrix effects are complex and are not fully understood. They are related to phenomena involved in plasma formation and/or sample ablation. Further work remains to be done to characterize these matrix effects for many types of samples under various conditions to develop correction methods to improve the quality of LIBS analysis. For GCS applications LIBS analysis of Na-Ca brine needs to be understood better. In particular it is advantageous to examine the effects of changes in sodium concentration on the spectral emission of laser-induced plasma in Na-Ca brines.

LIBS analysis of water composition has been studied by a number of researchers [52-55]. Although none of the cited research efforts specifically addresses carbon sequestration issues, they all have the potential to provide adequate measurements to complement and enhance existing methods for sequestration monitoring. For instance several research efforts examined the ion species and concentrations in and around under sea volcanic vents which are high temperature and high pressure zones rich in minerals and naturally discharging CO_2. Ion identification and concentrations were recorded in both the parts per million and parts per billion ranges and at pressures exceeding 4000 psi for the following atomic species: Magnesium, Sodium, Calcium, Lithium Potassium, Manganese, Zinc, and Hydrogen. The production of laser sparks in water can be a bit more difficult than on the surface of a solid or in a gas due to the lower availability of free electrons and quenching due to the mass and diffusion of the condensed phase (water). The primary reason for using the double laser pulse excitation is that the initial pulse creates a tiny bubble whereby the second pulse, timed accordingly, can initiate a larger and more energetic plasma in the gas that is inside of the initial bubble. This technique has the potential to be a significant source of information in the field of sequestration. A small laser spark generator can be submerged down hole to sample contaminants or CO_2 being displaced from one formation to another while possibly performing identification, concentration, and isotopic measurements. The experiment to conduct LIBS analysis of liquid samples in optically accessible pressure cells is in progress in our laboratory.

When injecting CO_2 into brine filled formations the brine will be displaced into other adjacent formations including those above and below depending on the degree of connectivity. The CO_2 will partially dissolve into the brine however most of the displaced brine will have little to no CO_2 content. When displacing this fluid it is important to understand where the fluid is moving, especially in inhabited areas. It is very important not to displace fluid into formations used for domestic or agricultural water production. The formations that are typically being targeted for CO_2 injection are very high in salinity and total dissolved solids (TDS). If a CO_2 injection site begins forcing fluid into the production formations for human use there will be a need to either treat the produced water or stop using it. In areas where water is scarce, the need for treatment, or additional treatment, will incur additional costs and liability. LIBS can be a useful technique for a long-term reliable water monitoring process across a wide range of analytess and a wide range of landscapes.

The experiments performed in our laboratory examined the effect of sodium on the spectral emission of calcium by varying NaCl molar concentrations from 0 to 3 M [56]. It is worth noting that instead of using the liquid-air interface, which is the most common configuration used in LIBS analysis of liquids, we opted for direct measurements within the bulk liquid for a specific underground application. The experimental apparatus used for this study is shown in Figure 11.

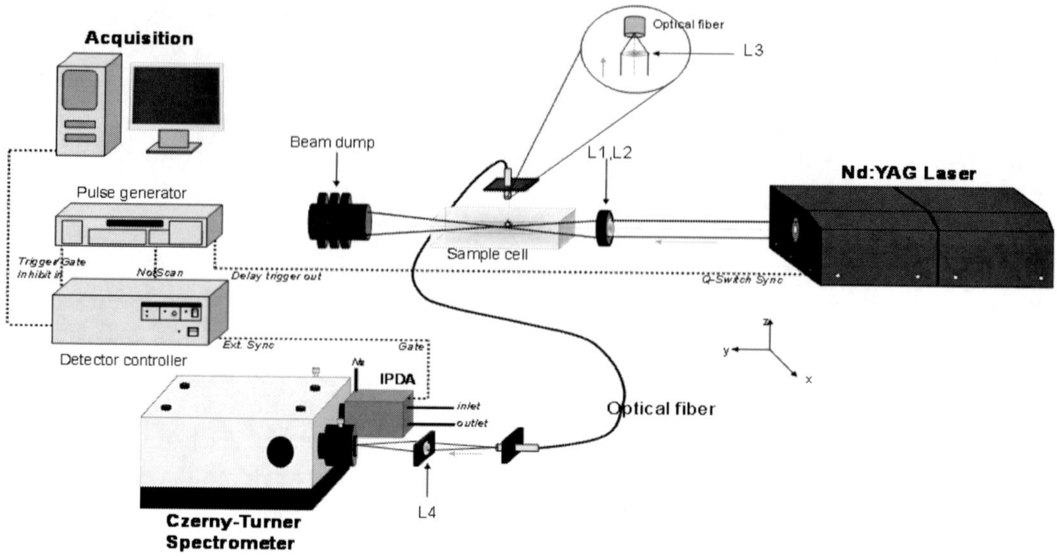

Figure 11. Schematic of the LIBS Experimental apparatus used for liquid sample [56].

A Q-switched Nd:YAG laser operated at 1064 nm with a 5 ns pulse duration and a 10 Hz repetition rate was used to generate the plasma. The laser was kept at constant pulse energy of 10 and 20 mJ and was monitored by a high-energy pyroelectric head detector connected to a hand-held optical power and energy display meter. The pulse-to-pulse laser energy fluctuation was typically less than 5%. The sample was contained in a rectangular (102 mm × 13 mm interior dimensions and 3.2 mm wall thickness) quartz liquid cell. In order to obtain a well-localized breakdown in the bulk liquid, the near-Gaussian beam (5 mm diameter at the laser output) was focused by a pair of fused-silica plano-convex lenses (75 mm focal length, 25.4 mm diameter). The lenses were separated by 1 mm and positioned as close as possible (i.e., few millimeters) to the wall of the sample cell. The resulting focal length was about 37 mm. The plasma breakdown occurred approximately 1.5 cm below the liquid surface. A total liquid volume of about 20 ml was used in each measurement. The plasma emission was then collected by a fused silica collimating lens of 10 mm focal length and 5 mm diameter. The fiber collection optic was mounted onto the end of an SMA 905-terminated 2-m optical fiber with a 400 μm core diameter. The optical fiber was mounted at a right angle to the laser beam on a micro xyz-translation stage. Light emitted at the other end of the optical fiber was focused by means of a plano-convex lens (50 mm focal length, 25.4 mm diameter,) at the entrance slit (30 μm aperture) of a Czerny-Turner spectrometer. The spectrometer focal length was 0.32 m while its effective aperture was f/4.2. It was equipped with a 1200 g/mm holographic grating with dimensions 58 mm × 58 mm. The spectral width of the acquisition window was therefore about 25 nm. The spectrometer was coupled to an intensified diode array (IDA) with 1024 linear elements. A detector controller with a thermo-electric cooling unit was used to reduce the temperature of the detector to -20 °C. The detector was controlled via a programmable pulse delay generator triggered by the laser Q-switch. In these experiments, each acquisition corresponds to an accumulation over 100 laser shots. Calibration curves were established with an average over 4 acquisitions to take into account any lack of signal reproducibility. Each error bar represents the associated standard deviation.

Brine solutions were laboratory-made by dissolving calcium chloride ($CaCl_2$) and sodium chloride (NaCl) in a known volume of deionized water. Twenty four solutions of Na-Ca brine were prepared by varying the amount of NaCl salt dissolved in the solution to 0 M, 0.5 M, 2 M and 3 M, while the concentration of Ca range from 45 to 900 ppm. The results obtained with each solution were systematically compared. Spectra of the 422.67 nm calcium emission line, in the 415 – 430 nm range are shown in Figure 12.

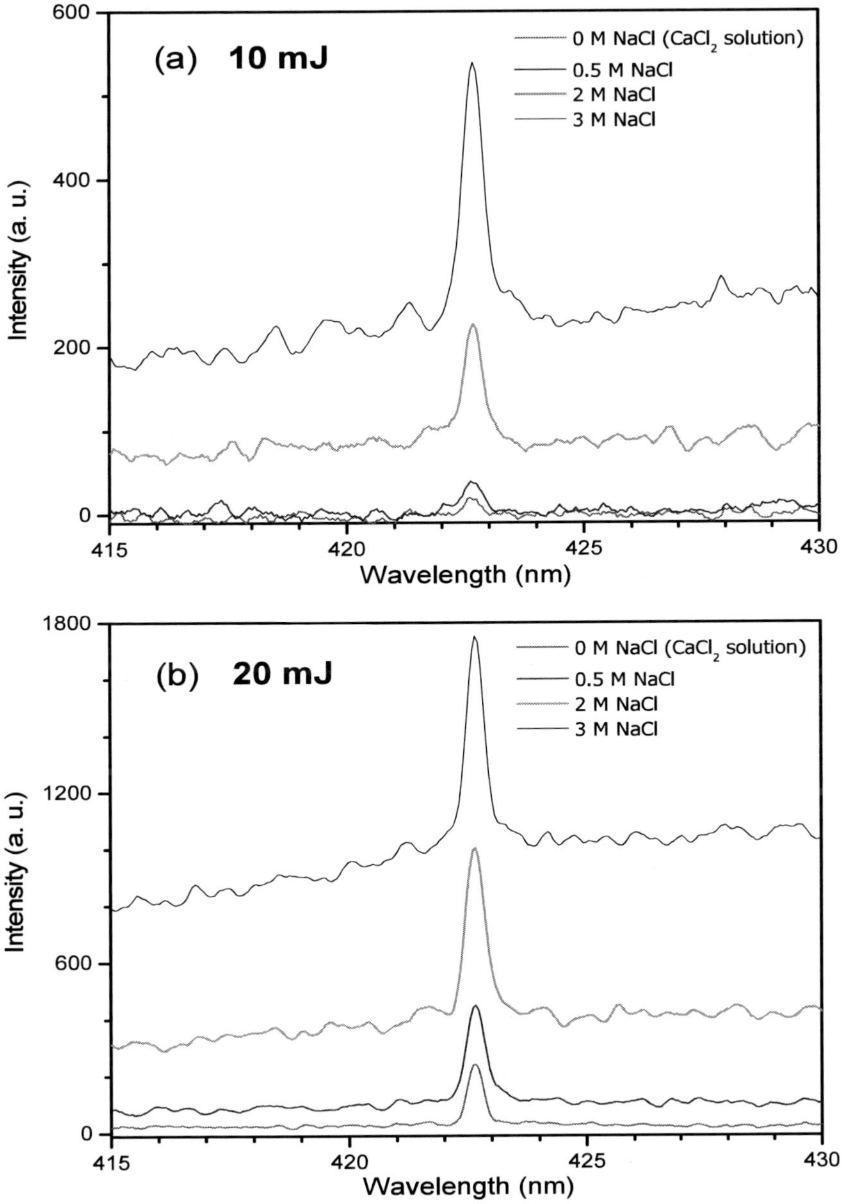

Figure 12. Emission spectra of the Ca I line at 422.67 nm observed at laser energy of (a) 10 mJ and (b) 20 mJ. Results are shown for solutions containing 0 M ($CaCl_2$ solution), 0.5 M, 2 M and 3 M NaCl. The time delay and gate width were 300 ns and 3 μs, respectively. Each solution contains 180 ppm of calcium [56].

Results are shown for laser energy levels of (a) 10 and (b) 20 mJ. For comparison, measurements were performed on Na-Ca brine containing 0 M (i.e., pure $CaCl_2$ solution), 0.5 M, 2 M, and 3 M NaCl. Each solution contains 180 ppm of calcium. Figure 12 shows that the spectral characteristics of Ca signal are strongly dependent on the amount of Na dissolved in the solution. At laser energy levels of 10 mJ (Figure 12a) the signal emitted by Ca was barely discernible from the background noise, when using 0 M NaCl solution ($CaCl_2$ solution). On the other hand the presence of Na causes an increase in both the intensity of the Ca line, and the continuum emission that is associated with electron-ion recombination and electron Bremsstrahlung emission. This likely indicates that the presence of Na increases the number density of free electrons, the efficiency of the laser energy absorption, and thus, the bremsstrahlung emission responsible for the continuum. In Figure 12a, one can also see that the spectrum intensity increases as the amount of dissolved Na increases. Indeed, compared to the $CaCl_2$ solution, the 0.5 M, 2 M, and 3 M NaCl mixtures yield an enhancement factor of 1.5, 6, and 11, respectively for the intensity of the Ca line and of 1, 88.95, and 239.08, respectively for the average background values (close to the Ca line). For the greater laser energy of 20 mJ (Figure 12b), similar behavior was observed except that the intensities of the Ca line and the continuum emission is higher than that observed for 10 mJ. As a result, the enhancement factor for the intensity of the Ca line and for the average background values decreased. Several explanations for these matrix effects have been suggested [48, 53, 54]. The increase of the observed LIBS signal, when increasing the amount of Na dissolved in the solution could be attributed to the increase in the number of free electrons and Na atoms in the focal volume of the laser beam (so that the absorption of incident photons by the plasma via Inverse Bremsstrahlung become more effective). These free electrons likely come from the ionization of Na atoms by thermal excitation, which produces an initial free electron density in the focal volume of the laser pulse. These "seed" electrons can subsequently gain sufficient kinetic energy from the laser pulse by inverse Bremsstrahlung absorption to produce a large amount of free electrons (the so-called avalanche ionization). Therefore the rapid ionization of the vapor/gas within the bubble leads to plasma formation. Some authors have also suggested that the higher plasma emission is likely due to easily ionized matrix species (Na has low ionization potential of 5.14 eV), which contribute to the first stages of the plasma formation [48, 53]. Eppler *et al.* [48] have investigated the matrix effects in sand/soil mixtures. They observed significant differences in the spectral intensity for trace elements at constant concentration in these materials. An explanation for the increase in the intensity of trace elements with sand percentage in the examined mixture was suggested. The authors attributed this matrix effect to the increase of the electron density, while the excitation temperature remains roughly constant. Another explanation proposed by Krasniker *et al.* [54], could be that sand particulates are responsible for a larger crater, which results in ablation of more material. The measurements were based on the amount of energy coupled into the plasma produced in the sand/soil mixtures by following the velocity of the shock waves. It's also worth mentioning that the authors in [48] have pointed out that higher electron density levels might also explain an increase in the population of neutral versus ionic species for trace elements. This is consistent with our observation in which, unlike $CaCl_2$ solutions, the Ca II lines at 390 and 393.40 nm were not observed for the mixtures solutions.

As illustrated in Figure 12a and 12b, the fact that these matrix effects seem to decrease with increasing laser energy might suggest that the plasma parameters (electron density and temperature) were less affected by changes in brine composition at higher energy levels. To correlate the observed signal enhancement with an increase of free electron in the plasma, we measured the electron density using Stark broadening of the Ca I line at 422.67 nm. This line has been extensively used in the literature for the determination of electron density in liquid plasmas [55]. The spectral lines were fitted using Lorentzian profiles. Instrumental broadening (0.32 nm at FWHM) was systematically subtracted from the observed line widths. Figure 13 shows the time-resolved evolution of the electron density for delays in the 250–850 ns range.

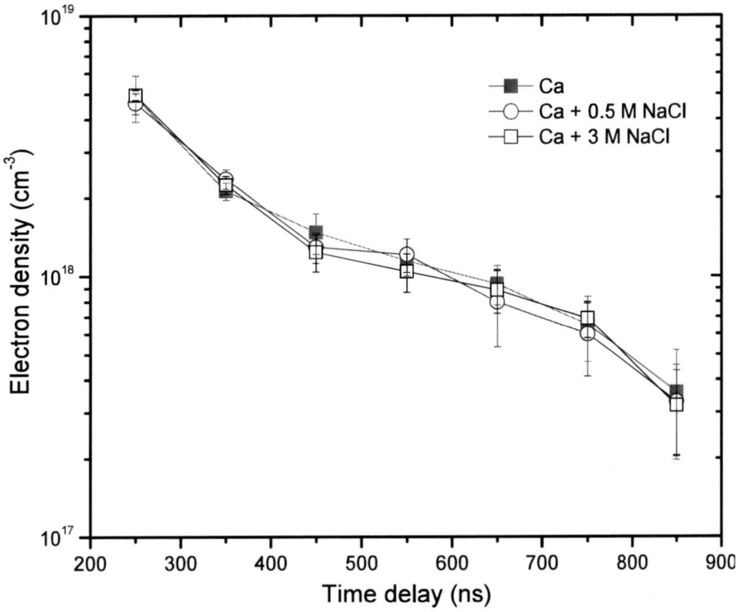

Figure 13. Time-resolved evolution of the electron density. Results are shown for solutions containing 0 M ($CaCl_2$ solution), 0.5 M and 3 M NaCl [56].

The laser energy was fixed at 20 mJ. Results obtained with the 0 M ($CaCl_2$ solution), 0.5 M and 3 M NaCl solutions are presented for comparison. The uncertainties of the electron density values were calculated by evaluating the error during the fitting procedure of the Ca line profile. One observes that Figure 13 exhibits no significant difference in electron density between the solutions. This result, which seems to be in contradiction with those of the previous paragraph, is probably due to larger uncertainties that arise from the Stark broadening method used to determine the electron density. A similar observation was made by Vrenegor *et al.* [50] in the analysis of high-alloy steel samples. In order to explain the origin of the matrix effects observed in steel samples, they measured the electron density and temperature in four samples with a high variation of the iron concentration, in the range of 43–99 wt.%. They found that despite a strong variation of the iron concentration, measurements of electron density and temperature showed no significant difference. These authors suggested that changes of the plasma state may occur but cannot be detected experimentally, which may explain the larger uncertainties in the determination of the plasma electron density and temperature. Figure 14 shows a comparison of calibration curves obtained for the Ca I line at 422.67 nm in Na-Ca brine.

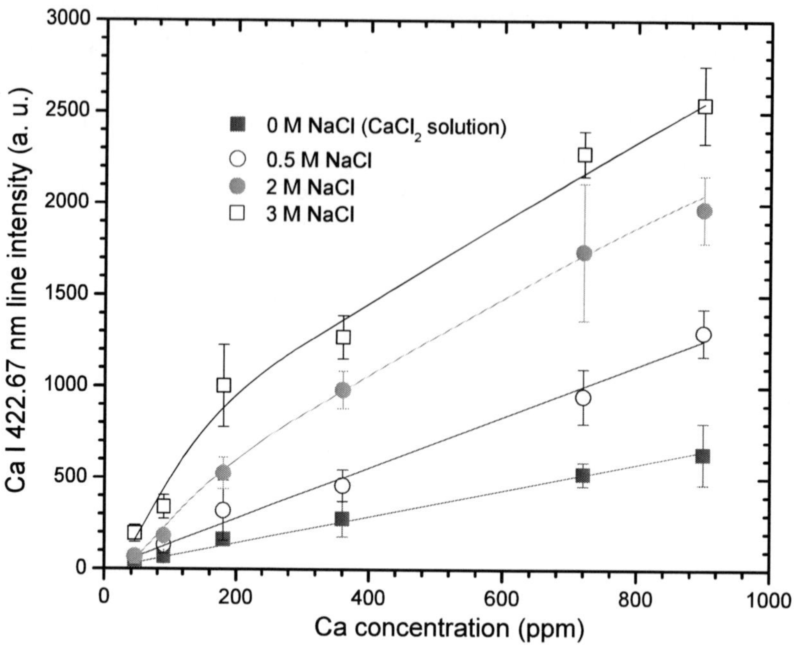

Figure 14. Calibration curves of the Ca I line at 422.67 nm obtained with solutions containing 0 M ($CaCl_2$ solution), 0.5 M, 2 M and 3 M NaCl. Laser energy was 20 mJ. Gate delay and gate width were 300 ns and 3 μs, respectively [56].

Results are shown for Na-Ca brine solutions containing 0 M, 0.5 M, 2 M and 3 M NaCl. The calibration curves shown in Figure 14 were established under experimental conditions of 20 mJ laser energy, for time delay and gate width of 300 ns and 3 μs, respectively. It can be observed in Figure 14 that as the NaCl concentration increases from 0 M to 3 M, a gradual increase of the Ca line intensity is observed over the whole range of calcium concentrations considered. Also, solutions with low NaCl concentration (0 and 0.5 M) exhibit a linear calibration curve, while solutions with high NaCl concentration (2 and 3 M) display a linear behavior for low calcium concentrations (up to about 180 ppm), and departs from linearity for higher concentrations due to self-absorption. Comparing the slope of these calibration curves (the linear part in case of solutions with high NaCl concentration), a greater slope is observed with increasing NaCl concentration. Some papers have shown similar results in different samples and matrices. For example, Chaléard *et al.* [51] observed a change in slopes of the calibration curves of copper in brass compared to aluminum and steel samples. By normalizing the intensity of the Cu line with the plasma acoustic signal, (which is proportional to the vaporized mass and the excitation temperature) the authors demonstrated an efficient correction of the observed matrix effects.

It is clear from this chapter that Na (in the 0–3 M NaCl range) plays a significant role in Ca signal emission and calibration curves. Enhancement of the Ca signal with increasing Na concentration is interesting.. The increase in continuum emission was also observed as Na concentration increased, which might be due to an increase in the number density of free electrons. Comparison of the calibration curves obtained with Na-Ca brine solutions containing 0 M, 0.5 M, 2 M and 3 M NaCl showed strong dependence of the slope of the curves on the amount of Na dissolved. For the lowest Ca concentrations, a systematic increase of the slope (the sensitivity) with increasing Na concentration was observed. For higher Ca concentrations, calibration curves obtained with 2 and 3 M NaCl solutions exhibited self-absorption; while those obtained with solutions containing lower NaCl concentrations (0 and 0.5 M) remain linear. Although a better understanding of these matrix effects still remain, these results allow us to conclude that LIBS analysis in brines is strongly dependent of changes in sodium concentration, and appears less affected as laser energy increases. Further experiments would be needed to fully understand the difference of spectral intensity observed between brine samples at different chemical composition and at temperature and pressure representative of the GCS condition.

5. New Directions

5.1. Isotopic Analysis

When suspected leakage from a CCS site is identified, measuring the carbon isotope ratios allows an investigator to determine the origin of the carbon. Natural deposits of CO_2 have significantly different carbon isotope ratios when compared to CO_2 that has been captured from a power plant to be injected underground [11]. Also depending on the feedstock to a particular power plant, the CO_2 produced may have an identifiable difference in its carbon isotope ratio as compared to that of another source. Therefore, a need arises to quickly and accurately measure key isotope ratios to aid in the identification of leakage sources.

Many conventional isotope measurement techniques are commercially available however they are typically lab based, not easily portable, and/or not suitable for the outdoors or harsh environments. This chapter highlights recent research efforts which show that LIBS is a feasible measurement technique for carbon sequestration leakage detection applications. With the introduction of new technologies, such as laser ablation molecular isotopic spectrometry (LAMIS) which makes LIBS based isotope measurement easier and more cost effective. The application of newly available miniaturization technologies for both lasers and spectrometers will make highly portable LIBS analysis system a reality in the near future.

The measurement of elemental isotopes via atomic emission relies on the spectral line discrimination of approximately one picometer, or less. The recent development of a technique that looks at the light emitted from the molecular recombination of atomic isotopes has managed to reduce the detection resolving requirements by several orders of magnitude [57-59]. This LAMIS technique measures the extreme temporal end of the laser induced plasma where the excited species are beginning to recombine. As the plasma continues to shrink and cool after the initial discharge the excited atoms and their isotopes begin to form oxides and nitrides. These recombination events are continuing to release spectral energy from molecular band structures. The addition of the mass during recombination produces isotopic spectral peak separation on the order of hundreds of picometers. The simple addition of like masses to the isotopes causes a vastly different degree of vibration and subsequent spectral emission which allows for a much easier detection. This technique has not yet been adapted for use with carbon sequestration monitoring however there is a great need for isotopic sensing as explained above. This technology is a great example of new techniques that are being introduced to simplify the carbon sequestration monitoring and make if far more cost effective.

5.2. Downhole Measurement

All of the measurement arrangements discussed thus far have been single point acquisitions and analysis. A much more useful approach to larger scale detection and measurement would be to connect the measurements from a wide array of locations and measurement modalities across the same injection basin. Rather than deploying a complete LIBS analysis system at each location, researchers could deploy a centrally located computer/controller and then connect it to a number of remotely energized laser spark sources which could be placed in the field, atmosphere, submerged in surface water systems, or down hole. The centrally located controller, shown in Figure 15, would then energize each sensor independently inducing sparks and either receive sample spectra to be analyzed with an on-board high resolution spectrometer or receive data from miniaturized spectrometers deployed with the spark generator.

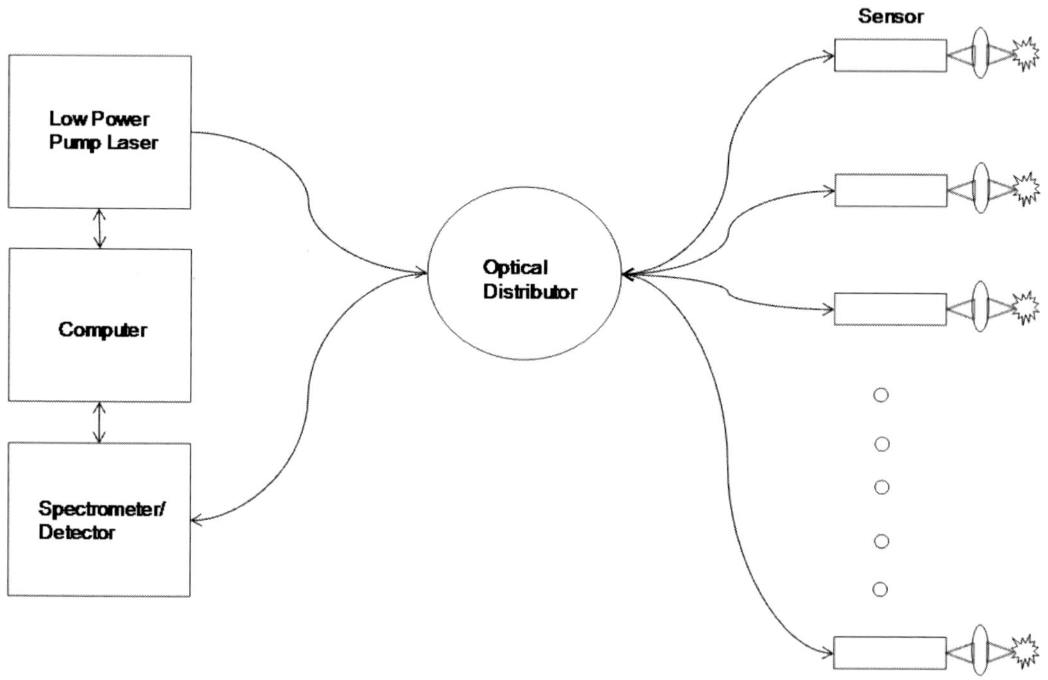

Figure 15. Optically distributed LIBS sensor network.

The centrally located controller would address and activate each sensor through an optical multiplexer and/or electrical distributor which would allow for the consolidation of the most expensive and/or sensitive pieces of equipment to be utilized for every sensor when activated. Recent advances in the areas of diode end pumped solid state passively Q-switched laser systems and miniaturized spectrometers show that the technologies are moving in the direction of smaller more compact and less costly equipment.

REFERENCES

[1] S. Solomon, D. Qin, M. Manning, Z. Chen, M. Marquis, K. B. Averyt, M. Tignor, H.L. Miller (eds.). *"Contribution of Working Group I to the Fourth Assessment Report of the Intergovernmental Panel on Climate Change,"* Cambridge University Press, Cambridge, United Kingdom and New York, NY, USA, Page 996 (2007).

[2] 2010 Carbon Sequestration Atlas of the United States and Canada – Third Edition (Atlas III) http://www.netl.doe.gov/technologies/carbon_seq/refshelf/atlasIII/

[3] P.L. Bondor, "Applications of carbon dioxide in enhanced oil recovery," Energy Conversion and Management, *Proceedings of the First International Conference on Carbon Dioxide Removal,* Vol. 33, Issues 5-8, May-August, Pages 579-586 (1992).

[4] Q.M. Malik, M.R. Islam, *"CO_2 Injection in the Weyburn Field of Canada: Optimization Of Enhanced Oil Recovery and Greenhouse Gas Storage With Horizontal Wells,"* SPE Paper 59327-MS, SPE/DOE Improved Oil Recovery Symposium, April 3-5, 2000, Tulsa, OK, DOI: 10.2118/59327-MS.

[5] R. Lal, *Forest Ecology and Management,* 220 (2005) 242-258.

[6] R. Lal, *Soil Tillage Res.* 43 (1997) 81–107.

[7] R. Lal, *Prog. Environ. Sci.* 1 (1999) 307–326.

[8] C. D. Farrar, M. L. Sorey, W. C. Evans, J. F. Howle, B. D. Kerr, B. M. Kennedy, C. Y. King, J. R. Southon, *Nature* 376 (1995) 675–678.

[9] K. A. McGee, T. M. Gerlach, *Geology* 26 (1998) 463-466,

[10] M. Z. Martin, N. Labbe, N. Andre, R. Harris, M. Ebinger, S. D. Wullschleger and A. A. Vass, *Spectrochim. Acta* Part B. 62 (2007) 1426-1432.

[11] G. Johnson, M. Raistrick, B. Mayer, M. Shevalier, S. Taylor, M. Nightingale, I. Hutcheon, *Energy Procedia* 1 (2009) 2315-2322.

[12] R. J. Gehl, C. W. Rice, *Clim. Change 80* (2007) 43–54.

[13] Z. Sulcek, P. Povondra, *Methods of Decomposition in Inorganic Analysis,* CRC Press, Boca Raton (1989).

[14] D. A. Cremers, L. J. Radziemski, *Handbook of laser-induced breakdown spectroscopy,* Wiley, New York, USA (2006).

[15] A. Miziolek, V. Palleschi, I. Schechter, Laser-Induced Breakdown Spectroscopy (LIBS): *Fundamentals and Applications,* Cambridge University Press (2006).

[16] J. P. Singh, S. N. Thakur, *Laser-Induced Breakdown Spectroscopy,* Elsevier Science B. V., Amsterdam, The Netherlands (2007).

[17] M. Holzinger, J. Maier, W. Sitte, *Solid State Ionics,* 94 (1997) 217-225.

[18] A.J. Ball, V. Hohreiter, D.W. Hahn, *Appl. Spectrosc.* 59 (2005) 348-353.

[19] M. Tran, B.W. Smith, D.W. Hahn, J.D. Winefordner, *Appl. Spectrosc.* 55 (2001) 1455-1461.

[20] E.D. McNaghten, A.M. Parkes, B.C. Griffiths, A. I. Whitehouse, S. Palanco, *Spectrochim. Acta* Part B. 64 (2009) 1111-1118.

[21] K.E. Eseller, F.Y. Yueh, J.P. Singh, *Appl. Phys* B 102 (2011) 963–969.

[22] F. Ferioli, P.V. Puzinauskas, S.G. Buckley, *Appl. Spectrosc.* 57 (2011) 1183-1189.

[23] V. Hohreiter, D.W. Hahn, *Anal. Chem* 77 (2005) 1118–1124.

[24] V. Dikshit, F.Y. Yueh, J.P. Singh, D. L. McIntyre, J.C. Jain, N. Melikechi, *Spectrochim. Acta* B 68 (2012) 65-70.

[25] D.A. Cremers, M.H. Ebinger, P.J. Unkefer, S.A. Kannerdiener, M.J. Ferris, K.M. Catlett, J.R. Brown, *J. Environ. Qual.* 30 (2001) 2202–2206.
[26] M.Z. Martin, S.D. Wullschleger, C.T. Garten, V.A. Palumbo, *Appl. Opt*. 42 (2003) 2072-2077.
[27] K. W. Thoning, P.P. Trans, W.D. Komhyr, *J. Geophys. Res.* 94 (1989) 8549-8565.
[28] R. M. Da Silva, D. M. B. P. Milori, E. C. Ferreira, E. J. Ferreira, F. J. Krug, L. Martin Neto, *Spec. Chem. Acta* Part B 63 (2008) 1221–1224.
[29] S. Pandhija, N. K. Rai, A. K. Rai, S. N. Thakur, *Appl. Phys.* B 98 (2010) 231–241.
[30] D. A. Cremers, M. H. Ebinger, D. D. Breshears, P. J. Unkefer, S. S. Kammerdiener, M. J. Ferris, K. M. Catlett, J. R. Brown, *J. Environ. Qual.* 30 (2001) 2202–2206.
[31] M. Z. Martin, N. Labbe, N. O. Andre, S. D. Wullschleger, R.H. Harris and M.H. Ebinger, *Soil Sci. Soc. of America J.* 74 (2010) 87-93.
[32] M. V. Belkov, V. S. Burakov, A. De Giacomo, V. V. Kiris, S. N. Raikov, N. V. Tarasenko, *Spectrochim. Acta* Part B. 64 (2009) 899–904.
[33] D. R. Hirmas, C. Amrhein, R. C. Graham, *Geoderma.* 154 (2010) 486–494.
[34] Y. Kuzyakov, *Soil Biology & Biochem.* 38 (2006) 425–448.
[35] M. Corsi, G. Cristoforetti, M. Hidalgo, S. Legnaiola, V. Palleschi, A. Salvetti, E. Tognoni, C. Vallebona, *Applied Geochemistry* 21 (2006) 748-755.
[36] K.K. Ayyalasomayajula, F.Y. Yueh, J.P. Singh, D.L. McIntyre*, J. Jain, Applied Optics,* 51 (2012) B149-B154.
[37] B. Lal, H. Zheng, F. Y. Yueh, J. P. Singh, *Appl. Opt*. 43 (2004) 2792–2797.
[38] S. Weisberg, *Applied Linear Regression*, 3rd ed. (John Wiley & Sons, 2005)
[39] K. H. Esbensen, Multivariate Data Analysis in Practice, 5th ed. (Camo Inc., 2004)
[40] K. K. Ayyalasomayajula, V. Dikshit, F. Y. Yueh, J. P. Singh, L. T. Smith, *Anal Bioanal. Chem.* 10 (2011) 3315−3322.
[41] Q. L. Zhang, J. P. Zhang, L. Zhang, *Chinese Sci. Bull.* 55 (2010) 386−389.
[42] B. Metz, O. Davidson, H. D. Coninck, M. Loos, L. Meyer, *Special Report on Carbon Dioxide Capture and Storage;* IPCC: Cambridge, England (2005).
[43] S. Bachu, *Energy Convers. Manage* 41 (2000) 953−970.
[44] *Monitoring, verification, and accounting of CO_2 stored in deep geologic formations,* U.S. Department of Energy, DOE/NETL-311/081508 (2009).
[45] D. W. Hahn, N. Omenetto, Appl. Spectrosc. 66 (2012) 347–419.
[46] Y. K. Kharaka, J. S. Hanor, Deep fluids in the continents: I. sedimentary basins. In *Treatise on Geochemistry,* J. I. Drever, Elsevier: Oxford, 5, 1−48 (2007).
[47] M. Lawrence-Snyder, J. Scaffidi, S.M. Angel, A.P.M. Michel, A.D. Chave, *Appl. Spectrosc.* 61 (2007) 171-176.
[48] A. S. Eppler, D. A. Cremers, D. D. Hickmott, M. J. Ferris, A. C. Koskelo, *Appl. Spectrosc., 50* (1996) 1175– 1181.
[49] J.A. Aguilera, C. Aragón, V. Madurga, J. Manrique, *Spectrochim. Acta* Part B 64 (2009) 993–998.
[50] J. Vrenegor, R. Noll, V. Sturm, *Spectrochim. Acta* Part B, 60 (2005) 1083–1091.
[51] C. Chaléard, P. Mauchien, N. Andre, J. Uebbing, J.L. Lacour, C. Geertsen, *J. Anal. At. Spectrom.* 12 (1997) 183–188.
[52] J.-S. Huang, C.-B. Ke, K.-C. Lin, *Spectrochim. Acta* Part B 59 (2004) 321–326.
[53] V. Bulatov, R. Krasniker, I. Schechter, *Anal. Chem.* 70 (1998) 5302–11.
[54] R. Krasniker, V. Bulatov, I. Schechter, *Spectrochim. Acta* Part B 56 (2001) 609–618.

[55] C. Aragón, J. A. Aguilera, *Spectrochim. Acta* Part B 63 (2008) 893–916.
[56] C. Goueguel, J. P. Singh, D. L. McIntyre, J. Jain, A. K. Karamalidis, *Appl. Spectro.* Submitted (2013).
[57] R.E. Russo, A.A. Bol'shakov, X. Mao, C.P. McKay, D.L. Perry, O. Sorkhabi, *Spectrochim. Acta* Part B 66 (2011) 99–104.
[58] X. Mao, A. Bol'shakov, D.L. Perry, O. Sorkhabi, R.E. Russo, *Spectrochimica Acta* Part B 66 (2011) 604-609.
[59] X. Mao, A.A. Bol'shakov, I. Choi, C.P. McKay, D.L. Perry, O. Sorkhabi, R.E. Russo, *Spectrochim. Acta* Part B 66 (2011) 767-775.

In: Spectroscopic Techniques for Security, Forensic ... ISBN: 978-1-63117-404-9
Editors: Y. Dwivedi, S. B. Rai and J. P. Singh

Chapter 3

LASER SPECTROSCOPY FOR NUCLEAR APPLICATIONS

M. L. Shah, A. K. Pulhani, G. P. Gupta and B. M. Suri*
Laser and Plasma Technology Division, Bhabha Atomic Research Centre, Trombay, Mumbai, India

ABSTRACT

Laser spectroscopy plays a vital role for applications related to nuclear science and technology. Here, we have reviewed our work on development of some of these techniques including laser-induced fluorescence, resonance ionization and laser-induced breakdown spectroscopy, keeping some of these applications in mind. The development of these techniques provided extensive data on different species of interest to nuclear science and technology. At the same time the techniques developed here have universal applicability wherever complex spectra are involved. The techniques for measurement of branching fractions and absolute transition probabilities of ground state to excited state and excited state to excited state are much simpler and yet quite accurate. The accurate quantitative analysis is possible using the calibration free laser-induced breakdown spectroscopy technique by choosing appropriate experimental conditions, which can enable online elemental composition analysis for the materials of nuclear interest.

1. INTRODUCTION

Nuclear technology, right from its inception has been intimately linked with high purity of materials (aimed at neutron poisons), study of isotopes and isotope effects and production and detection of isotopes of interest to nuclear industry. Each of these is enabled by classical spectroscopic techniques. Analytical spectroscopic techniques have been quite useful for facilitating production and characterization of high purity materials. Optical isotope shift and hyperfine structure studies have given a major boost to generating database on nuclear charge radii and nuclear magnetic moments and electric quadrupole moments of chain of stable and

* Email: profbmsuri@gmail.com.

unstable isotopes. With the development of tunable narrow linewidth dye lasers, spectroscopic techniques witnessed a renaissance. There was improvement of resolution and detection sensitivity by several orders of magnitude. New range of spectroscopic techniques appeared on the scene, besides the classical absorption and emission spectroscopy. Development of high power dye lasers gave rise to possibility of altering atomic (or molecular) population structure in a drastic manner. This resulted in techniques to purify or separate different components from a mixture using laser spectroscopy, evolving a major new technique of laser isotope separation. As is well known, in nature, uranium fissile isotope exists with abundance unfavorable to production of nuclear power. The fissile isotope U-235 is only 0.7 % abundant and needs to be enriched to at least 3 % for it to be employed in nuclear power reactor.

It is often required to detect radio-isotopes (fissile or otherwise) with very high sensitivity and isotope selectivity for reasons of environment surveillance or otherwise. Alternately, there is need to detect many of these radioactive species in complex matrices like glass or ceramics coming from a radioactive waste vitrification plant – preferably from a remote location for sake of minimum dose to operating personnel. Normally many of these isotopes of interest to nuclear industry are detected by radiometric techniques. But there are various situations where these techniques have limitations, either in terms of sensitivity or isotope selectivity or the possibility of remote detection. It is here that many laser spectroscopic techniques offer themselves to be a suitable candidate for such applications.

Most of the techniques meant for sensitive and selective detection of species or laser isotope separation are centered around multi-step excitation and photoionization of atomic vapor. Most often the species of interest are lanthanides and actinides. The processes being spectroscopic in nature, there is need for characterizing these multi-step photoexcitation/photoionization schemes and thus measurement of several atomic parameters accurately. We have developed in our laboratory the techniques of multi-step Laser Induced Fluorescence (LIF), Resonance Ionisation Spectroscopy (RIS) and also simultaneous LIF and RIS techniques in atomic beams. These were aimed at understanding photoexcitation/ photoionization dynamics and search of highly excited energy levels which could yield efficient and isotope selective photoionization. We also report in this chapter development of new techniques for measurement of radiative lifetimes, branching ratios, photoexcitation and photoionization cross-sections. This has not only yielded large new database but has also shown possibilities of its applications in general.

We have also been working on developing techniques on characterizing multi-component systems like alloys, glasses and ceramics. In this regard, we have developed both calibration based and calibration free Laser Induced Breakdown Spectroscopy (LIBS) and employed them to investigate samples like brass, stainless steel, nickel based alloys, and uranium vitrified glasses. This was aimed at both qualitative and quantitative detection of species of interest in these matrices.

In this chapter, we have reviewed some of our work related to the development of laser spectroscopic techniques like laser-induced fluorescence, resonance ionization and laser-induced breakdown spectroscopy, both in gaseous phase as well as in condensed phase. Besides the development of these techniques, extensive data were generated on different species of interest to nuclear science and technology.

2. Laser-Induced Fluorescence

Most of spectroscopic information on elements particularly lanthanides and actinides had been obtained using classical emission spectroscopy of hollow cathode discharges and absorption spectroscopy in furnaces [1]. However, there are limitations of these classical techniques for exploring energy range or parity of excited levels. These limitations are overcome by use of laser spectroscopic techniques, such as RIS and LIF. In RIS technique, atomic levels in the energy region of interest are pumped by multi-step excitation and detected by further photoionization. In LIF spectroscopy the fluorescence is collected from the high-lying atomic levels populated by multi-step excitation. Both the RIS and LIF techniques have been carried out for different elements by various authors [2-9]. By carefully reviewing the literature, it has been noted that RIS technique was considered as most sensitive technique for identification of energy levels and thus the experimentalists developed and used it more extensively than LIF. Moreover also the LIF detection is difficult because fluorescence photons per decay channel were less as the density of energy levels to which excited atoms can decay are much more. A case in illustration is of uranium, where Miron *et al.* [10] have carried out energy level search in the energy region 32267.65 cm $^{-1}$ to 34803.73 cm $^{-1}$, using two-color laser-induced fluorescence technique. Several researchers [2-5] have also explored the same energy region of atomic uranium using the technique of two-color three-photon photo ionization. Comparison of these works in atomic uranium, it is noted, that many energy levels observed by one technique are not observed by the other and vice versa, which otherwise should be detectable from spectroscopic considerations.

2.1. Identification of Energy Levels

To ascertain this point, and for sake of laser spectroscopy of atomic beams, LIF spectrometer was developed indigenously in our laboratory. We could observe several new high-lying even-parity energy levels of atomic samarium, which could not be observed by more sensitive technique of RIS. The experiment and the results are published in detail in our research paper [11]. It has been confirmed that two-color laser induced fluorescence and two-color photoionization are complementary in nature and, therefore, to have a complete knowledge of energy position of atomic resonances, both the techniques have to be employed. Such studies can be extended for other elements and the experimental data may be useful for identifying optimal ionization schemes for efficient and isotope selective laser excitation. This observation was explained physically as excitation strength of a two-color LIF feature is directly related to $(\sigma_1(\lambda_1)\ P_1)\ (\sigma_2(\lambda_2)\ P_2)$, while ionization signal in case of two-color three-photon ionization is directly related to $(\sigma_1(\lambda_1)\ P_1)\ (\sigma_2(\lambda_2)\ P_2)\ (\sigma_3(\lambda_2)\ P_2)$, when the powers of the corresponding lasers are in the non-saturated regime of excitation. In these expressions, σ_1 and σ_2 are excitation cross-sections of first-step and second-step transitions at λ_1 and λ_2 respectively, $\sigma_3\ (\lambda_2)$ is the cross-section for the photoionization from the second excited state at λ_2 and P_1 and P_2 are power densities of dye lasers. In case of photoionization, it is quite possible that due to rich autoionization spectrum, which is in general true for lanthanides and actinides, $\sigma_3\ (\lambda_2)$ could be very high in a coincidental way and raise the detection sensitivity for some of the second-step resonances having weak σ_2, making them observable in a two-color three-photon ionization experiment, but could not be observed in two-color laser induced fluorescence spectrum because of very little fluorescence yield at these excitation wavelengths. Moreover in LIF experiment, the upper state may decay by infrared (IR) cascades or via ultraviolet transition and avoid detection. Hence, the resonances, which could be observed in two-color three-photon ionization and not observed in the present LIF experiment, fall in this category. These resonances are either fairly weak in σ_2 values or produce weak fluorescence in the detection region (decay by other channels) and/ or have relatively long lifetimes (low amplitudes). The other category of resonances, where σ_2 is very large, is easily detectable in present two-color LIF experiment but their non-observance in photoionization experiment can be attributed to the absence of autoionization resonance in the continuum at those wavelengths. We believe these could also become observable in photoionization experiment if substantially much larger P_2 is employed. This indicate the fact that, in order to have a complete knowledge of atomic resonances, both the techniques have to be employed, thus confirming the complementary nature of two-color LIF and two-color three-photon ionization techniques in normal circumstances.

2.2. Measurements of Radiative Lifetimes, Branching Fractions and Absolute Transition Probabilities

Radiative lifetimes, branching fractions and absolute transition probabilities are required in many areas of science such as atomic physics, plasma physics, laser chemistry, atmospheric sciences and are also of importance in the determination of solar and stellar abundances of elements and in the search for new laser materials. These parameters also play a crucial role in laser isotope separation.

Commonly employed methods for determining the absolute transition probability include measurements of line strengths from absorption and emission spectra, measurement of saturation energy, measurement of Rabi frequency and combined measurements of lifetimes and branching fractions. The difficulty in measuring the absolute transition probabilities employing most of the conventional techniques is due to either the uncertainties in measurement of absolute number densities of the interacting atoms or the intensity of laser photons used for excitation. The potentials and limitations of different methods are discussed in a review article by Petit *et al.* [12]. Laser-based methods for measurement of absolute transition probability use either laser-induced fluorescence (LIF) or laser-induced photoionization (LIP). Carlson *et al.* [13] have described a branching-fraction-lifetime (β - τ) method for determining absolute transition probabilities by independently measuring branching fractions (β) and radiative lifetimes (τ) using time-resolved optical pumping, probed by time-resolved laser-induced photoionization (LIP). This method has a distinctive advantage of being independent to laser beam quality, such as temporal profile, intensity and linewidth, unlike most of the other methods. They have used this method to obtain accurate values of transition probability for atomic transitions in uranium. Hackel and Rushford [14] have refined this method and measured the branching fraction and the accurate absolute transition probability in atomic uranium. In the above works the β-τ method is applied for optical transitions between excited and ground states.

Measurement of absolute transition probability for atomic transitions between excited-level-to-excited-level is complicated. The β-τ method using time-resolved LIP technique was used by Bisson *et al.* [15] for measurement of absolute transition probabilities of excited-level-to-excited-level optical transitions in atomic cerium and gadolinium. They have discussed advantages and limitations of their method. Petit *et al.* [12] and Avril *et al.* [16] have used LIP to measure the lifetimes and LIF to measure the branching fractions for determining the absolute transition probabilities of uranium atomic transitions originating from the 34659 cm^{-1} excited energy level. Radiative lifetime measurements are commonly conducted by monitoring time-resolved LIF [10, 17-23] spectrum from an atomic beam with a fast detector like photomultiplier tube (PMT) and branching fraction measurements by either monitoring the LIF spectrum in an atomic beam or the emission spectrum of a hollow cathode discharge lamp with a Fourier-transform spectrometer (FTS) [12]. Komarovskii and Smirnov [24] have used electron beam for excitation and measured the branching fractions of different decay channels in atomic samarium. Using these measurements and lifetimes reported earlier [17, 18], they have determined absolute transition probabilities for excited-state-to-ground-state Sm I atomic transitions in the wavelength range 350-650 nm.

We developed and reported an experimental method [25] for the measurements of radiative lifetimes, branching fractions and absolute transition probabilities of excited-level-to-ground-level and excited-level-to-excited-level transitions in atomic Sm by carrying out simultaneous measurements of radiative lifetimes and branching fractions based on spectrally and temporally resolved LIF using intensified charge coupled device (ICCD) in conjunction with a monochromator. Availability of tunable narrowband lasers in combination with advanced sensitive detection systems like ICCD provided the possibility to detect weak atomic fluorescence spread over wide spectral regions. This method is intrinsic in nature and the value of transition probability does not depend on accurate measurement of number density of atoms and photons but on the measurement of lifetimes of the excited levels and the branching fractions. The accuracy of the measurement of atomic transition probability using this method depends on how accurately the lifetime and branching fraction measurements are done. For detection of temporally and spectrally resolved laser-induced fluorescence, we have used ICCD camera of high temporal resolution (< 1 ns) and moderate gains ($>10^4$). ICCD camera when coupled with a monochromator has the capability of simultaneous temporal and spectral measurements. The experimental setup is shown in Figure 1.

For measurement of excited-level-to-ground-level transition probabilities, radiative lifetimes and branching fractions of low-lying odd-parity energy levels were measured using single-colour laser excitation. Figure 2 shows a typical time-resolved single-color LIF decay signal from the 18225.13 cm^{-1} energy level with an exponential fit. Laser was scanned in the wavelength region 556 – 572 nm to resonantly excite Sm atoms to the excited odd-parity energy levels in the energy region 17769.71-19990.25 cm^{-1}. We have confirmed that the right energy level is populated by observing the known decay channels from that level. Table 1 shows the measured radiative lifetimes of odd-parity energy levels in atomic samarium.

The absolute transition probabilities of 21 excited-level-to-ground- level transitions of Sm I were determined from the measured lifetimes of the first-excited level and the measured branching fractions of the fluorescence decay channels terminating on one of the levels of the $4f^6 6s^2$ ^7F ground septet. Table 2 shows the measured branching fractions and absolute transition probabilities for optical transitions from excited level to ground level. Figure 2 shows the different decay channels from the odd-parity energy level 18475.3 cm^{-1}.

Reasonably good agreement is observed between our measured values and the earlier reported values for these transitions [25].

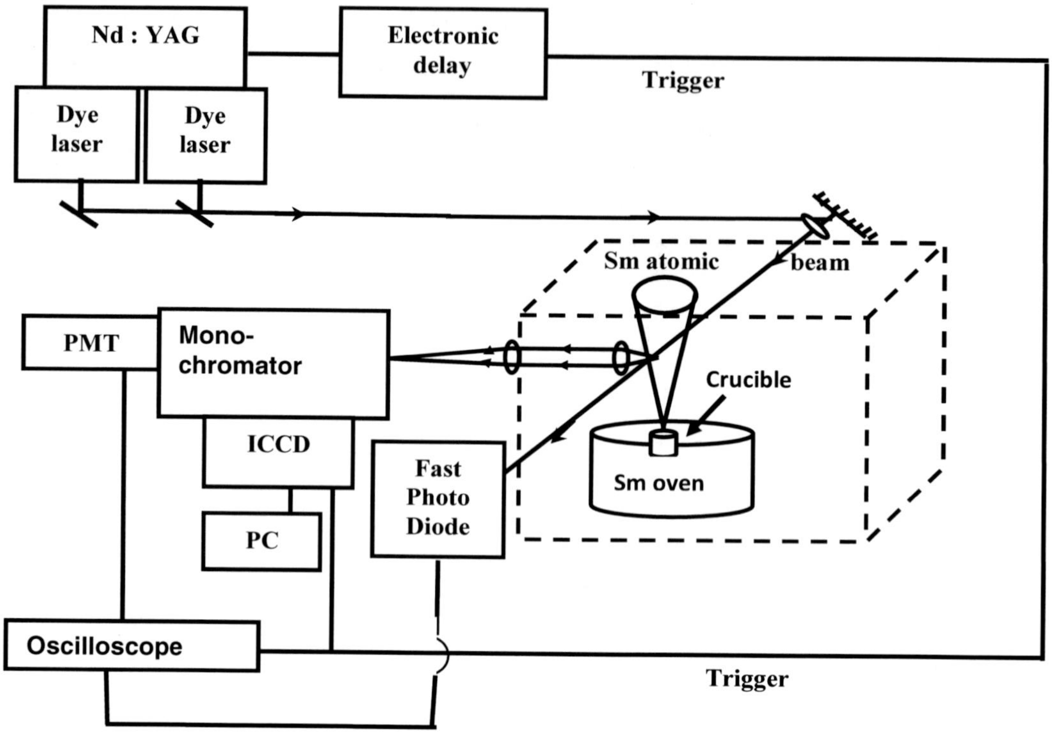

Figure 1. Schematic diagram showing the experimental setup for laser-induced fluorescence.

Table 1. Measured radiative lifetimes of odd-parity energy levels in atomic samarium using single-colour laser-induced fluorescence

Energy level (cm^{-1})	Excitation wavelength (nm)	Fluorescence wavelength (nm)	Lifetime (ns)	
			This work	Reported value*
17769.71	562.60	562.60	136 ± 11	157 ± 5
18075.67	562.18	562.18	471 ± 49	450 ± 50
18225.13	557.49	548.54	143 ± 10	146 ± 6
18475.28	565.99	549.82	63 ± 5	71 ± 2
18788.08	556.14	540.52	121 ± 18	128 ± 5
19009.52	570. 62	549.37	46 ± 3	51 ± 2
19990.25	564.27	540.37	259 ± 38	259 ± 10

*Ref. [18]

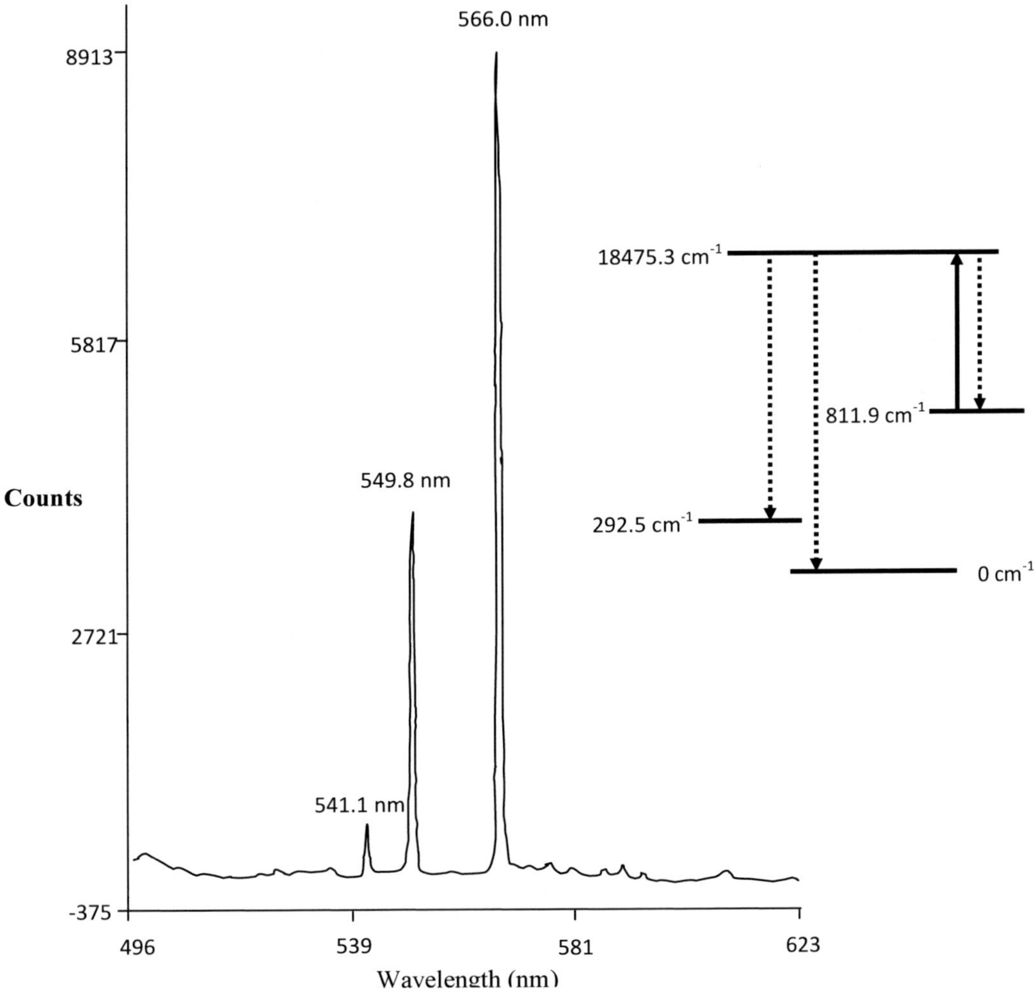

Figure 2. Spectrally resolved LIF showing three different decay channels form the excited level at 18475.28 cm^{-1}. In the inset, the solid line represents laser-induced excitation and the dotted lines represent the observed fluorescence.

For measurement of excited-state-to-excited-state transition probabilities, two-color, two-step resonant excitation have been used. For excited-level-to-excited-level transition probabilities measurements, the first-step excitation laser was tuned to 591.6 nm to excite the samarium atoms from 293 cm^{-1} to 17190 cm^{-1} and the second-step excitation laser was tuned from 558 – 568 nm to further excite Sm atoms from the first excited level to different second-excited high-lying even parity energy levels. One of the resonant or non-resonant decay channels was chosen to measure the lifetime of the excited state. The fluorescence counts were recorded as a function of delay between the trigger pulse of ICCD and the opening of ICCD shutter for the exposure. Exposure time and delay of gate pulse with respect to trigger can be varied from 1 ns – 80 sec and 0 ns – 80 sec, respectively with minimum step of 0.1 ns. For the lifetime measurements, delay between the trigger pulse of ICCD and the opening of ICCD shutter for the exposure was varied till signal falls to just above the noise level. Background of ICCD camera was suitably subtracted from the actual fluorescence counts.

The plot of the corrected fluorescence counts versus this delay was fitted exponentially to get the radiative lifetime of the excited level. Figure 3 and 4 depicts a typical time-resolved single-color and two-color LIF signals from the energy levels 18225.13 and 34814.4 cm^{-1} respectively with an exponential fit. The ICCD gain is set such that the maximum intensity during predetermined gate exposure time after excitation laser pulse does not saturate the ICCD camera. We have used the exposure time of 10 ns. With these conditions we could get counts corresponding to an average of 25 laser pulses per data point.

Experiment was repeated many times for each excited level of interest under different experimental conditions of laser pulse energy, atomic beam number density, etc. to overcome any systematic error. The average of these measurements was adopted as the final radiative lifetime of the excited upper level. The statistical uncertainty in the lifetime values is estimated from the spread in the fitted lifetime values. Radiative lifetime values were also checked with other decay channels but the values remained same within the experimental errors. The measured radiative lifetimes of the six stepwise-excited even-parity energy levels along with the excitation and fluorescence wavelengths are shown in Table 3.

Table 2. Measured branching fractions of different decay channels from odd-parity energy levels to lower even-parity energy levels of the ground septet and determined absolute atomic transition probabilities of Sm I using single-colour LIF

Upper level k (cm^{-1})	Lower level i (cm^{-1})	Excitation Wavelength (nm)	Fluorescence Wavelength (nm)	Intensity (arb. units)	Branching fraction	Transition probability A_{ki} (10^6 s^{-1})	
						This work	Reported value*
17769.71	0.0	572.02	562.59	3284	0.755	5.6	3.9
	292.58		572.02	632	0.146	1.1	1.2
	811.92		589.54	431	0.099	0.7	0.4
18075.67	292.58	562.17	562.17	2799	0.813	1.7	1.7
	811.92		579.08	236	0.069	0.1	0.29
	1489.55		602.7	407	0.118	0.2	0.21
18225.13	0.0	557.49	548.54	2924	0.575	4.0	4.4
	292.58		557.49	1966	0.387	2.7	2.2
	811.92		574.11	192	0.038	0.3	0.2
18475.28	0.0	565.98	541.11	396	0.032	0.5	---
	292.58		549.82	3425	0.281	4.5	6.0
	811.92		565.98	8379	0.687	10.9	8.2
18788.08	292.58	556.14	540.52	991	0.525	4.3	5.7
	811.92		556.14	462	0.245	2.0	0.65
	1489.55		577.92	433	0.230	1.9	1.5
19009.52	292.58	570.62	534.13	1110	0.289	6.3	4.9
	811.92		550.09	1527	0.398	8.6	11.0
	1489.55		570.62	1200	0.313	6.8	---
19990.25	1489.55	564.26	540.37	1499	0.633	2.4	2.7
	2273.09		564.26	362	0.153	0.6	2.9
	3125.46		592.78	506	0.214	0.8	---

*Ref [24].

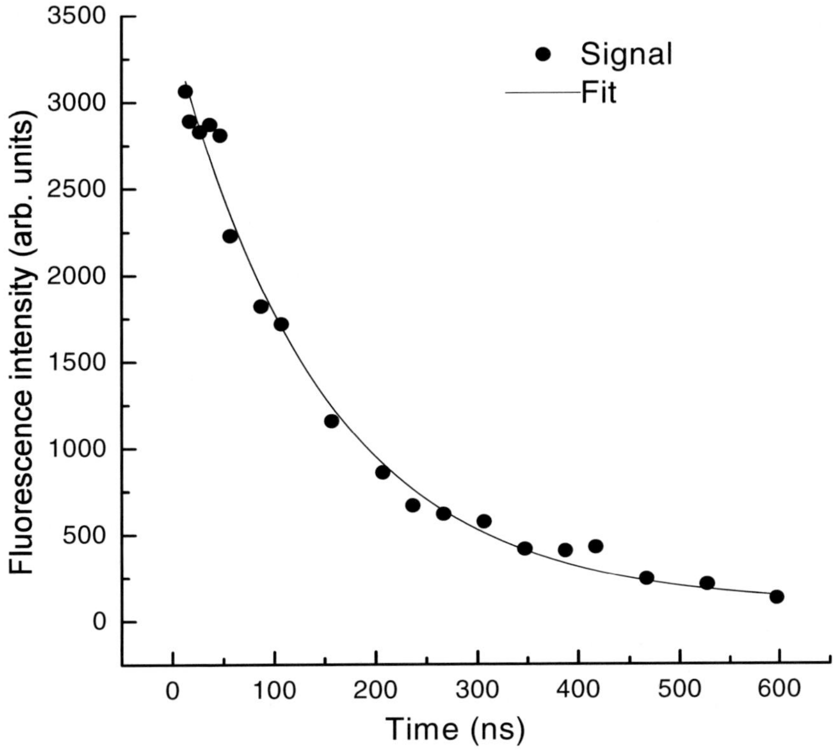

Figure 3. Experimental time-resolved single-color LIF decay signal with an exponential fit. The lifetime for this level (18225.13 cm-1) is measured to be 143 ± 10 ns.

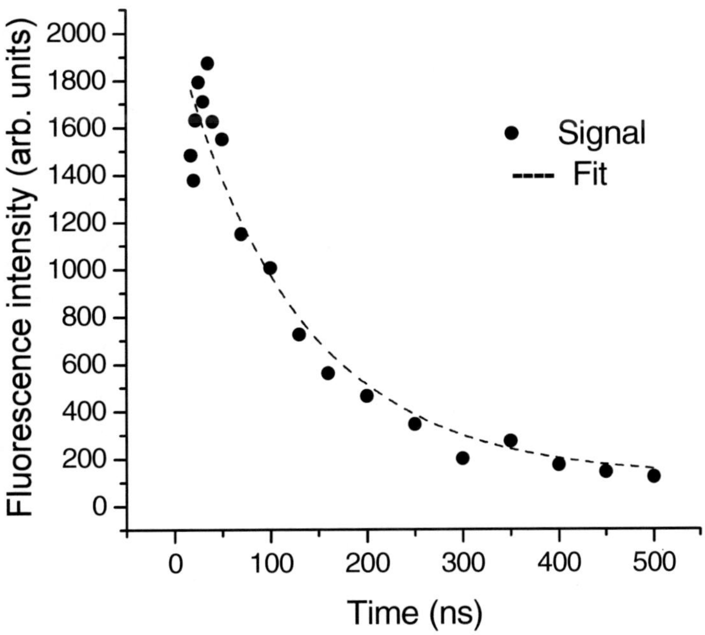

Figure 4. Experimental time-resolved two-color LIF decay signal with an exponential fit. The lifetime for this level (34814.4 cm^{-1}) is measured to be 127 ± 13 ns.

Table 3. Measured radiative lifetimes of stepwise-excited even-parity energy levels in atomic samarium using two-colour laser-induced fluorescence

Energy level (cm^{-1})	First excitation wavelength (nm)	Second excitation wavelength (nm)	Fluorescence wavelength (nm)	Lifetime (ns)
34814.4	591.64	567.22	553.4	127 ± 13
34924.0	591.64	563.73	531.4	106 ± 9
34935.5	591.64	563.36	516.2	78 ± 10
34972.1	591.64	562.32	571.0	44 ± 3
35072.6	591.64	559.04	543.9	28 ± 2
35092.1	591.64	558.44	494.2	80 ± 7

Branching fractions of optical transitions from each excited level were determined from LIF light intensities of different decay channels to lower energy levels. Branching fraction was obtained for each decay channel from line intensities through $\beta_{ki} = I_{ki} / \Sigma_i I_{ki}$, where I is intensity of the spectral line and sum is taken over all lower levels (i) to which upper level (k) decays. The spectra showing the fluorescence decay channels were recorded after the delay of 25 - 30 ns with respect to laser trigger pulse in order to cut down the contribution of laser scattering to resonant fluorescence. Experiment was repeated many times under different experimental conditions as in the case of lifetime measurements. The average of these measurements was used to determine the branching fractions. The observed decay channels from stepwise-excited even-parity energy level at 34924.4 cm^{-1}, measured branching fractions and absolute transition probabilities are listed in Table 4.

Table 4. Measured branching fractions and transition probabilities of the transitions from the stepwise-excited even-parity energy level at 34924.4 cm^{-1} to various lower odd-parity levels using two-colour laser-induced fluorescence with λ_1 = 591.64 nm and λ_2 = 563.73 nm

Energy of lower level (cm^{-1})	J-value of lower level	Fluorescence wavelength (nm)	Intensity (arb. Units)	Branching fraction	Transition probability (10^6 s^{-1})
14380.50	2	486.6	202	0.039	0.4
14863.85	1	498.4	157	0.031	0.3
15039.59	2	502.8	153	0.030	0.3
15567.32	2	516.5	121	0.024	0.2
16116.42	2	531.5	1442	0.284	2.7
16748.30	3	550.0	268	0.053	0.5
17190.20	2	563.7	1476	0.291	2.7
17243.55	3	565.4	319	0.063	0.6
17830.80	3	584.9	561	0.110	1.0
18225.13	1	598.7	381	0.075	0.7

Fluorescence decay channels, which are due to two-step resonant excitation (involving λ_1 and λ_2), are used for obtaining the values of branching fractions. We determined absolute transition probabilities for 51 excited-state-to-excited-state transitions in atomic Sm. Transition probability A_{ki} was obtained from the measured values of lifetime and branching fraction using the expression $A_{ki} = \beta_{ki} / \tau$, where τ is the radiative lifetime of the excited level and β_{ki} is the branching fraction for the k $\rightarrow$ i transition.

The statistical uncertainty in measurements of lifetime is ~ ± 10 %, based on the spread in the fitted lifetime values to the time-dependent LIF signals. Apart from this statistical uncertainty, there could be various effects causing systematic uncertainties in the lifetime measurements, such as radiation trapping, collisional depopulation, atomic motion, cascade repopulation and depopulation, Zeeman and hyperfine quantum beats [20]. Due to the reasons mentioned above and considering experimental conditions, the systematic uncertainties have been neglected, because any possible small contribution due to systematic errors would not change the uncertainty beyond statistical value of 10%.

The statistical uncertainty in our β measurements estimated from the spread of β-values measured under different experimental conditions is ≤ 10%. The spread is minimum ~5% for the strong transitions and maximum 10% for weak transitions. The maximum value of 10% is taken for estimation of uncertainties. The uncertainty of calibration of the lamp with the uncertainty of the response of spectrometer using this lamp is about 6 %. These two uncertainties are combined in quadrature with the uncertainty in lifetime measurements to give the uncertainty of ~15 % in the measurement of transition probabilities. This estimate of uncertainty is without considering the infrared (IR) transitions. The non-observation of allowed transitions in the IR region is a systematic uncertainty in the branching fraction measurement, limiting its accuracy. Biemont *et al.* [26] have observed in the spectrum of Sm III that the IR transitions are mostly weak transitions and do not give a significant contribution. But, if there are many transitions of weak intensity - a distinct possibility that they could add up to as much as 20% or more of total decay as mentioned by Wickliffe *et al.* [27] in the nearby spectrum of Dy I. Although the contribution of IR transitions to uncertainty is dependent on element and transition, this could lead to a systematic uncertainty ranging from negligibly small value to 20 % or more in the branching fraction measurements of Sm transitions considered, in addition to the statistical uncertainties. Similarly overall uncertainties in transition probability measurements will increase due to contribution from IR transitions.

Table 5. Assigned values of the total angular momentum J to each of the stepwise-excited even-parity energy levels in atomic samarium based on examination of J-values of the lower odd-parity levels to which a stepwise-excited even-parity level decays, removing the ambiguity from the proposed J-values in Ref [11]

Even-parity energy level (cm^{-1})	J -value
34814.4	2
34924.0	2
34935.5	2
34972.1	3
35072.6	1
35092.1	3

This method has also been exploited to uniquely assign J values to high-lying even-parity energy levels earlier identified by us [15] as new energy levels by identifying known lower levels (odd parity) with specific total angular momentum (J) values to which a stepwise-excited even parity energy level decays via fluorescence decay channels. Table 5 lists the stepwise-excited even-parity energy levels and their unique angular momentum values thus obtained. This technique of J identification is relatively simpler and does not require excitation to the same energy level by different excitation routes.

3. Laser-Induced Breakdown Spectroscopy

Laser-induced breakdown spectroscopy (LIBS) [28, 29] is an elemental analytical technique which has been evolving since last five decades. It has become very popular in the last decade owing to its numerous advantages, such as rapidity, multi-elemental analysis, minimal sample preparation, minimal sample destruction, low cost, versatility of being applied to a wide range of materials without any pre-processing and flexibility of the experimental set-up. It has a strong potential for in situ and real time analysis. LIBS plasma is generated when a high-power pulsed laser is focused on a solid, liquid or gaseous sample with irradiance > 1×10^8 Wcm^{-2} and the light emitted from the plasma is used to determine the elemental composition of the sample. In the recent past, it has been extensively used for multi-element analysis of metals and alloys [30], art works [31], cultural heritage targets [32], environmental samples [33], and geological [34], biomedical [35] and nuclear materials [36]. LIBS of nuclear materials is both interesting as well as challenging. Specific interest is to exploit remote handling (and analysis) capability of this technique-enabling working upon radioactive materials. Also many of the species of interest are heavy atoms with complex spectra. In this section we deal with development of calibration free LIBS of metallic alloys and uranium vitrified glasses. Many of these are exploratory studies for the final goal of compositional analysis of radioactive glasses.

Quantitative analysis of unknown samples is conventionally performed using calibration-based LIBS (CB-LIBS) method in which the composition of an unknown sample is determined by comparing the analyte line signals with spectral intensities from the calibration curves [37-42] of emission line intensity ratios versus elemental concentrations using many matrix-matched certified reference samples with known compositions. We have utilized the CB-LIBS method and carried quantitative analysis of nickel alloys [43]. This method requires reference samples with a matrix having composition similar to the unknown sample. The CB-LIBS method however is inappropriate in more complex cases, such as multi-elemental analysis in environmental and inaccessible samples. Ciucci *et al.* [44] have developed an alternative method, calibration-free LIBS (CF-LIBS) method, for quantitative elemental analysis of a given sample. In this method, the elemental composition of a sample is determined from the LIBS spectrum using computational methods in analyzing the basic physics of the plasma process through the estimation of the plasma temperature and electron number density, assuming that the plasma composition represents exactly the composition of the sample, i.e., stoichiometric ablation and the plasma is optically thin and is in local thermodynamic equilibrium (LTE). Since its inception, the CF-LIBS method has been applied by several research groups across the world to the analysis of metallic alloys [44-54], as well as non-metallic samples, such as soils, rocks and glasses [47,52,55-57], human hair [58], coral skeletons [59] and environment samples [60]. Several CF-LIBS techniques have been developed and reported in the literature in the past.

Tognoni *et al.* [61] and Hann and Omenetto [62] have discussed the applications of CF-LIBS method and the measurements of temperature and electron number density in their review articles respectively. The distance, which is used as a monitor of global quality of the CF-LIBS results, is being defined as [61],

$$\text{distance} = \sum_{i=1}^{N} abs\ |M_i - C_i|,$$

where the index i runs over all the N elements present in a sample. M_i and C_i represent the measured and certified elemental concentrations respectively.

The advantage of the CF-LIBS method developed by Ciucci *et al.* [44], which is used in the above mentioned works, is that it can be applied to any samples without the matrix problem. But, this method requires detecting at least one line of each element in the plasma with known atomic data. Gomba *et al.* [63] have developed a CF-LIBS procedure different from that of Ciucci *et al.* [44] to quantify the contents of the elements through the estimation of plasma temperature, electron number density and relative number densities of the neutral and singly ionized ionic species in the LIBS plasma by making use of the experimental spectral line intensities in the time window where the plasma is optically thin and is in LTE. CF-LIBS may be a very important tool for many analytical problems, especially in rapid screening of unknown samples. It can only provide a qualitative estimation of the trace components. This is an intrinsic limit of CF-LIBS because even a small uncertainty in major component is translated into a huge relative error affecting the minor component. However, this limit can be bypassed if the internal standard is available.

Owing to the economic and technical importance of steel in a wide field of industrial applications, home appliances and medical technology, the steel manufacturers are bound to provide customers the chemical composition of steel. The stainless steel is used in nuclear industry for a variety of applications ranging from the reactor containment building to internal reactor components. Because of the growing demand for real-time, in-situ analytical results in steel industry, LIBS has received increasing attention of several research groups for compositional analysis of steel [37-39, 45, 47, 48, 64-69].

In the following text, we are summarizing our work on steel sample by employing a Q-switched Nd:YAG laser (532 nm wavelength) using the CF-LIBS technique for its quantitative analysis. We have extended the procedure developed by Gomba *et al.* [63] for quantitative analysis of two-element Li-Al alloy to multi-element steel (Fe, Cr, Ni, Mg and Si) sample. For the quantification of elemental contents in the sample, we have developed an algorithm relating the experimentally measured spectral intensity values at a time delay where the LIBS plasma is optically thin and in LTE with the basic physics parameters of the plasma. Using the algorithm we have determined the concentrations of Fe, Cr, Ni, Mg and Si in the steel sample. The analytical results obtained from the CF-LIBS method agree well with the certified values of the elements in the sample with a relative uncertainty < 5 %.

3.1. Experimental

A schematic diagram of the experimental set-up used for LIBS study is shown in Figure 5. A flash lamp pumped Q-switched Nd:YAG laser (Quantel YG 980) at the second harmonic wavelength λ=532 nm with pulse duration τ_L=7 ns and repetition rate ν_{rep}=20 Hz was focused normally onto a polished and cleaned steel sample, placed in air at atmospheric pressure using a planoconvex lens with a focal length f=100 mm. The focal spot of laser beam was always located on the surface of the sample so that the distance Δs between the position of the beam waist and the surface of the sample was always zero.

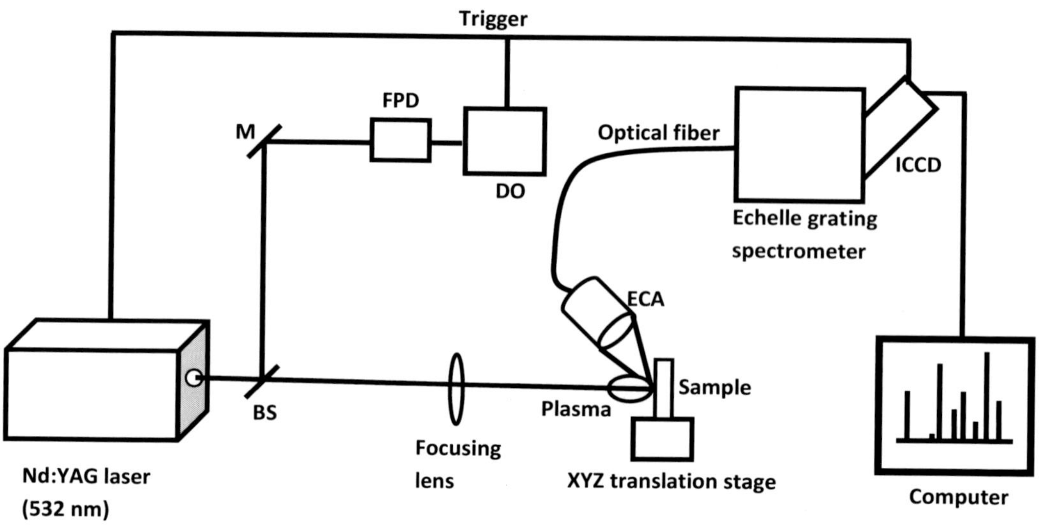

Figure 5. Schematic diagram of the experimental set-up for LIBS studies. BS - Beam splitter, DO - Digital oscilloscope, ECA - Emission collection assembly, FPD - Fast photo diode, M – Mirror.

We have used 532 nm visible laser radiation for the sake of ease in the alignment. The manufacturer certified elemental compositions of the steel sample used are: Fe (72.4%), Cr (18.3%), Ni (8.03%), Mn (0.82%) and Si (0.33%). The sample, mounted vertically on a X-Y-Z translation stage, was moved manually to perform each measurement on a new part of the sample surface. The laser beam with pulse energy E_L=70 mJ was focussed to a focal spot diameter (beam waist width $d_{\sigma 0}$) of about 1 mm measured from the observed ablation spot size on the steel sample. This yields the average laser irradiance at the sample surface equal to ~ 1×10^9 Wcm^{-2}, sufficient for inducing the plasma on the steel surface but insufficient for air breakdown in front of the sample. An Echelle grating spectrometer coupled with a gateable ICCD camera (Andor Mechelle ME 5000) was used to record the emission spectrum. The spectrometer covers a wide wavelength range 200-975 nm in a single spectrum with a good wavelength resolution (0.05 nm). An Hg-Ar lamp, which provides sharp lines from 200-1000 nm, was used for wavelength calibration of this system. Intensity calibration of the Echelle spectrometer-ICCD system was done using NIST certified Deuterium-Quartz-Tungsten-Halogen lamp. The spatially integrated plasma light emission was collected and imaged onto the spectrograph slit using an optical-fibre-based collection system. This emission collection assembly (ECA) was positioned at a distance of about 20 cm from the plasma, making an angle 45° to the laser beam. The detector integration time was limited to 1000 ns to avoid saturation of the detector whereas four delay times of 800, 900, 1000 and 1200 ns were chosen for recording the plasma emission spectra, avoiding the high intensity continuum emission at the early times of the plasma formation. One LIBS spectrum at a given delay time is derived from an integration of 120 laser pulses to minimize the effects of pulse-to-pulse variations from run to run and for improving the signal-to-noise ratio. In order to obtain enough reproducibility of the results, we have analyzed four sets of such emission spectra recorded at a given delay time. The average line intensity values are calculated as the mean of line intensities form the four spectra, which has been used in the present work for the LIBS analysis. The details can be found in our earlier work [70].

3.2. Plasma Characterization

For quantitative elemental analysis from the LIBS spectral line intensities, it is essential to characterize the time-evolving LIBS plasma in terms of its temperature and electron number density and find out the time window where the LIBS plasma is optically thin and in LTE. Based on plasma spectroscopy, the Boltzmann plot method yields the temperature T and the Saha-Boltzmann equation method yields the electron number density n_e of plasma, which is optically thin as well as in LTE [71]. We briefly present below the relevant equations for characterizing the LIBS plasma using line intensities of the LIBS spectrum.

3.2.1. Determination of Plasma Temperature T

The spatially integrated spectral line intensity of the k-i electronic transition of the species in charge state Z (Z = 0 for neutral atoms, Z = 1 for singly-ionized atoms, etc.) of element α in optically thin and LTE plasma is given as [71]

$$I_{ki,Z}^{\alpha} = Fn_Z^{\alpha} \frac{g_{k,Z}^{\alpha} A_{ki,Z}^{\alpha}}{\lambda_{ki,Z}^{\alpha} P_Z^{\alpha}} \exp(-\frac{E_{k,Z}^{\alpha}}{k_B T}), \quad (1)$$

where $F = \dfrac{hcL}{4\pi}$, L is the characteristic length of the plasma, h is the Planck constant, c is the speed of light, k_B is the Boltzmann constant, $E_{k,Z}^{\alpha}$ and $g_{k,Z}^{\alpha}$ are the energy and degeneracy of the upper energy level k respectively, $A_{ki,Z}^{\alpha}$ is the transition probability and $\lambda_{ki,Z}^{\alpha}$ is the transition wavelength of the k-i transition, P_Z^{α} is the partition function and n_Z^{α} is the number density of the species in ionization stage Z. By taking the natural logarithm, Eq. (1) can be rewritten as

$$\ln\left(\frac{I_{ki}^{\alpha}\lambda_{ki}^{\alpha}}{g_k^{\alpha}A_{ki}^{\alpha}}\right) = -\frac{E_k^{\alpha}}{k_B T} + \ln\left(\frac{Fn_Z^{\alpha}}{P_Z^{\alpha}}\right). \quad (2)$$

If the term on the left-hand side is plotted against E_k for two or more transition lines, the linear plot obtained is called the Boltzmann plot where the integral intensity of each spectral line is represented by a point in the Boltzmann plane. This plot yields straight line with a slope of $(-1/k_BT)$. Thus, the slope of the Boltzmann plot yields the plasma temperature T.

3.2.2. Determination of Electron Number Density n_e

The electron number density using atom and ion spectral lines emitted from the plasma is determined employing the Saha-Boltzmann equation for the line intensities of the species in two consecutive charge states Z and Z+1of a particular element as [63]

$$n_e = \frac{I_Z'}{I_{Z+1}'} 6.04 \times 10^{21} (T_{eV})^{3/2} \times \exp[(-E_{k,Z+1} + E_{k,Z} - \chi_Z)/k_B T] \text{ cm}^{-3}, \quad (3)$$

where $I_Z' = \frac{I_Z \lambda_{ki,Z}}{g_{k,Z} A_{ki,Z}}$ and χ_Z is the ionization energy of the species in the ionization stage Z. T_{eV} is the plasma temperature in eV. Electron number density is experimentally determined using Eq. (3).

3.2.3. *Determination of Density Ratio of the Species of a Particular Element*

The Saha Equation relating the concentrations in the two consecutive charge states Z and Z+1 of a particular element α is written as [63]

$$\frac{n_e n_{Z+1}^{\alpha}}{n_Z^{\alpha}} = 6.04 \times 10^{21} T_{eV}^{3/2} \frac{P_{Z+1}^{\alpha}}{P_Z^{\alpha}} \exp(-\frac{\chi_Z^{\alpha}}{T_{eV}}) \text{ cm}^{-3}, \quad (4)$$

This equation determines the value of the density ratio of the species in two consecutive charge states of a particular element, using experimentally determined values of T and n_e.

3.2.4. *Determination of Density Ratio of the Species of Different Elements*

The density ratio of the species in different charge states of two elements is determined from the Saha-Boltzmann equation for the line intensities of the species in different charge states of two elements α and β as [63]

$$\frac{n_Z^{\alpha}}{n_{Z+1}^{\beta}} = \frac{I_{\alpha,Z}'}{I_{\beta,Z+1}'} \frac{P_{\alpha,Z}(T)}{P_{\beta,Z+1}(T)} \exp(-\frac{E_k^{\beta,Z+1} - E_k^{\alpha,Z}}{k_B T}). \quad (5)$$

3.2.5. *Criterion for Optically Thin and LTE Plasma*

The condition for the validity of the LTE is often checked with the McWhirter's [72,73] criterion,

$$n_e(cm^{-3}) \geq 1.6 \times 10^{12} T(K)^{1/2} [\Delta E(eV)]^3, \quad (6)$$

which provides the lowest limit of the electron number density for a homogeneous plasma to be in LTE. Here ΔE is the largest energy gap of the transition lines. There are articles demonstrating the existence of regions of different temperatures and densities in the LIBS plasma [74]. The criterion whether the LIBS plasma is optically thin is obtained from the intensity ratio of two lines of a particular element in the same charge state Z which can be expressed using Eq. (1) as

$$\frac{I_1}{I_2} = \left(\frac{\lambda_{nm,Z}}{\lambda_{ki,Z}}\right)\left(\frac{A_{ki,Z}}{A_{nm,Z}}\right)\left(\frac{g_{k,Z}}{g_{n,Z}}\right)\exp\left(-\frac{E_{k,Z} - E_{n,Z}}{k_B T}\right), \quad (7)$$

where I_1 is the line intensity from the *k-i* transition and I_2 is that from the *n-m* transition. If we consider two emission lines having the same upper level or as close as possible, the temperature effect of the Boltzmann factor on the reproducibility of the line intensity ratio is minimized and at the same time the consideration of the efficiency factor of the collecting system is avoided. Neglecting the exponential factor in that condition, one can find out the theoretical value of the intensity ratio of the two lines by using the atomic parameters of the transitions. By matching this theoretical intensity ratio with the measured values at different delay times, one finds out the time window where the plasma is optically thin.

3.3. Quantitative Elemental Analysis

For quantitative elemental analysis Using the CF-LIBS method, we have followed the CF-LIBS procedure developed by Gomba *et al.* [63] where the concentrations of the elements are determined by matching theoretically obtained values of n_e and relative number densities of the neutral and singly ionized ionic species of a particular element as well as different elements in the plasma with those experimentally determined from Eqs. (3), (4) and (5) using the spectral line intensities in a time window where the LIBS plasma is optically thin and in LTE. The used CF- LIBS procedure is mentioned in the following.

The total number density of element α is written as

$$n_{tot}^{\alpha} = n_0^{\alpha} + n_1^{\alpha} + n_2^{\alpha} + \ldots\ldots\ldots n_Z^{\alpha} = n_0^{\alpha}[1 + \sum_{Z=1}^{Z_N}\left(\frac{n_Z^{\alpha}}{n_0^{\alpha}}\right)], \quad (8)$$

where n_Z^{α} is the number density of the species of element α in charge state Z (Z = 0 for neutral atoms, Z=1 for singly-ionized atoms, etc.). The element α contributes to the total n_e with n_e^{α} electrons from the atoms in different charge states, equal to the sum of all ionic state contributions, i.e.,

$$n_e^{\alpha} = n_1^{\alpha} + 2n_2^{\alpha} + 3n_3^{\alpha} + \ldots\ldots\ldots = \sum_{Z=1}^{Z_N} Z n_Z^{\alpha} \quad (9)$$

Let us define the functions S_Z^α and R_Z^α as

$$S_Z^\alpha = \frac{n_e n_{Z+1}^\alpha}{n_Z^\alpha} = 6.04\times10^{21} T_{eV}^{3/2} \frac{P_{Z+1}^\alpha}{P_Z^\alpha} \exp(-\frac{\chi_Z^\alpha}{T_{eV}})\ \text{cm}^{-3} \tag{10}$$

and $R_Z^\alpha = \dfrac{n_Z^\alpha}{n_0^\alpha}$ (11)

The function R_Z^α can be rewritten as

$$R_Z^\alpha = \frac{n_Z^\alpha}{n_0^\alpha} = \frac{n_1^\alpha}{n_0^\alpha}\frac{n_2^\alpha}{n_1^\alpha}\cdots\cdots\frac{n_Z^\alpha}{n_{Z-1}^\alpha} = \frac{S_1^\alpha}{n_e}\frac{S_2^\alpha}{n_e}\cdots\cdots\cdots\frac{S_Z^\alpha}{n_e} = \prod_{j=1}^{Z}\frac{S_j^\alpha}{(n_e)^Z}. \tag{12}$$

n_e^α , n_Z^α and $n_Z^\alpha / n_{Z+1}^\beta$ can be expressed in terms of R_Z^α as

$$n_e^\alpha = \frac{n_{tot}^\alpha \sum_{Z=1}^{Z_N} Z R_Z^\alpha}{1+\sum_{Z=1}^{Z_N} R_Z^\alpha} \tag{13}$$

$$n_Z^\alpha = \frac{n_{tot}^\alpha}{1+\sum_{Z=1}^{Z_N} R_Z^\alpha} R_Z^\alpha \tag{14}$$

$$\frac{n_Z^\alpha}{n_{Z+1}^\beta} = \frac{n_0^\alpha}{n_0^\beta}\frac{R_Z^\alpha}{R_{Z+1}^\beta} \tag{15}$$

The total electron number density n_e of the plasma is obtained by summing up of electron contributions from all the elements:

$$n_e = \sum_{\alpha'=1}^{M} n_e^{\alpha'} , \tag{16}$$

where M is the total number of elements in the sample. Using T obtained from the Boltzmann plot and a proposed (initial) set of $n_{tot}^{\alpha'}$ of all the elements, n_e is calculated, incorporating electron contributions from all the elements, starting from its default value until convergence is reached. The convergence is accepted when the ratio of difference of the last and previous values is ≤ 10^{-4} %. The converged value is taken as the theoretical value of n_e. Then, the values of $\frac{n_{Z+1}^{\alpha}}{n_Z^{\alpha}}$ and $\frac{n_Z^{\alpha}}{n_{Z+1}^{\beta}}$ are calculated. These calculations are carried out for a variable set of $n_{tot}^{\alpha'}$ of all the elements to find out an appropriate set of $n_{tot}^{\alpha'}$ of all the elements that results in the theoretical estimates of n_e, $\frac{n_{Z+1}^{\alpha}}{n_Z^{\alpha}}$ and $\frac{n_Z^{\alpha}}{n_{Z+1}^{\beta}}$ very close (≤ 1%) to the experimentally measured values. These elemental concentrations are then stored. The relative concentration of element α is then obtained as

$$c^{\alpha} = \frac{n_{tot}^{\alpha}}{\sum_{\alpha'} n_{tot}^{\alpha'}}, \tag{17}$$

in terms of molar fractions or as

$$c^{\alpha} = \frac{n_{tot}^{\alpha}\mu_{\alpha}}{\sum_{\alpha'} n_{tot}^{\alpha'}\mu_{\alpha'}}, \tag{18}$$

in terms of mass abundance, where μ_α is the atomic weight of element α. Based on the above procedure, we have developed a CF-LIBS algorithm for the quantification of elemental contents in a steel sample. A schematic diagram of the algorithm is shown in Figure 6.

4. Results and Discussion

4.1. Emission Spectra

In order to arrive at the proper time window for quantitative analysis, where the LIBS plasma is optically thin and in LTE, we have recorded four sets of LIBS spectra of the steel sample at each of the four delay times of 800, 900, 1000 and 1200 ns respectively relative to the laser pulse. The average line intensity values, calculated as the mean of line intensities from the four spectra, have been used for the LIBS analysis. Figure 7 shows typical experimental LIBS spectra of a steel sample at a delay time of 1000 ns. We have chosen the wavelength range from 250 to 550 nm in Figure 7, covering the atomic lines of Fe, Cr, Ni, Mn and Si chosen for the plasma characterization and the compositional analysis. In Figure

7B, we have chosen the wavelength range from 232 to 242 nm to show the ionic lines of Fe used in the analysis. Since Fe is a major element with a content of more than seventy percent, its rich atomic and ionic transitions make the spectrum much crowded, overlapping with lines of the other elements and making the spectral analysis much tedious.

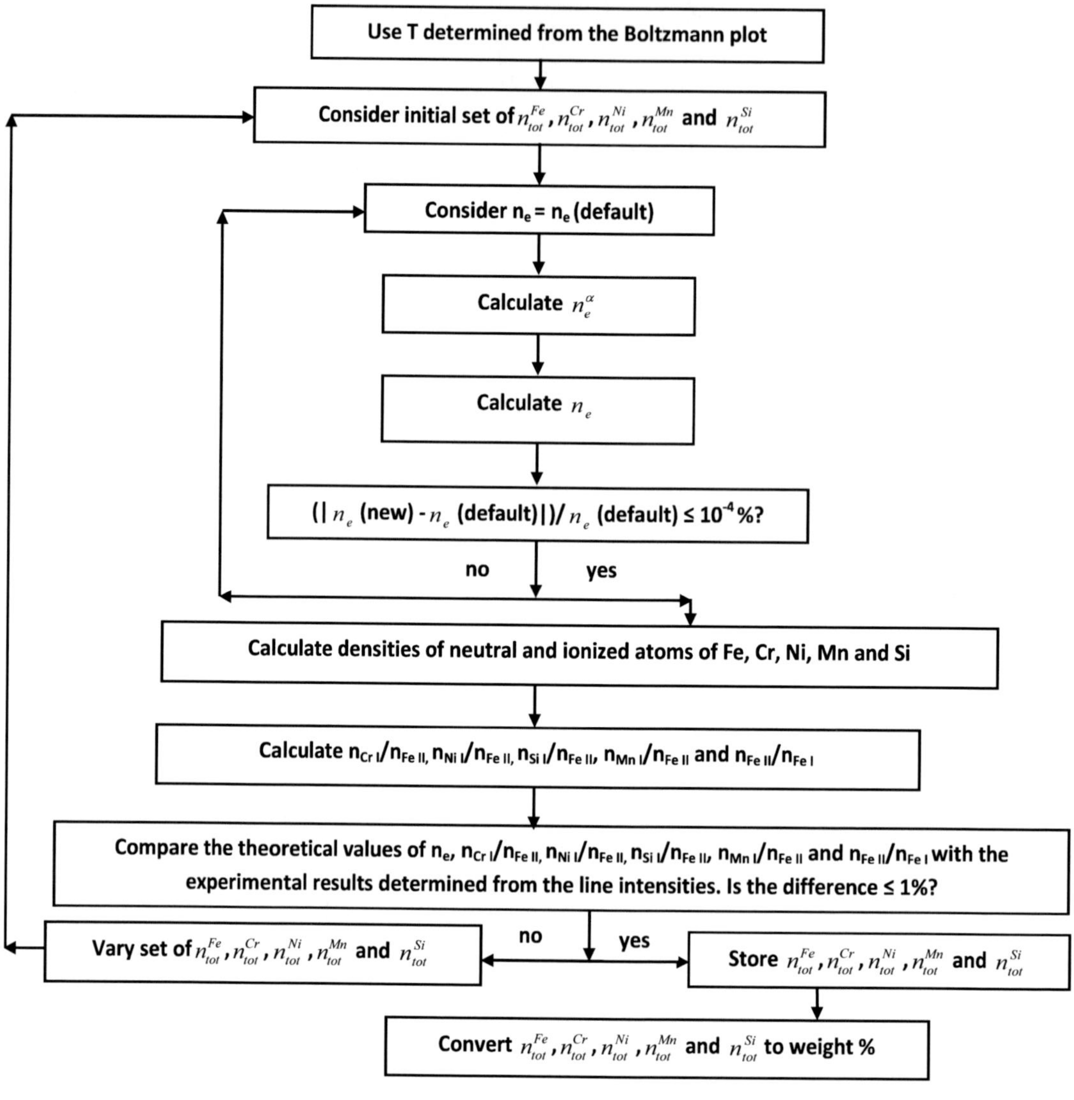

Figure 6. Schematic diagram of the CF-LIBS algorithm.

As mentioned in Figure 7, eight Fe I and three Fe II emission lines, which are well resolved, non-resonant and have minimal spectral interference with other spectral lines in a complex steel spectrum, are chosen for the plasma characterization in the present work. Aydin *et al.* [75] have discussed line selection criterion in detail. These lines along with their spectroscopic parameters, taken from the NIST atomic database [76], are shown in Table 6. The non-resonant atomic emission lines of other elements (Cr, Ni, Mn and Si), which are used in the quantitative analysis, along with their spectroscopic parameters are also given in the table.

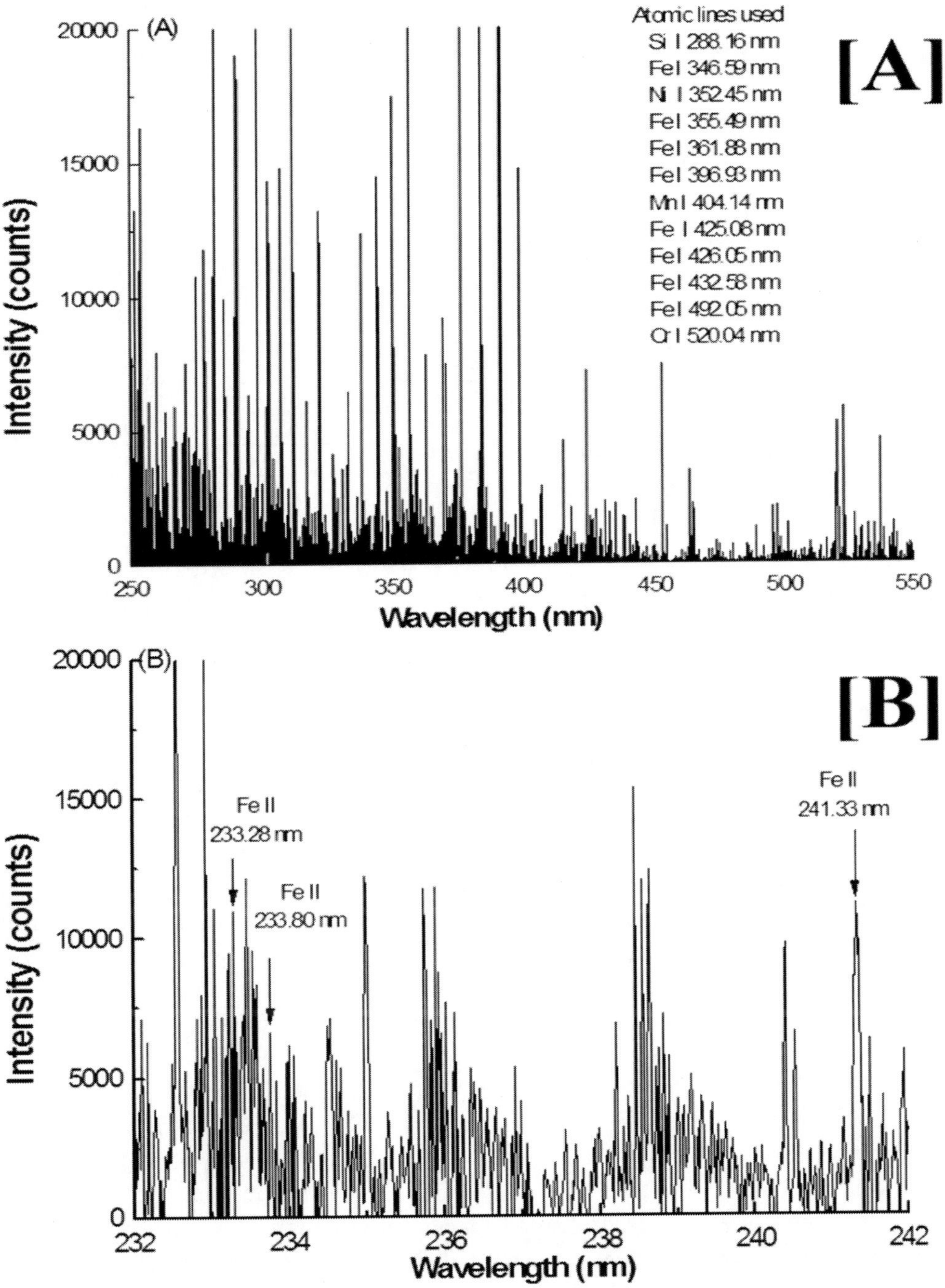

Figure 7. LIBS spectra of a steel sample at a delay time of 1000 ns, covering (A) atomic lines of Fe, Cr, Ni, Mn, and Si (B) ionic lines of Fe used in the analysis.

Table 6. Wavelength, lower and upper energy levels, upper level degeneracy, transition probability for the atomic and ionic emission lines used in the analysis

Atom/ion	Wavelength (nm)	Upper level energy (eV)	Lower level energy (eV)	Upper level degeneracy	Transition probability (s^{-1})
Fe I	346.59	3.686	0.110	3	1.19×10^7
Fe I	355.49	6.319	2.832	13	1.40×10^8
Fe I	361.88	4.415	0.990	7	7.22×10^7
Fe I	396.93	4.608	1.485	7	2.26×10^7
Fe I	425.08	4.473	1.557	7	1.02×10^7
Fe I	426.05	5.308	2.399	11	3.99×10^7
Fe I	432.58	4.473	1.608	7	5.16×10^7
Fe I	492.05	5.352	2.832	9	3.58×10^7
Fe II	233.28	5.361	0.048	6	1.31×10^8
Fe II	233.80	5.408	0.107	4	1.13×10^8
Fe II	241.33	5.257	0.121	4	$1.02x10^8$
Cr I	520.84	3.321	0.941	7	$5.06\ x10^7$
Ni I	352.45	3.542	0.025	5	$1.00\ x10^8$
Mn I	404.14	5.181	2.114	10	$7.87\ x10^7$
Si I	288.16	5.082	0.781	3	$2.17\ x10^8$

4.2. Time Delay for Optically Thin and LTE Plasma

To infer the time delay where the plasma is optically thin, we have considered two Fe I emission lines, one at 425.08 nm and the other at 432.58 nm. The first emission line at 425.08 nm is chosen owing to its upper energy level (4.473 eV) and its degeneracy (7) being same as that of the second emission line. The transition probability of the first emission line is $1.02\times10^7\ s^{-1}$ and that of the second emission line is $5.16\times10^7\ s^{-1}$ as provided in NIST data [49, 50]. The theoretical value of the intensity ratio for this couple of lines is calculated from Eq. (7) to be equal to 0.20. Since the values of the transition probabilities chosen are uncertain in range of 10% [77], the overall uncertainty in the theoretical line intensity ratio of two lines is in the range 20%, owing to two transition probabilities involved in Eq. (7). Thus, the theoretical intensity ratio of the two lines chosen is 0.20 ± 0.04. From the measured average intensities of these Fe I lines with relative error in the measurement of each of the line intensities in the range of 10%, the experimentally measured intensity ratios of this couple of lines are 0.17 ± 0.03, 0.23 ± 0.05, 0.25 ± 0.05 and 0.13 ± 0.03 at delay times of 800, 900, 1000 and 1200 ns respectively. Thus, the LIBS plasma produced in this work is optically thin (within the theoretical and experimental errors) at all the four delay times considered.

We have applied the McWhirter's criterion [Eq. (6)] to obtain the lowest limit of n_e for which the plasma will be in LTE. In our case ΔE is ~ 5.0 eV (Table 1). Using $T \sim 10^4$ K (as mentioned in section 2.4.3) at 1000 ns delay time, the lowest limit of n_e for the plasma to be in LTE, as obtained from Eq. (6), is $\sim2\times10^{16}\ cm^{-3}$. The measured value of n_e equal to $9.8\times10^{16}\ cm^{-3}$ (as mentioned in section 2.4.4) is more than the lowest limit, satisfying the assumption of LTE prevailing in the plasma at 1000 ns delay time. Thus, the LIBS plasma in the present work is optically thin as well as in LTE at 1000 ns delay time. Hence, we have carried out further LIBS analysis at a delay time of 1000 ns.

4.3. Plasma Temperature

We have generated the Boltzmann plot (Figure 8) at a delay time of 1000 ns using seven Fe I lines mentioned in Table 6. The 425.08 nm line is not used because it has the same upper energy level as that of 432.58 nm line. The slope of the plot gives T = 0.875 eV (10150 K). The relative error in the measurement of T corresponding to a relative error in the measurement of line intensities and a relative error in the measurement of the transition probabilities may be estimated form the error analysis of Eq. (2) as

$$\frac{\delta T}{T} = \frac{k_B T}{E}\left(\frac{\delta I}{I} + \frac{\delta A}{A}\right) \tag{19}$$

The error in our measurement of line intensities is in the range of 10% and the reported errors in the transition probabilities are in the range 10%. Considering the lowest upper energy level used in the present work, for which E=3.7 eV and T= 0.875 eV (measured), the resultant relative error in the temperature measurement is about 5%. Thus the value of T obtained from the Boltzmann plot is 0.875 ± 0.044 eV (10150 ± 510 K).

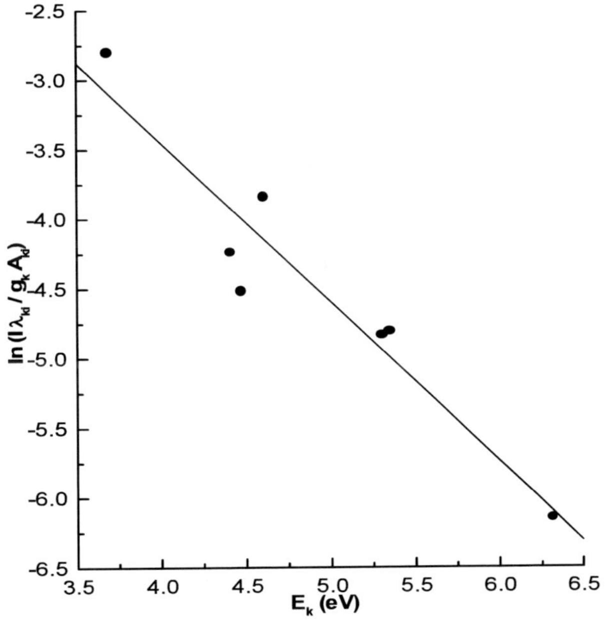

Figure 8. Boltzmann plot made using seven Fe I transitions, considering the intensities at a delay time of 1000 ns. The continuous line represents the results of a linear best fit. The slope gives the temperature equal to 0.875 ± 0.044 eV.

4.4. Electron Number Density

Using the measured intensity ratio of Fe I and Fe II lines at a given delay time, the electron number density n_e is determined from Eq. (3). We have considered five intensity ratios, Fe I 346.59 nm and Fe II 233.28 nm, Fe I 355.49 nm and Fe II 233.28 nm, Fe I 361.88 nm and Fe II 233.28 nm, Fe I 346.59 nm and Fe II 233.80 nm and Fe I 346.59 nm and Fe II 241.33 nm and determined five values of n_e at a given delay time. The arithmetic mean of the five values of n_e is taken as the average value of n_e at that particular delay time. The average value of the electron number density at a delay time of 1000 ns is thus determined to be 9.8×10^{16} cm^{-3}. The spread in the five values of n_e gives a relative uncertainty of ~10% in its measurement. Thus the value of n_e is $(9.8 \pm 1.0)\times10^{16}$ cm^{-3}.

4.5. Density Ratios of the Species of Particular and Different Elements

We have determined the density ratio $n_{Fe\,II}/n_{Fe\,I}$ from Eq. (4) using the measured values of T and n_e at a delay of 1000 ns and the corresponding partition function from the NIST atomic database [76]. Its value is determined as 6.7. Using Eq. (5), the intensities of Cr I, Ni I, Mn I, Si I and Fe II lines at a delay of 1000 ns and their spectroscopic parameters, taken from the NIST database [76], the density ratios $n_{Cr\,I}/n_{Fe\,II}$, $n_{Ni\,I}/n_{Fe\,II}$, $n_{Mn\,I}/n_{Fe\,II}$ and $n_{Si\,I}/n_{Fe\,II}$ are determined from Eq. (5). These experimental results are given in Table 7. The experimental values of T and n_e determined are also given in the table.

Table 7. Values of T, n_e, $n_{Cr\,I}/n_{Fe\,II}$, $n_{Ni\,I}/n_{Fe\,II}$, $n_{Si\,I}/n_{Fe\,II}$, $n_{Mn\,I}/n_{Fe\,II}$ and $n_{Fe\,II}/n_{Fe\,I}$ in the LIBS plasma determined from the LIBS spectral line intensities as experimental results at a delay time of 1000 ns

T (eV)	n_e (cm^{-3})	$n_{Cr\,I}/n_{Fe\,II}$	$n_{Ni\,I}/n_{Fe\,II}$	$n_{Si\,I}/n_{Fe\,II}$	$n_{Mn\,I}/n_{Fe\,II}$	$n_{Fe\,II}/n_{Fe\,I}$
0.875	9.8×10^{16}	3.24×10^{-2}	3.01×10^{-2}	4.67×10^{-3}	5.80×10^{-3}	6.7

Table 8. Comparison of the elemental contents of the steel sample obtained from the CF-LIBS analysis with their certified values and the corresponding relative uncertainty

Element	Certified concentration (wt %)	LIBS determined concentration (wt %)	Relative uncertainty (%)
Fe	72.40	72.66	0.4
Cr	18.30	17.88	2.3
Ni	8.03	8.25	2.7
Mn	0.82	0.86	4.9
Si	0.33	0.34	3.0

4.6. Quantitative Elemental Analysis

Like many reported articles [61], we have studied one steel sample and applied the CF-LIBS algorithm developed in this work (Figure 6) for its quantitative elemental analysis. In the CF-LIBS algorithm, we have varied the values of n_{tot}^{Fe}, n_{tot}^{Cr}, n_{tot}^{Ni}, n_{tot}^{Mn} and n_{tot}^{Si} each in the range 10^{14}-10^{17} cm^{-3} and calculated the theoretical values of n_e, $n_{Cr\,I}/n_{Fe\,II}$, $n_{Ni\,I}/n_{Fe\,II}$, $n_{Si\,I}/n_{Fe\,II}$, $n_{Mn\,I}/n_{Fe\,II}$ and $n_{Fe\,II}/n_{Fe\,I}$ in the LIBS plasma, which match well (≤ 1%) with the experimental results when n_{tot}^{Fe} = 8.7 x 10^{16} cm^{-3} , n_{tot}^{Cr} = 2.3 x 10^{16} cm^{-3}, n_{tot}^{Ni} = 9.4 x 10^{15} cm^{-3}, n_{tot}^{Mn} = 1.1 x 10^{15} cm^{-3}, and n_{tot}^{Si} = 8.1 x 10^{14} cm^{-3} are considered. Using these density values we have determined the fractional molar concentrations of Fe, Cr, Ni, Mn and Si in the steel sample. The values of Fe, Cr, Ni, Mn and Si in the steel sample obtained from the CF-LIBS analysis along with their certified values are shown in Table 8.

The residual deviation (i.e., the deviation of CF-LIBS determined values from the certified values) for Fe, Cr, Ni, Mn, Si in our measured values are 0.26, 0.42, 0.22, 0.04 and 0.01 wt % respectively. The distance as defined by Tognoni *et al.* [61] for our measurements is 0.95. For the appraisal of LIBS performance, the relative uncertainty of the elemental content value is determined by dividing the residual deviation with the certified value. The relative uncertainties for all the elements are also given in the table. It is noted from the table that the analytical results obtained from the CF-LIBS method agree well with the certified values of the elements with a relative uncertainty < 5 %. The relative uncertainties (or distance) for our measurements are comparable to those reported in the literature on steel samples [38, 45, 47, 48, 61, 66, 69] and other metallic alloys [61]. Aiming the analysis of nuclear materials in mind, we have further explored the same technique to analyze the uranium vitrified glass sample. Figure 9 and 10 shows the spectra of uranium vitrified glass sample and Boltzmann plot respectively. After carrying out the entire CF-LIBS analysis procedure as discussed above, the elemental concentration of uranium in the uranium vitrified glass sample was determined. The experimental results matches well within 10% of certified values. Hence, the technique can be further explored for other nuclear materials. It can be used for online monitoring of uranium concentration in uranium waste management and reprocessing plants.

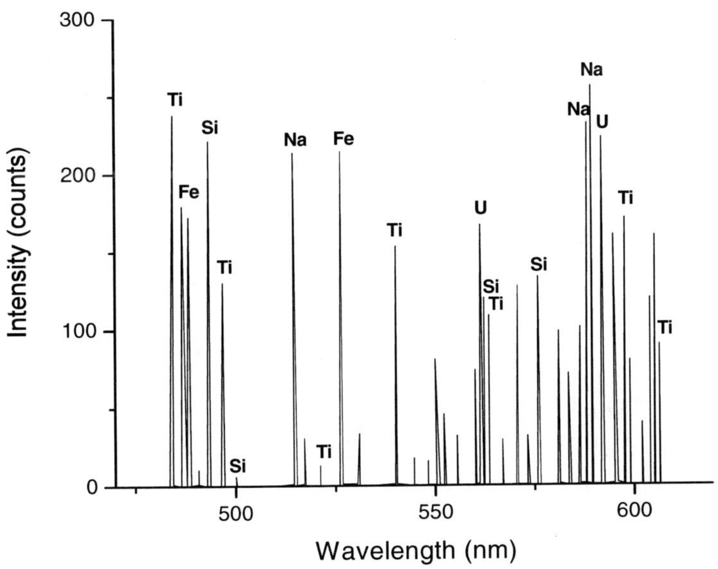

Figure 9. LIBS spectra of a uranium vitrified glass sample covering atomic lines of U, Fe, Si, Ti, and Na.

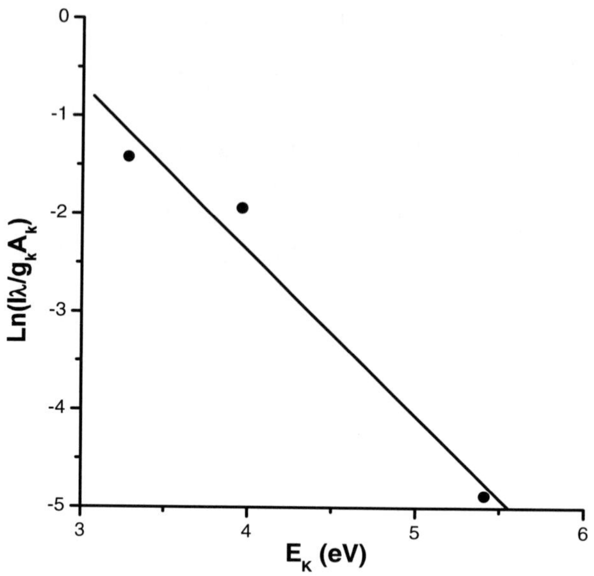

Figure 10. Boltzmann plot made using Fe I transitions. The continuous line represents the results of a linear best fit. The slope gives the temperature equal to 0.595 ± 0.029 eV.

Conclusion

As can be seen from the above reported results, laser spectroscopy has ample scope for applications to nuclear science and technology. Most of the materials studied are relevant for nuclear applications. At the same time the techniques developed here have universal applicability wherever complex spectra are involved. The techniques for measurement of branching fractions and absolute transition probabilities of ground state to excited state and excited state to excited state are much simpler and yet quite accurate. The accurate quantitative analysis can be done using the CF-LIBS technique by an appropriate choice of experimental conditions for materials including the nuclear materials of interest for online elemental composition analysis.

Acknowledgement

The authors are thankful to Dr. L.M. Gantayet, Director, Beam Technology Development Group and Dr. A.K. Das, Head, Laser and Plasma Technology Division, Bhabha Atomic Research Centre for their encouragement and support.

References

[1] W.C. Martin, R. Zalubas, and L. Hagan, "*Atomic Energy Levels – The rare earth elements*" *Natl. Bur. Stand.* (U.S. Rep. NSRDS-NBS 60 (1978)).

[2] B.M. Suri, K. Dasgupta, P. N. Bajaj, K.G. Manohar, R. Talukdar, P. K. Chakraborti, P.R. K. Rao, *J. Opt. Soc. Am.* B 4 (1987) 1835 - 1836.

[3] V.K. Mago, A.K. Ray, B. Lal, P.R.K. Rao, *J. Phys.* B, 21 (1988) 955 – 961.

[4] L.R. Carlson, J.A. Paisner, E.F. Worden, S.A. Johnson, C.A. May, and R.W. Solarz, *J. Opt. Soc. Am. 66* (1976) 846 – 853.

[5] M. Miyabe, M. Oba, I. Wakaida, *J. Phys.* B, 33 (2000) 4957 - 4972.

[6] M. Oba, K. Akaoka, M. Miyabe, I. Wakaida, *Eur. Phys. J. D,* 21 (2002) 255 - 260.

[7] S. Rochester, C. J. Bowers, D. Budker, D. DeMille, M. Zolotorev, *Phys. Rev.* A 59 (1999) 3480-3494.

[8] K. Song, H. Cha, J. Lee, *J. Anal. At. Spectrom,* 13 (1998) 1207 - 1210.

[9] A. Gomonai, V. Kelman, O. Plekan, *Radi. Phys. Chem.* 68 (2003) 137 - 141.

[10] E. Miron, R. David, G. Erez, S. Lavi, L.A. Levin, *J. Opt. Soc. Am.* 69 (1979) 256 - 264.

[11] A. K. Pulhani, M.L. Shah, Vas Dev, B.M. Suri, *J. Opt. Soc. Am.* B 22 (2005) 1117-1122.

[12] A. Petit, R. Avril, D. L'Hermite, A. Pailloux, *Physica Scripta* T 100 (2002) 114-119.

[13] L. R. Carlson, S. A. Johnson, E.F. Worden, C. A. May, R.W. Solarz, J. A. Paisner, *Opt. Commun.* 21 (1977) 116-119.

[14] L.A. Hackel, M.C. Rushford, *J. Opt. Soc. Am.* 68 (1978) 1084-1087.

[15] S. E. Bisson, B. Comaskey, E. F. Worden, *J. Opt. Soc. Am.* B, 12 (1995) 193-202.

[16] R. Avril, A. Petit, J. Radwan, E. Vors, *Proc. SPIE* 1859 (1993) 38-48.

[17] B. Blagoev, V. A. Komarovskii, N.P. Penkin, *Opt. Spectrosk.* 42 (1977) 424-426.

[18] P. Hannaford, R. M. Lowe, *J. Phys.* B, 18 (1985) 2365-2370.

[19] C. J. Bowers, D. Budker, E. D. Commins, D. Demille, S. J. Freedman, A.-T. Nguyen, S.-Q. Shang, M. Zolotorev, *Phys. Rev.* A 53 (1996) 3103-3109.

[20] M. Song, Y. P. Li, W. X. Peng, Z. K. Jiang, C. Guo Y. N. Yu, *Eur. Phys. J.* D 2 (1998) 115-116.

[21] S. Rochester, C. J. Bowers, D. Budker, D. DeMille, M. Zolotorev, *Phys. Rev.* A, 59 (1999) 3480-3494.

[22] H. Xu, Z. Jiang, H. Lundberg, *J. Opt. Soc. Am.* B 23 (2006) 2597-2600.

[23] V. Fivet, P. Palmeri, P. Quinet, E. Biemont, H. L. Xu, S. Svanberg, *Eur. Phys. J.* D 37 (2006) 29-35.

[24] V. A. Komarovskii, Yu. M. Smirnov, *Opt. Spectrosk.* 80 (1996) 357-361.

[25] M. L. Shah, A. K. Pulhani, G.P. Gupta, B.M. Suri, *J. Opt. Soc. Am.* B 27 (2010) 423-431.

[26] E. Biemont, H. P. Garnir, U. Litzen, K Nielsen, P. Quinet, S. Svanberg, G. M. Wahlgren, Z. G. Zhang, *Astronomy and Astrophysics* 399 (2003) 343-349.

[27] M. E. Wickliffe, J. E. Lawler, G. Nave, *J. Quant. Spectos. Radiat. Transfer* 66 (2000) 363-404.

[28] D. A. Cremers, L. J. Radziemski, *Handbook of Laser-Induced Breakdown Spectroscopy* (John Wiley & Sons Ltd., England (2006).

[29] A. W. Miziolek, V. Pallesschi, I. Schechter, eds., *Laser-Induced Breakdown Spectroscopy* (LIBS), Cambridge University Press, New York (2006).

[30] Y. N. Ledney, A. V. Yakovlev, T. A. Labutin, A. M. Povov, N. B. Zorov, *J. Anal. Chem.* 62 (2007) 1151-1155.

[31] V. Burakov, V. Kiris, A. Klyachkovskaya, N. Kozhukh, S. Rainkov, *Microchim. Acta* 156 (2007) 337-342.

[32] A. Giakoumaki, K. Melessanaki, D. Anglos, *Anal. Bioanal. Chem.* 387 (2007) 749-760.

[33] F. Capitelli, F. Colao, M. R. Provenzano, R. Fantoni, G. Brunetti, N. Senesi, *Geoderma* 106 (2002) 45-62.

[34] B. Salle, D. A. Cremers, S. Maurice, R. C. Wiens, *Spectrochim. Acta* B 60 (2005) 479-490.

[35] O. Samek, D. C. S. Beddows, H. H. Telle, G. W. Morris, M. Liska, J. Kaiser, *Appl. Phys.* A 69 (1999) S179-S182.

[36] J. P. Singh, F. Y. Yueh, H. Zhang, K. P. Karney, *Rec. Res. Dev. Appl. Spectrosc.* 2 (1999) 59-67.

[37] R. Sattmann, V. Sturm, R. Noll, *Phys. D: Appl. Phys.* 28 (1995) 2181-2187.

[38] C. Aragon, J. A. Aguilera, F. Penalba, *Appl. Spectrosc.* 53 (1999) 1259-1267.

[39] I. Bassiotis, A. Diamantopoulou, A. Giannoudakos, F. R. Kalantzopoulou, M. Kompitsas, *Spectrochim. Acta* B 56 (2001) 671-683.

[40] J. D. Winefordner, I. B. Gornushkin, T. Correll, E. Gibb, B. W. Smith, N. Omenetto, *J. Anal.At. Spectrom* 19 (2004) 1061-1083.

[41] B Salle, D Cremers, S Maurice, R Wiens, P Fichet, *Spectrochim. Acta* B 60 (2005) 805-815.

[42] G. Cristoforetti, S. Legnaioli, V. Palledchi, A. Salvetti, E. Tognoni, P. A Benedetti, F. Brioschi, F. Ferrario, *J. Anal. At. Spectrom.* 21 (2006) 697-702.

[43] G. P. Gupta, B. M. Suri, A. Verma, M. Sunderaraman, V. K. Unnikrishnan, K. Alti, V. B. Kartha, C. Santhosh, *J. Alloys and Compd.* 509 (2011) 3740-3745.

[44] A. Ciucci, M. Corsi, V. Palleschi, S. Rastelli, A. Salvetti, *E. Tognoni, Appl. Spectro.* 53 (1999) 960-964.

[45] E. Tognoni, G. Cristoforetti, S. Legnaioli, V. Palleschi, A. Salvetti, M. Mueller, U. Panne, Gornushkin, *Spectrochim. Acta* B 62 (2007) 1287-1302.

[46] K. K. Herrera, E. Tognoni, N. Omenetto, Y. B. Gornushkin, B. W. Smith, J. D. Winefordner, *J. Anal. At. Spectrom.* 24 (2009) 426-438.

[47] K. K. Herrera, E. Tognoni, N. Omenetto, B. W. Smith, J. D. Winefordner, *J. Anal. At. Spectrom.* 24 (2009) 413-425.

[48] D. Bullajic, M. Corsi, G. Christoforetti, S. Legnaioli, V. Palleschi, A. Salvetti, E. Tognoni, *Spectrochim. Acta* B 57 (2002) 339-353.

[49] V. S. Burakov, V. V. Kiris, P. A. Naumenkov, S. N. Raikov, *J. Appl. Spectrosc.* 71 (2004) 740-746.

[50] F. Colao, R. Fantoni, V. Lazic, I. Caneve, A. Giardini, V. Spizzichino, *J. Anal. At. Spectrom.* 19 (2004)502-504.

[51] I. Fornarini, F. Colao, R. Fantoni, V. Lazic, V. Spizzichino, *Spectrochim. Acta* B 60 (2005) 1186-1201.

[52] V. S. Burakov, S N. Raikov, *Spectrochim. Acta* B 62 (2007) 217- 223.

[53] J. A. Aguilera, C.Aragon, G.Cristoforetti, E. Tognoni, *Spectrochim. Acta* B 64 (2009) 685-689.

[54] M. V. Belkov, V. S. Burakov, V. V. Kiris, N. M. Kozhukh, S. N. Raikov, *J. Appl. Spectrosc.* 72 (2005) 376-381.

[55] F. Colao, R. Fantoni, V. Lazic, A. Paolini, F.Fabbri, G. G. Ori, L. Marinangeli, A. Baliva, *Planet. Space Sci.* 52 (2004) 117-123.

[56] B. Salle, J. I. Lacour, P. Mauchien, P. Fichet, S. Maurice, G. Manhes, *Spectrochim. Acta* B 61 (2006) 301-313.

[57] A. De Giacomo, M. Dell'Aglio, O. De Pascale, S. Longo, M. Capitelli, *Spectrochim. Acta* B 62 (2007) 1606-1611.

[58] M. Corsi, G. Christoforetti, M. Hidalgo, S. Legnaioli, V. Palleschi, A. Salvetti, E. Tognoni, C. Vallebona, *Appl. Opt.* 42 (2003) 6133-6137.

[59] S. Pandhija, A. K. Rai, *Appl. Phys.* B 94 (2009) 545-552.

[60] S. Pandhija, N. K. Rai, S. N. *Thakur, Appl. Phys.* B 98 (2010) 231-241.

[61] E. Tognoni, G. Cristoforetti, S. Legnaioli, V. Palleschi, *Spectrochim. Acta* B 65 (2010) 1-14.

[62] D. W. Hahn, N. Omenetto, *Appl. Spectrosc.* 64 (2010) 335A-366A.

[63] J.M. Gomba, C.D' Angelo, D. Bertuccelli, G. Bertuccelli, *Spectrochim. Acta* B 56 (2001) 695-705.

[64] C. M. Davies, H. H. Telle, D. J. Montgomery, R. E. Cobert, *Spectrochim. Acta* B 50 (1995) 1059-1075.

[65] J. A. Aguilera, C. Aragon, F. Penalba, *Appl. Sur. Sci.* 127 (1998) 309- 314.

[66] S. Palanco, J. J. Laserna, J. *Anal. At. Spectrom.* 15 (2000) 1321-1327.

[67] M. A. Ismail, H. Imam, A. Elhassan, H. Imran, A. Elhassan, W. T. Youniss, M A. Harith, *J. Anal. At. Spectrom.* 19 (2004) 489-494.

[68] C. Lopez-Moreno, K. Amponsah-Manager, B. W. Smith, I. B. Gornushkin, N. Omenetto, S. Palanco, J. J. Laserna, J. D. Winefordner, *J. Anal. At. Spectrochim.* 20 (2005) 552-556.

[69] J. Vrenegor, R. Noll, V. Sturm, *Spectrochim. Acta* B 60 (2005) 1083-1091.

[70] M. L. Shah, A. K. Pulhani, G. P. Gupta, B. M. Suri, *Applied Optics* 51 (2012) 4612-4621.

[71] H. R. Griem, *Principles of Plasma Spectroscopy* (Cambridge University Press, Cambridge, 1997).

[72] R. W. P. McWhirter, in *Plasma Diagnostic Techniques,* R. H. Huddlestone and S. L. Leonard, Eds. (Academic Press, New York, 1965), Chap. 5, pp. 201-264.

[73] G. Cristoforetti, A. D. Giacomo, M. Dell'Aglio, S. Legnaioli, E. Tognoni, V. Palleschi, N. Omenetto, *Spectrochim. Acta* B 65 (2010) 86-95.

[74] R. Wester, R. Noll, *J. Appl. Phys.* 106 (2009) 1233021-10.

[75] U. Aydin, P. Roth, C. D. Gehlen, R. Noll, *Spectrochim. Acta* B 63 (2008) 1060-1065.

[76] NIST Atomic Spectra Database, http://physics.nist.gov.

[77] T. R. O'Brian, M. E. Wickliffe, J. E. Lawler, W. Whaling, J. W. Brault, *J. Opt. Soc. Am* B 8 (1991) 1185-1201.

In: Spectroscopic Techniques for Security, Forensic ... ISBN: 978-1-63117-404-9
Editors: Y. Dwivedi, S. B. Rai and J. P. Singh

Chapter 4

PHOTOACOUSTIC AND PHOTOTHERMAL DETECTION OF EXPLOSIVES AND HARMFUL CHEMICALS

Surya Narayan Thakur*

Ex-Professor, Physics Department, Banaras Hindu University, India

ABSTRACT

Photoacoustic spectroscopy involves measurements of optical absorption, due to molecules as well as condensed phase-atoms, by its subsequent conversion into heat and sound due to nonradiative transitions in gaseous, liquid and solid samples. Present chapter discusses various aspects of this technique and its applications for detection of explosives and harmful chemicals.

1. INTRODUCTION

The origin of highly sensitive Photoacoustic spectroscopy technique lies in the discovery of the photoacoustic effect by Graham Bell [1] in 1880 and covers the wide range from the ultraviolet to the infrared of electromagnetic spectrum. Bell could make qualitative determination of the optical characteristics of materials by listening to their spectral response [2] in his spectrophone as shown in Figure 1. Substances replaced the crosswire of the spectrometer and were put in contact with the ear by means of a hearing tube. Bell found, with sunlight entering the spectrometer, good sound in all parts of the visible and invisible spectrum. The frequency of sound was the same as that of the mechanical chopper used for periodic interruption of the light beam but its intensity varied from one wavelength to other depending on the material content of the sample.

* Email: snthakur1@yahoo.com.

Figure 1. Spectrophone of Graham Bell with eyepiece of spectrometer replaced by a hearing tube.

The photpacoustic technique, however, remained dormant for nine decades for want of powerful optical sources and suitable electromechanical transducers. The advent of tunable lasers provided a different kind of source of electromagnetic radiation and Kreuzer was the first to revive photoacoustic spectroscopy [3] for detection of pollutant gases [4]. In recent times this technique has come into prominence for homeland and defense applications in rapid detection of hazardous materials (biological, energetic) and illicit drugs deposited on surfaces or nerve agents in a crowded environment during a terrorist attack. Each molecular vapor has its characteristic infrared fingerprint spectrum and in a photoacoustic gas sensor the molecule to be detected acts as its own selective filter. Laser is uniquely capable in the standoff detection of explosives and the availability of quantum cascade lasers in recent years has enhanced their role in photoacoustic and photothermal spectroscopy techniques.

Photoacoustic and photothermal spectroscopy encompass a number of highly sensitive methods of detecting trace levels of optical absorption in transparent as well as optically opaque samples. The underlying principle, that connectcts these spectroscopic techniques, involves the measurement of pressure, density or temperature due to nonradiative transitions in the sample resulting from the optical absorption. Quartz tuning fork (QTF) has emerged as a sharply resonant acoustic transducer [5-7] in recent years providing many advantages over the use of microphone and piezoelectric detectors in photoacoustic spectroscopy. Photothermal spectroscopy, on the other hand, involves the techniques of photothermal deflection (PTD) [8, 9] and photothermal lensing (PTL) [10, 11]. Safety and security concerns, during the past one decade, have attracted a great deal of innovations in laser based techniques for the detection of explosives, illicit drugs and harmful chemicals to stop their passage through international boundaries and their presence in crowded environments [12]. Some of these techniques, like Laser Ion Mobility Spectrometry (LIMS) [13], Raman Spectroscopy [14], Laser Induced Fluorescence Spectroscopy (LIF) [15], Photo-fragmentation-Laser Induced Fluorescence Spectroscopy (PF-LIF) [16, 17], Laser Induced Breakdown Spectroscopy (LIBS) [18-20], have been used with great success in special cases. This may be illustrated by giving a simple example. Triacetone triperoxide (TAPT) is a powerful explosive that is easy to synthesize using commonly available chemicals and it has high vapor pressures even at room temperature which facilitates vapor phase detection unlike the case of trinitrotoluene (TNT) which has very low vapor pressure. It has been found that femtosecond laser pulses are more effective in the detection of the parent molecular ion of TAPT using the technique of laser photo-ionization [21] whereas the technique of laser photofragmentation (PF-LIF) is more effective for TNT using nanosecond pulsed lasers [22].

But none of the laser-based techniques have the universal appeal of Photoacoustic (PA) and Photothermal (PT) spectroscopy due to the variety of samples - solid, liquid, gel, powder, vapor etc. that can be handled in the laboratory as well as in the field by using appropriate lasers in the wavelength region of 200 nm to 1600 nm.

2. BASIC PROCESSES IN PA & PT SPECTROSCOPY

The various processes involved in the creation of photoacoustic (PA) and photothermal (PT) signals follow the sequence illustrated in Table 1 where the former are detected by a microphone whereas the latter are detected by an optical probe. Monochromatic optical radiation of certain bandwidth is incident on the sample from a tunable source. When the excitation wavelength overlaps with an absorption band of the sample the atoms or molecules of the sample get transferred into a higher electronic or vibrational energy state which relaxes to a lower energy state via radiative as well as nonradiative transition. The nonradiative relaxation in the sample is accompanied by conversion of the absorbed energy into heat which results in a temperature change as well as changes in thermodynamic parameters of the sample which are related to temperature. The thermal diffusion in the sample removes the heat from its initial region to other parts leading to changes in density and pressure in the sample and also in a fluid in intimate contact of the sample. The subsequent heat induced thermal expansion in the sample and adjacent media produces a photoacoustic (or photothermal) signal when the incident beam intensity is modulated at a frequency in the acoustic range. The magnitude of PA/PT signal is linearly dependent on the sample's optically absorbing analyte concentration. Measurement of the temperature, pressure or density changes that occur due to optical absorption result from the photo heating of sample and so PA/PT signals are directly dependent on light absorption. Scattering and reflection losses do not produce photothermal signals.

Table 1. Basic processes and detection of optical heating

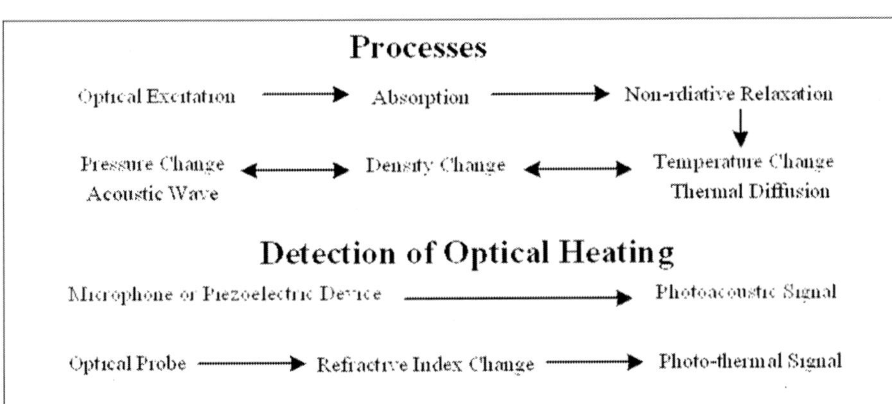

PA spectroscopy is employed in studies on gaseous sample or a solid sample in contact with a non-absorbing gas in which the sound waves are generated by periodic pressure changes and detected by a microphone. In the case of pulsed optical excitation of a condensed sample the pulsed pressure fluctuation is detected by a piezo-electric transducer in intimate contact with the sample. The sound pressure is proportional to the light power and the absorption coefficient of the constituent atomic/molecular species in the sample at the wavelength of light. For gas sensing applications the wavelength of the monochromatic light has to be tuned to a strong absorption line of the molecule to be detected. Tunable lasers are the best optical sources for photoacoustic applications due to their high light power and narrow emission linewidth.

PT spectroscopy refers to methods that monitor the temperature dependent refractive index changes with a probe laser. PT signal depends on the optical power as well as the geometry used to excite the sample which can be arranged to enhance the signal to be large, even for samples with relatively poor optical and thermal properties. There are two reasons for using laser light sources in PT spectroscopy. The first reason is the fact that lasers can deliver high power or energy in a very narrow optical wavelength enhancing the PT signal which is proportional to temperature change. The second reason is the spatial coherence of laser which allows the light to be focused to small diffraction limited volumes. The temperature change is not only proportional to the optical power or energy, but it is inversely proportional to the volume over which light is absorbed since the heat capacity increases with the amount of substance. Thus the small volumes used in PT spectroscopy permits microscopic analysis of heterogeneous materials. Continuous (cw) as well as pulsed lasers are used as optical source for PT and PA spectroscopy studies but the mode of generation of the resulting signals and their subsequent detection is different as will be explained in the next section.

3. PA & PT Spectroscopy Methods

A continuous source of optical excitation produces signals that increase in magnitude with the irradiating time duration. The heat flux produced by diffusion is proportional to the temperature gradient in the beginning and heat removal is slower than that produced by optical excitation. As time increases there is increase in temperature gradient leading to a stage when the optical heating flux equals the flux due to thermal conduction. This leads to a steady state for the spatially dependent temperature change. Thus the PA/PT signals eventually reach a steady-state value dependent on the thermal conductivity of the sample and the duration for signal to develop may vary from milliseconds to seconds.

While using CW laser the light beam intensity is periodically modulated, sometime using a mechanical chopper. The magnitude of resulting oscillating signal depends on the absorption coefficient and thermal conductivity of the medium as well as on the modulation frequency of the optical beam. The signal magnitude is proportional to the optical absorption in the sample but it decreases with increase in the modulation frequency. A phase sensitive lock-in analyzer is used in this case to also provide the signal-to-excitation phase-shift information. This frequency dependent phase-shift information is equivalent to that contained in the time-dependent signal transients produced using a pulsed laser.

Pulsed laser source produces a transient PA/PT signal which has maximum value immediately after the excitation and decays as the sample approaches equilibrium through thermal diffusion. The duration of the transient signals varies from a few microseconds in the gas phase to several milliseconds in condensed samples. In this case a boxcar averager is employed for electronic processing of the detector output.

3.1. Microphone Detection in PA Spectroscopy

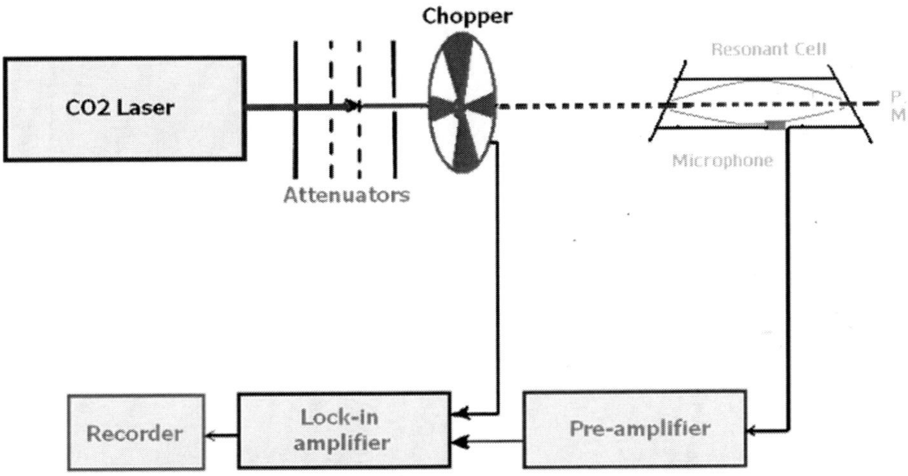

Figure 2. Microphone detection of signal for gaseous sample using a resonant PA cell.

The experimental set up for gas phase PA spectroscopy is schematically shown in Figure 2. The length of the longitudinal PA cell is acoustically resonant for a particular frequency of the mechanical chopper corresponding to the stationary wave formation with a single anti-node in the middle of the cell. The microphone is located near the center of the PA cell for maximizing the signal. The CO_2 laser beam is attenuated to a few milliwatt power before meeting the chopper and its power through the evacuated cell is measured at each rotational-line emission from the laser using a power meter at the exit end of the PA cell. The PA signal from the microphone is pre-amplified before its processing. The output of the lock-in amplifier was manually recorded at each wavelength of the line-tunable CO_2 laser. The PA signals were normalized at each wavelength using the corresponding power meter readings and the spectral features are plotted in Figure 3. It is found that characteristic vibrational bands of CH_3OH and C_2H_5OH vapors are located in the 9.6 μm region whereas those of C_2HCl_3 are found in the 10.6μm region only. This information has been shown to be useful to estimate the relative concentrations of CH_3OH and C_2H_5OH in a mixture of the two organic vapors [23].

The experimental arrangement for measurement of PA signal for solid samples using a non-resonant PA cell is shown in Figure 4 which differs from that of Figure 3 only in the shape of the PA cell and transverse incidence of the laser beam on the sample cell. In this case the PA cell was made in the form of an oval cavity in an aluminum block with the sample chamber at its wider end and the microphone at the narrow end. The cavity could be covered with an airtight flat ZnSe window with about one millimeter thin layer of air between the sample surface and the window. The sample chamber of circular cross-section was of larger depth than the microphone chamber and the upper ends of both were flush with the bottom surface of the ZnSe window. The PA spectra for powder samples of explosives and hazardous drugs were recorded manually and normalized as described above for the vapor samples and some of them are shown in Figure 5a and 5b.

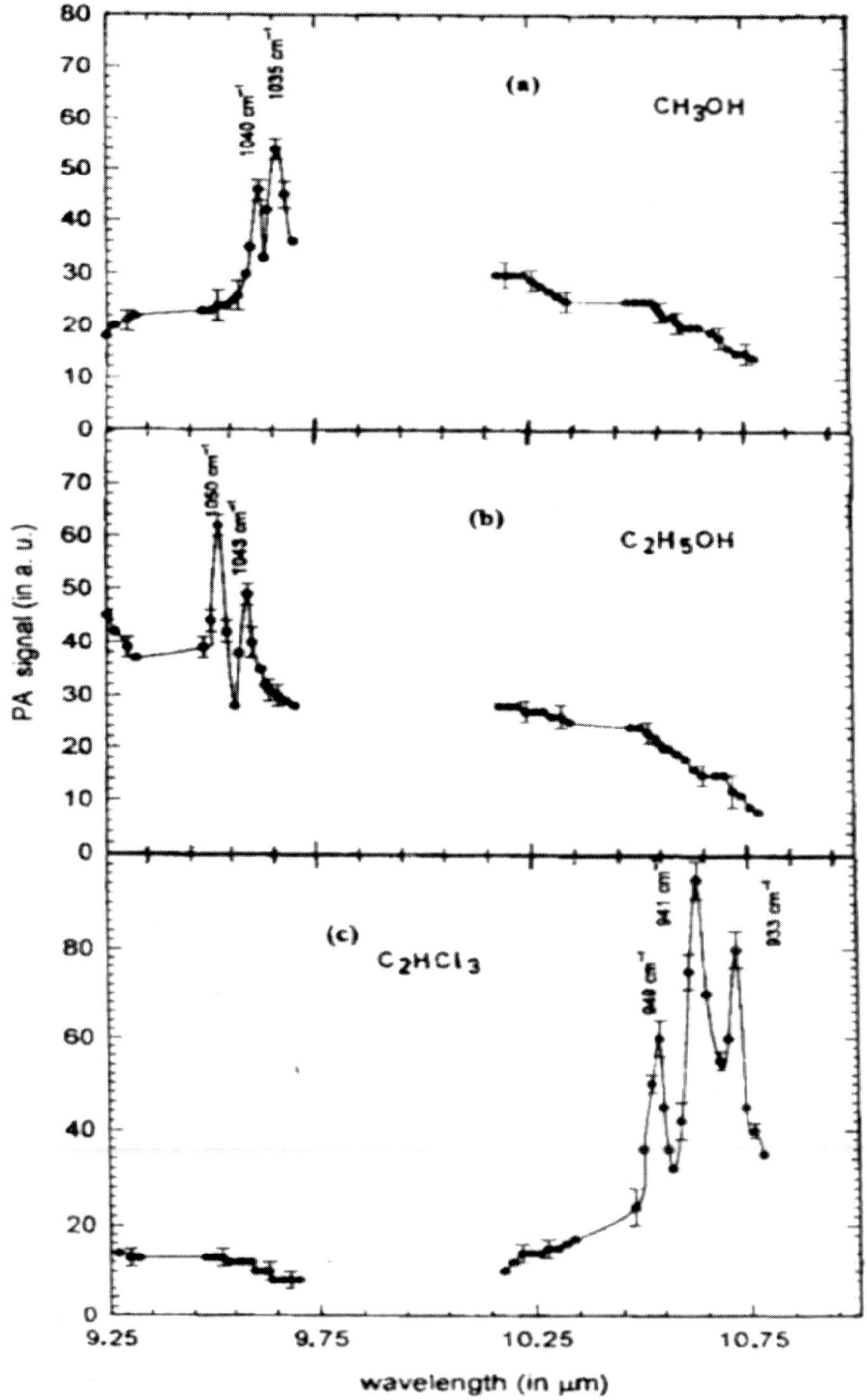

Figure 3. PA spectra of some organic vapors using a cw CO_2 laser.

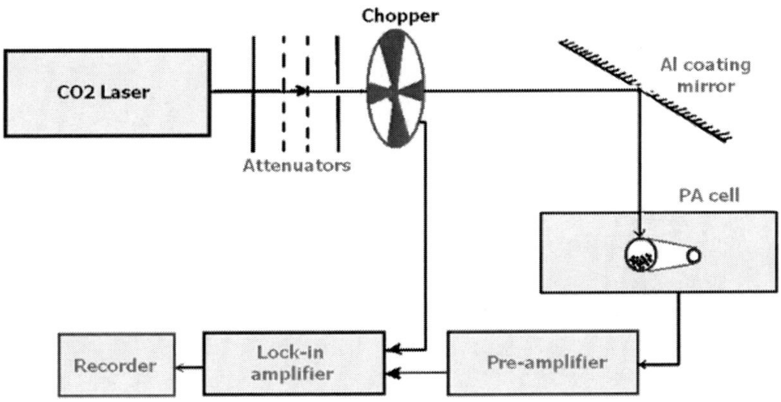

Figure 4. Microphone detection of PA signal from a solid sample using a non-resonant PA cell.

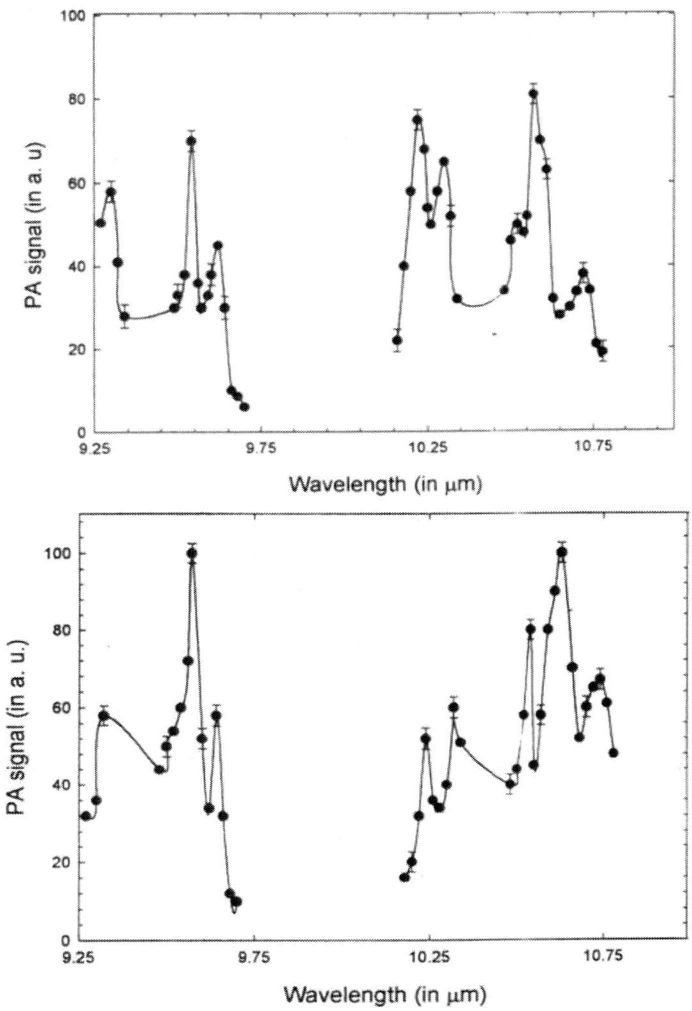

Figure 5a. PA spectral features of TNT (top) and RDX (bottom) in the 9.6μm and 10.6μm regions of CO_2 laser.

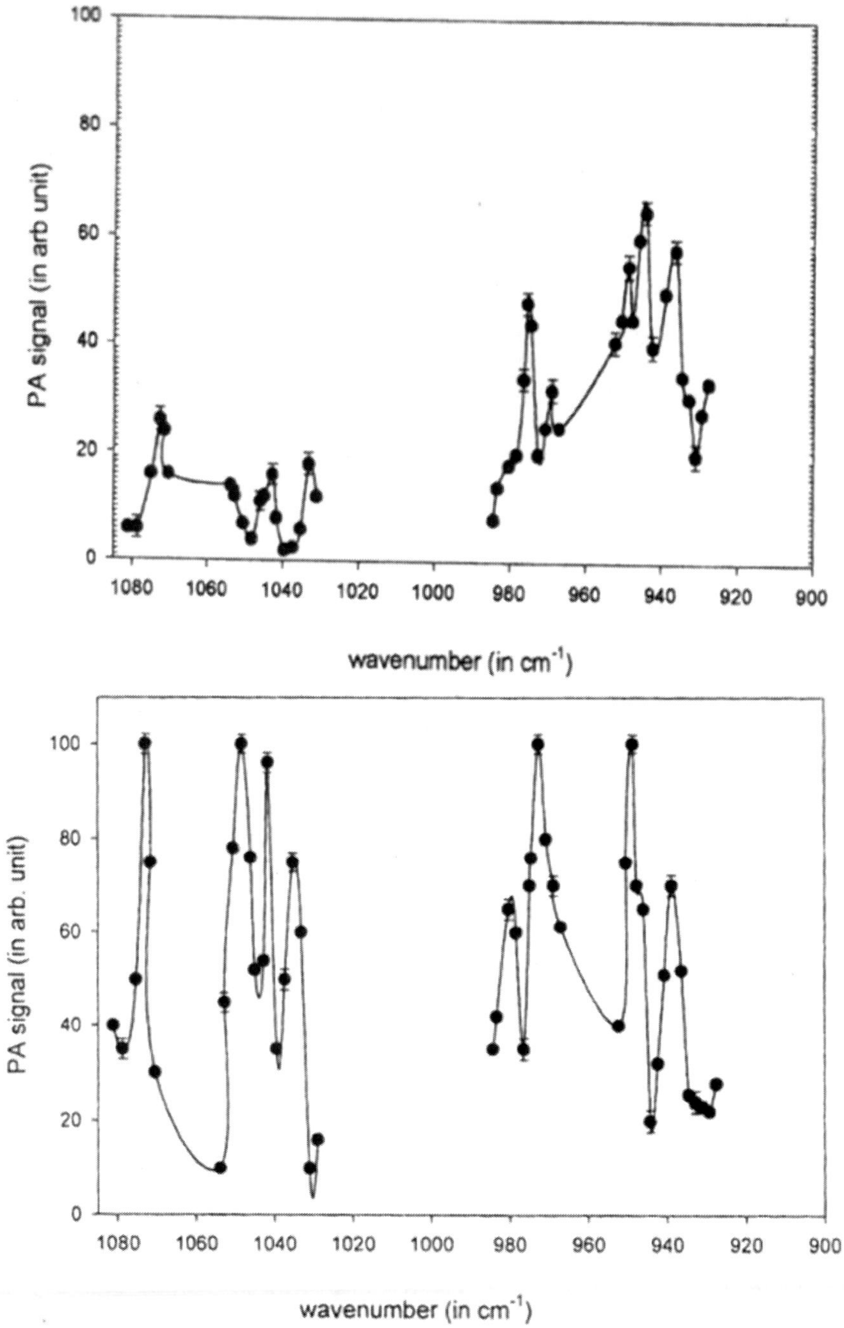

Figure 5b. PA spectral features of heroin (top) and morphine (bottom) in the 9.6 μm and 10.6 μm regions of CO_2 laser.

Some of the spectral peaks in the PA spectra of the explosives and the hard drugs shown in Figure 5 are found to persist at extremely low concentrations and it has been suggested that these features can be used in detection of trace amounts of these hazardous chemicals [24, 25].

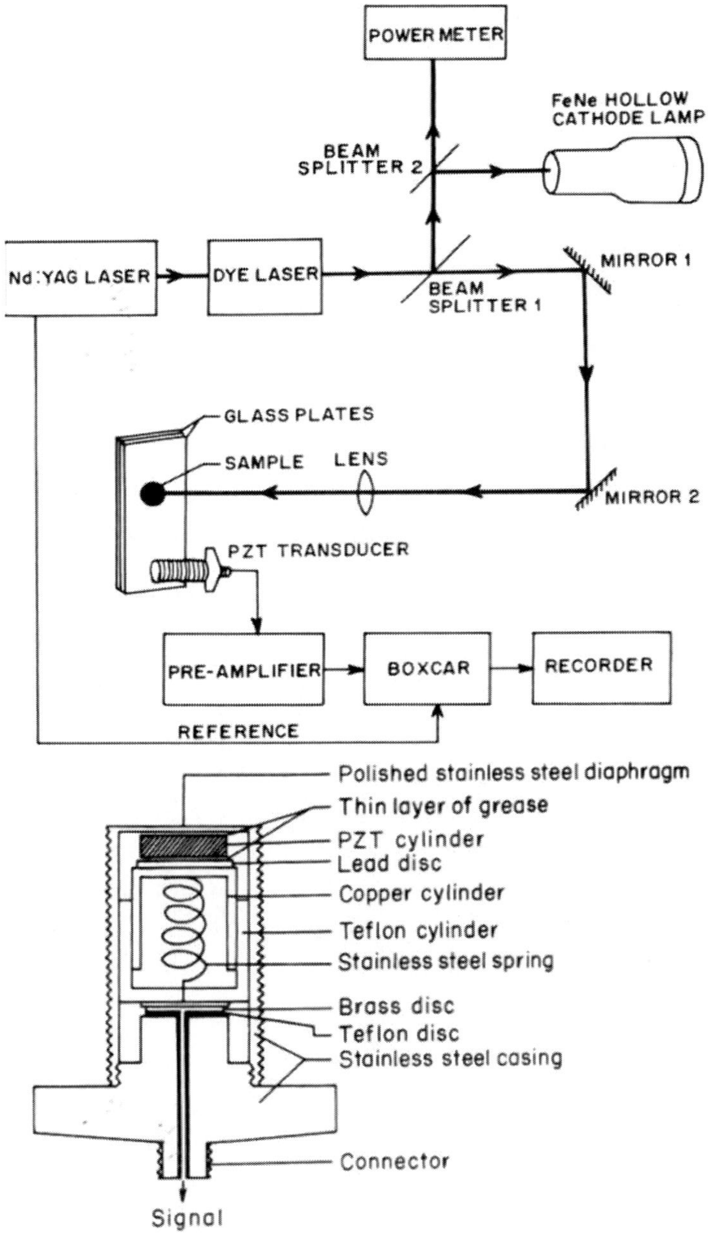

Figure 6. Experimental set up for PA spectroscopy with tunable dye laser (top) using a piezoelectric device (bottom).

3.2. PA Spectroscopy with Piezoelectric Transducer

Pulsed laser generated PA signals from powder samples have been recorded with the experimental arrangement shown in Figure 6 and a home-made piezoelectric transducer whose details are shown at the bottom of the experimental setup [26]. A thin circular layer of the powder sample was sandwiched between two plane glass plates and the PZT transducer was tightly held with its front face in firm contact with one of the glass plates.

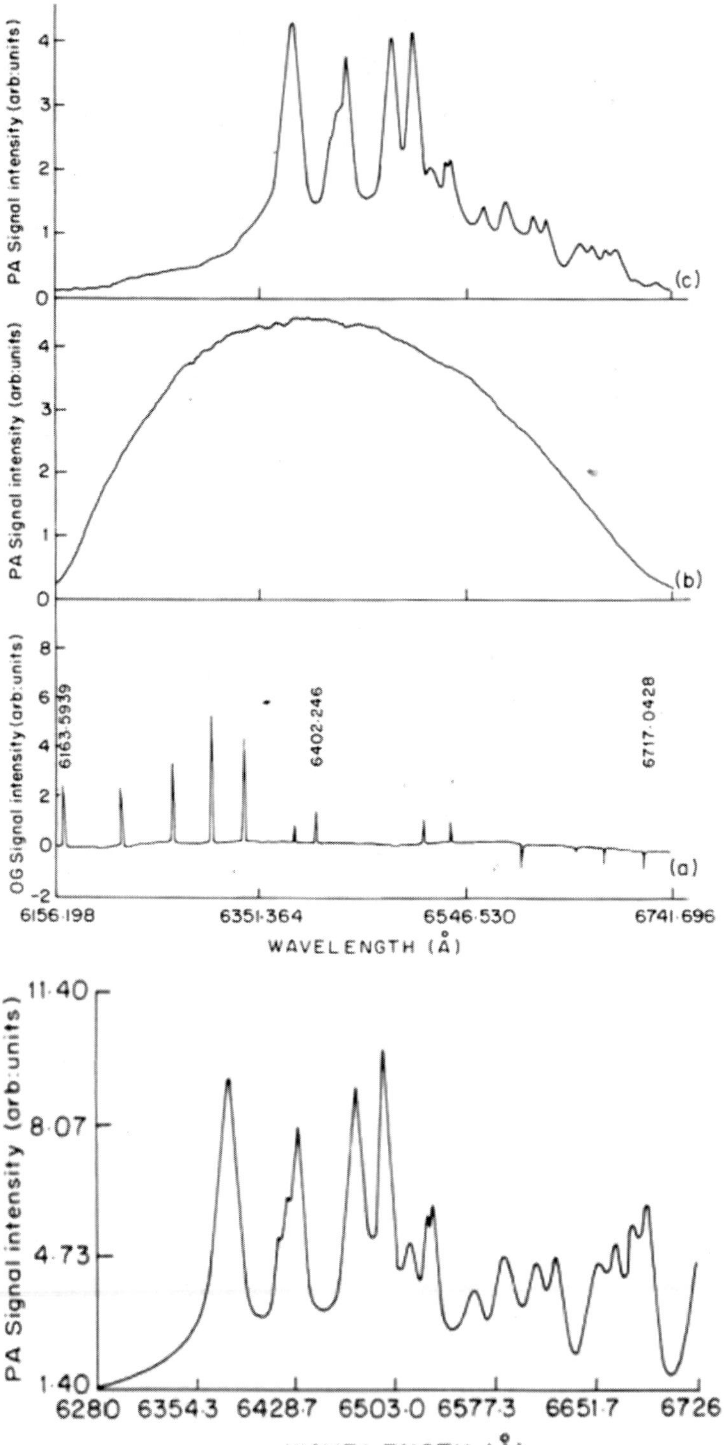

Figure 7. Un-normalized PA spectrum of Ho_2O_3 powder (top-c), spectral power profile of the dye laser (upper middle-b), optogalvanic signals of neon for wavelength calibration (lower middle-a) and normalized PA spectrum (bottom).

The linear separation between the sample spot and the face of the transducer was earlier determined by maximizing the PA signal, while recording the spectral profile of the dye laser output, by using carbon black in place of the sample. The wavelength of the dye laser output was calibrated using the opto-galvanic signals generated from a Fe-Ne hollow cathode lamp. The powder sample consists of micro-crystals of Ho_2O_3 and the light absorbing species are Ho^{3+} ions in these micro-crystals. The acoustic pressure wave generated in the glass plate, following the pulsed optical absorption of the sample, propagates to the transducer where these are converted into transient electrical signals as a result of the piezoelectric effect. The transient PA signals are then processed by the boxcar averager and the spectrum recorded by tuning over the emission wavelengths of the dye laser. A comparison between the raw PA spectrum at the top of Figure 7 and the normalized PA spectrum at the bottom clearly shows that the optical power dependent influence on the PA signal strength is removed by the process of normalization.

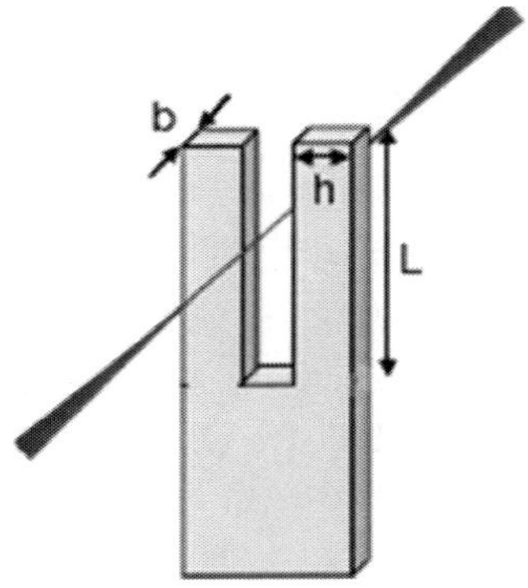

Figure 8. Quartz Tuning Fork located in a gas sample with the laser beam focused at its center.

3.3. PA Spectroscopy with Quartz Tuning Fork (QTF)

A novel approach to photoacoustic spectroscopy of gases utilizing a quartz tuning fork (QTF) as a sharply resonant acoustic transducer was first reported in 2002 [5]. The basic idea of quartz-enhanced photoacoustic spectroscopy (QEPAS) is to accumulate the acoustic energy not in a gas-filled cell but in a sharply resonant acoustic transducer. An easy material for such a transducer is crystal quartz, because of its low-loss piezoelectric property. Quartz tuning forks (QTF) intended for use in electronic clocks as frequency standards resonate at 32768 Hz in vacuum.

The detection of PA signal by a quartz tuning fork is also based on the piezoelectric effect. The two tines of the QTF shown in Figure 8 may be regarded as two identical beams of length L, width ‘b’ and thickness ‘h’. We may approximate the QTF as a system of two weakly coupled beams each of which is a cantilever with a fixed end. The tuning fork has two vibrational modes each with a different natural frequency. One is symmetric in which the two tines move in opposite directions with respect to the central plane in the gap and the other is anti-symmetric with one tine approaching the central plane while the other going away from it. The QTF is so designed that only the symmetric mode induces an electrical signal via the piezoelectric effect. When the periodically modulated laser beam is focused at the center of the gap (as shown in Figure 8) with the tuning fork placed in a gaseous sample, the optical energy absorbed by the gas results in a periodic expansion which generates a weak acoustic pressure wave. This pressure wave excites the symmetric vibration of the tuning fork thereby generating an electrical signal via the piezoelectric effect.

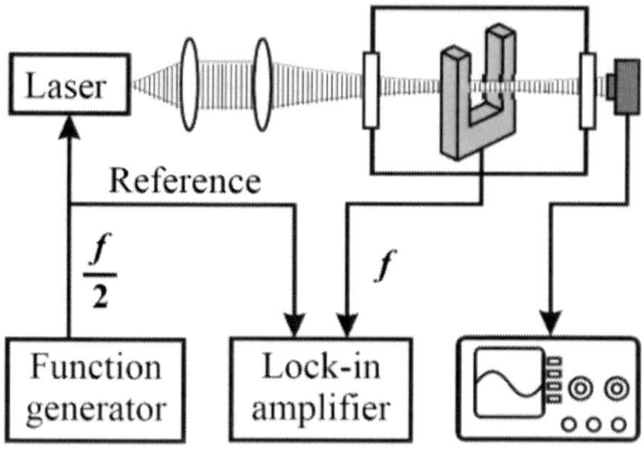

Figure 9. Experimental setup for gas phase PA detection. The DFB diode laser current modulated at half QTF frequency [ref. 5].

The experiment was carried out for methane gas using a DFB diode laser at 1.66 μm which could be wavelength tuned from 5997 to 6001 cm^{-1}. The diode laser temperature was set to 22.8°C so that its frequency coincided with an absorption peak of CH_4 at 5999.5cm^{-1}. The absorption line of the gas was crossed twice during each modulation period which resulted in acoustic wave generated at QTF frequency. PA signals at two different partial pressures of methane [5] are shown in Figure 10.

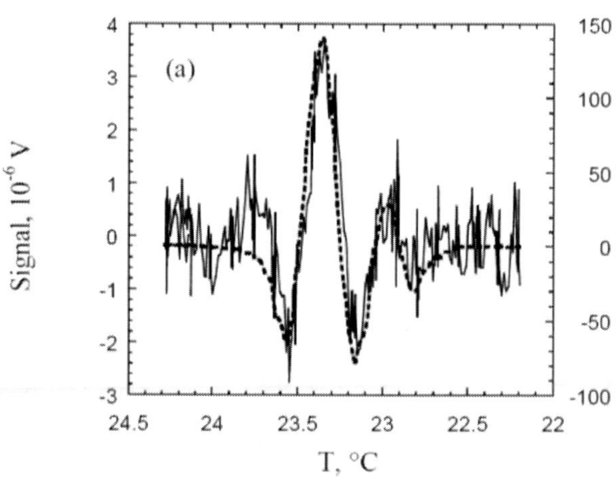

Figure 10. PA signals from QTF in gas cell at 375 Torr with CH_4 concentrations 6.7% (dotted line) and 0.17% (solid line).

The commercial availability of quantum cascade laser (QCL) has provided compact narrow-linewidth mid-IR source that combines single-frequency operation with powers up to tens of milliwatts [27]. The large wavelength coverage available coupled with their room temperature operation makes it possible to monitor trace amounts of numerous molecular species. The output of a typical QC laser operating around 7.35 μm at liquid nitrogen temperature is shown in Figure 11 with respect to the spectrum of TNT and the atmospheric window in the mid-IR region [28].

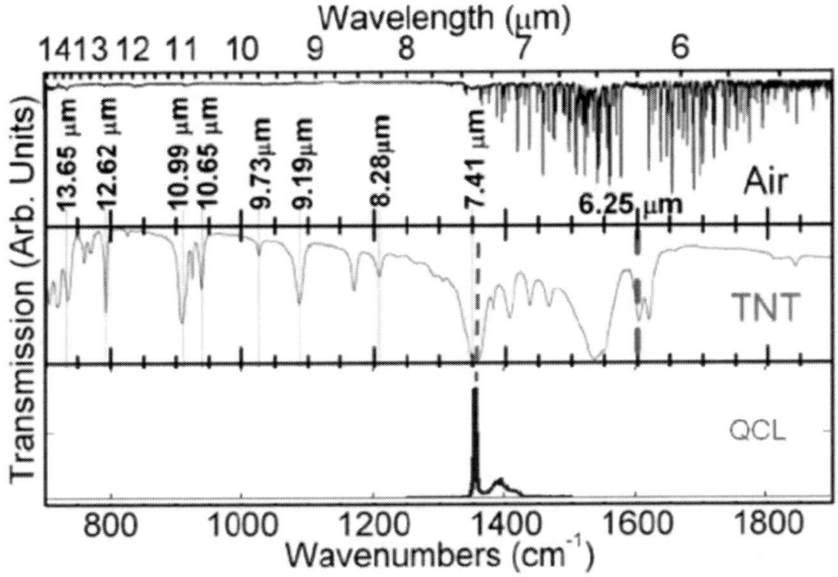

Figure 11. QCL lasing spectrum (bottom), TNT absorption spectrum (middle) and mid-IR atmospheric window (top).

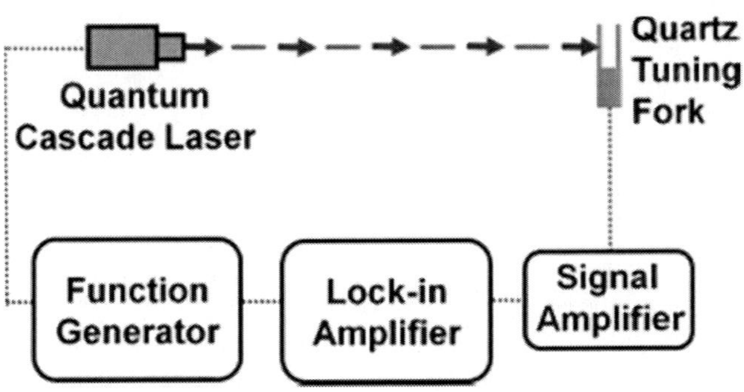

Figure 12. QTF setup with a QC Laser for PA spectroscopy of adsorbed chemicals on the tine surface. The function generator controls the pulse frequency of the QCL.

A pulsed QCL in Figure 12 illuminates adsorbed RDX sample on the surface of one of the tines of QTF with the laser pulse frequency equal to the mechanical resonant frequency of QTF generating acoustic waves at the air-surface interface at the tuning fork. This process gives rise to oscillating localized pressure and sets the QTF into resonance. The amplitude of this vibration and the resulting piezoelectric voltage is proportional to the amount of heat produced by optical absorption on the surface. A plot of quartz piezoelectric voltage thus created as a function of illumination wavelength yields the IR spectrum of the sample in the tunable range of QCL. This is illustrated in Figure 13 by the PA spectrum of RDX using tunable QCL in the wavelength range 7.4 to 8.0 μm.

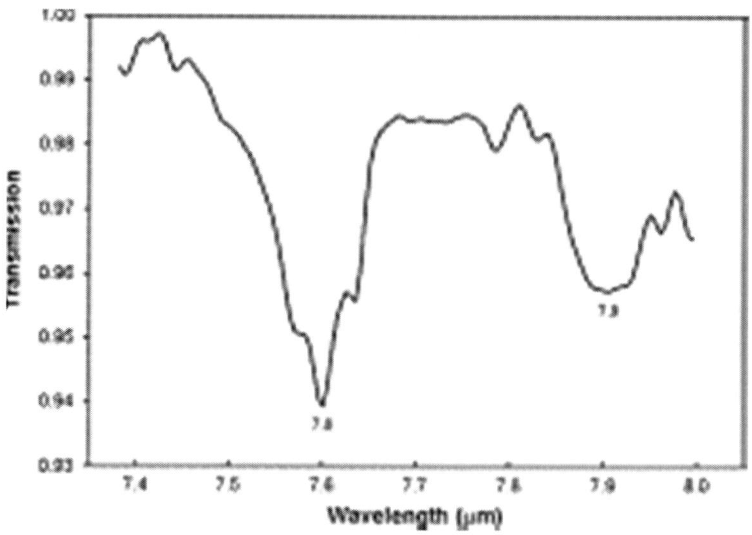

Figure 13. PA spectrum of 2ng adsorbed RDX on QTF in the 7.4- 8.0 µm region. Reprinted with permission from ref. [29].

3.4. Photothermal (PT) Lens Spectroscopy

A photothermal lens is formed due to optical absorption followed by heating of the sample in a localized region that depends on the excitation source. The lens is created because of the temperature dependence of the sample refractive index. This thermal lens has a negative focal length since most of the materials expand upon heating and their refractive index is proportional to the density. If a probe light beam is passed through the medium before and after the optical absorption then it suffers divergence in the latter case. If the absorption causing laser beam and the probe laser are collinear in medium then the formation of the negative lens generates a PT signal detected as time dependent decrease in the power at the center of the probe beam. This process is illustrated in Figure 14 where the excitation laser beam is periodically incident on the sample and the collinear probe laser is cw. The PT spectrum is obtained with a wavelength tunable excitation laser by plotting the PT signal against the excitation wavelength for the optically transparent sample.

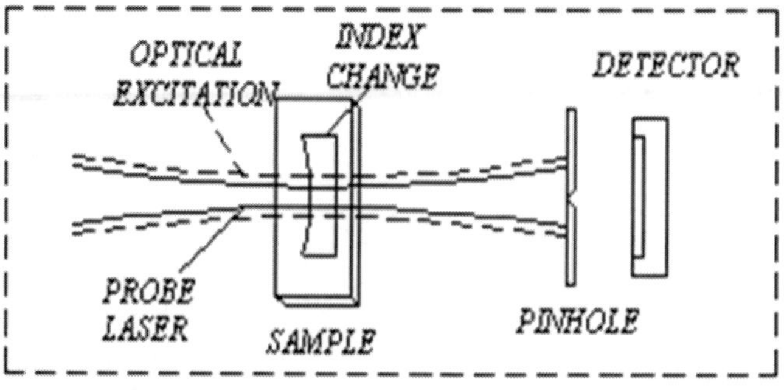

Figure 14. Formation of thermal lens. Wavelength tunable excitation beam is shown by dotted lines and the probe beam by solid lines. The probe laser power reaches the detector through a pinhole.

3.5. Photothermal Deflection Spectroscopy

As described earlier the temperature dependent refractive index measurement is possible in the case of transparent samples only. For opaque or scattering surfaces the temperature dependent changes in the refractive index of a fluid that couples heat out of a solid sample are measured. The optical absorption at sample surface heats the gas or liquid above the surface as shown in Figure 15. The heated fluid acts like a prism and deflects the probe laser incident tangent to the sample surface. A position sensitive detector monitors the change in angle of the probe laser beam. The linear displacement of the laser spot is directly proportional to the deflection angle for small angle. In the process of prism formation the temperature diffusion from the surface creates a thermal wave in the fluid causing the time dependent deflection signal to initially rise and then fall. Thus the time and the magnitude of the maximum signal are both functions of the separation of the probe laser beam from the sample surface.

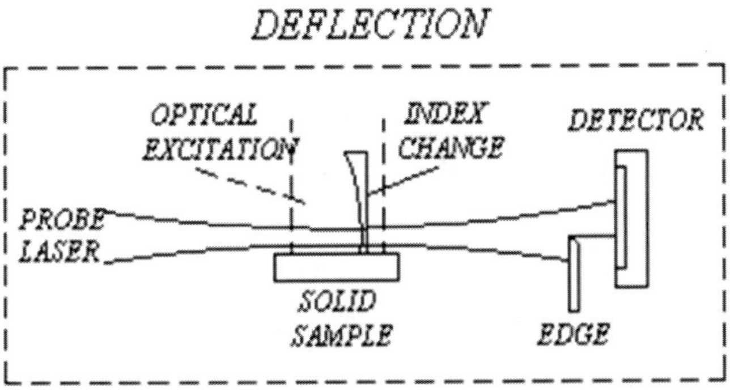

Figure 15. Wavelength tunable excitation beam is vertically incident on the surface of solid sample. The probe laser beam reaches straight to the detector in absence of excitation. Optical heating of sample creates a gradient of refractive index in the fluid in contact with sample surface. This leads to probe laser beam deflection that changes due to optical absorption with the wavelength of tunable excitation beam.

4. Standoff and Point Detection of Hazardous Chemicals in the Field

In the previous sections of this article we have tried to explain the principles of PA/PT spectroscopy with some illustrations of experimental systems being used in the laboratory set up for trace detection of chemicals. In this section we describe the equipments for PA/PT detection of molecules in environments outside of a sophisticated laboratory. The increase in the acts of terrorist incidents in crowded public places like airports, rail and bus terminals, temples and government offices etc during the past decade have attracted a great deal of appropriate security measures. The first step in this direction is the ability to detect the explosives or contraband chemicals concealed in the baggage or on the person or hidden in a place. Laser based detection of chemicals depends upon several factors including the physical state of the sample (solid, liquid or gas), its vapor pressure and knowledge of its spectral characteristics. The spectral signature of an explosive is obtained from its condensed-phase such as the residue left over from a fingerprint or the vapor emanating from the concealed explosive. As discussed earlier PA/PT techniques are based on localized heating and expansion in a gas, liquid or solid subjected to pulsed or modulated monochromatic light when it is absorbed by the sample. In recent times there has been a rapid development in PA/PT systems for remote sensing of explosives, biological agents and harmful chemicals and in the following sections we present examples of some of these advanced systems.

4.1. Photothermal Cantilever Deflection Spectroscopy (PCDS)

PCDS combines the high selectivity of mid-IR spectroscopy with the thermal sensitivity of a bi-material cantilever. A microcantilever used in photothermal spectroscopy is composed of two materials with different thermal expansion coefficients. The microcantilever becomes bent upon heating and this bending is measured optically by the deflection of a probe laser beam reflected on to a position sensitive detector as shown in Figure 16(top). It is possible to obtain molecular signatures of trace amounts of adsorbed molecules on the cantilever surface. The target molecules are first allowed to absorb on a bi-material cantilever which is then exposed to a wavelength tunable beam from an IR source. Upon resonant excitation of target molecules due to IR radiation, the bi-material cantilever undergoes deflection, the magnitude of cantilever deflection as a function of IR wavelength resembles the IR absorption spectra of the adsorbed molecules as shown in Figure 16(bottom).

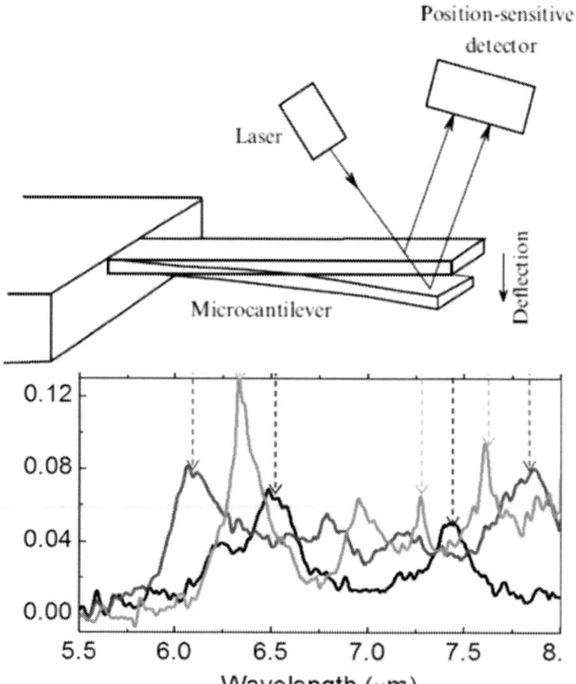

Figure 16. (top) Probe laser beam deflection monitoring the bending of a microcantilever (whose surface contains traces of explosive) when irradiated by a wavelength tunable IR source. (bottom) The PCDS spectra of TNT (black), RDX (green) and PETN (red) using a focused beam from IR monochromator on the microcantilever surface [30].

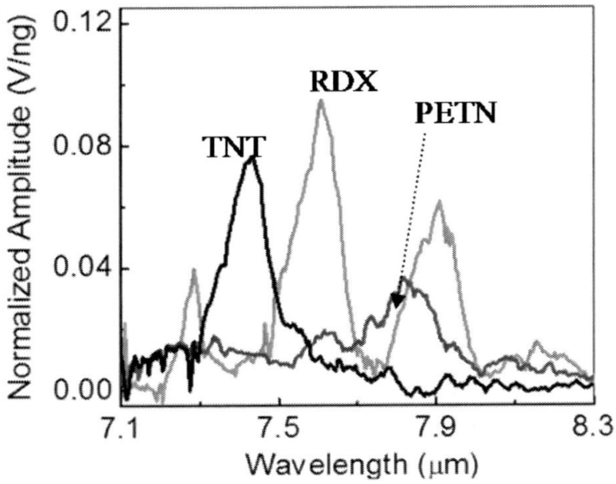

Figure 17. The Point detected PCDS spectra of TNT, RDX and PETN using a tunable QCL source and a few nano-grams sample adsorbed on the surface of the microcantilever [30].

Unlike conventional IR absorption spectroscopy, in which a small intensity change is measured by cryogenically cooled mid-IR detectors, the PCDS measures the photothermal effect of a small number of adsorbed molecules with a high photon flux taking full advantage of the high brightness of QCL light sources. Using high power tunable QCL the sensitivity of detection has been increased by one order of magnitude as shown in Figure 17 for the PCDS spectra of TNT, RDX and PETN recorded using a tunable QCL source [30].

The methods for detecting explosives in the standoff and point modes are radically different. As shown in Figure 18, a standard IR monochromator was used to tune the wavelengths in the range 2-14 μm focused on the distant target having trace residue of TNT on its surface. The scattered IR beam from the target was in turn focused on the clean surface of a microcantilever to record the photothermal spectrum of TNT. The microcantilever was bent as a result of heating by the IR radiation and this bending was recorded optically by deflection of the probe laser beam as shown in Figure 18. It has been found that the sensitivity to detecting TNT traces was about100 ng/cm^2 [31]. The distance from the target can be significantly increased by using a tunable QCL source.

4.2 Photoacoustic Quartz Tuning Fork Spectroscopy (PQTFS)

In the method of point detection the target molecules are allowed to be absorbed on the surface of the sensor and hence the detector and the source of hazardous chemicals have to be in close proximity. We have seen example of point detection by QTF in Figure 13. We will now discuss the method of photoacoustic signal detected in a standoff mode where both the detector and the operator are far removed from the source of the explosive or the harmful chemical [32- 37]. We have presented examples of PCDS for standoff detections where the sensitivity is proportional to the cantilever's ability to bend which in turn is directly related to the stiffness of the cantilever. Although lower stiffness yields greater bending but it reduces the quality factor of the cantilever and as a consequence there is increase in signal noise when exposed to an open environment. The QTF detection circumvents some of these problems and has proved to be a success in its field experiments in open environments.

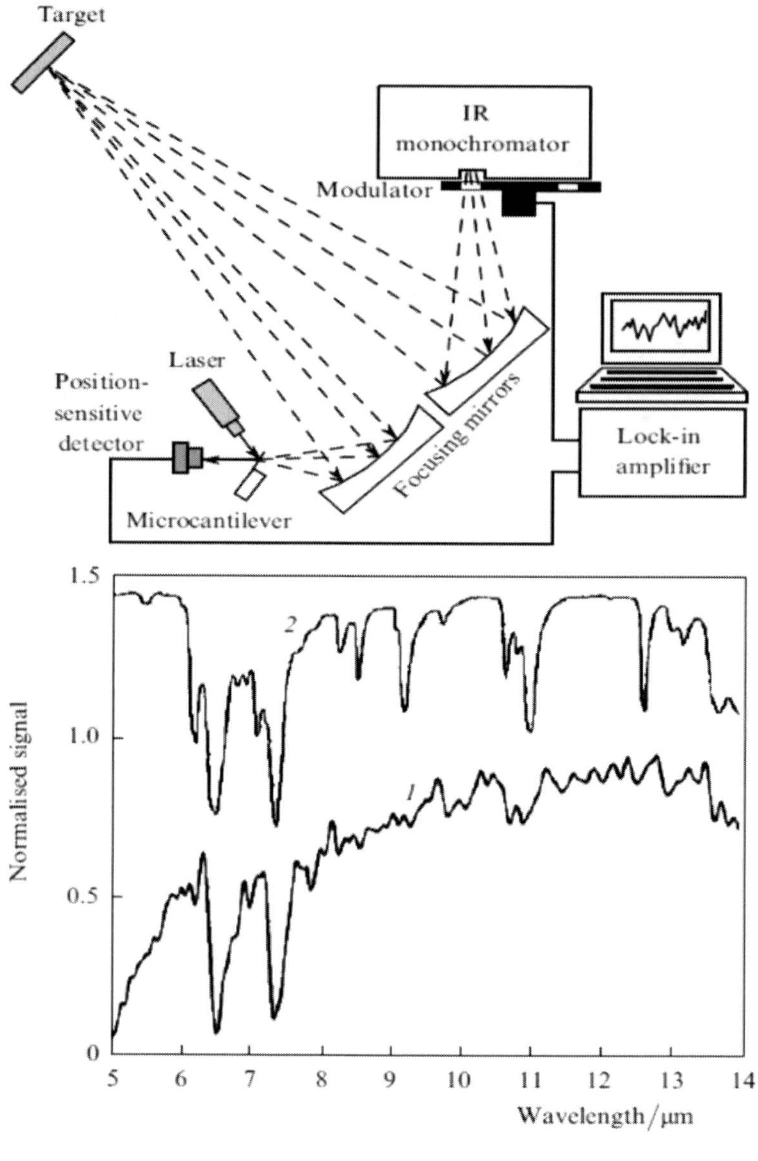

Figure 18. The Standoff detection system for recording the PCDS spectrum of TNT residue on a distant target (top) and the PCDS spectrum of TNT (bottom) compared with its IR absorption spectrum (middle) [31].

When a target surface containing traces of hazardous chemicals is illuminated with a wavelength tunable light source, the light scattered from the surface exhibits variation of light intensity at different wavelengths due to selective absorption by the chemicals. In the PQTF detection system a pulsed tunable QCL source is used to illuminate the target surface as shown in Figure 19. The scattered light from the target is focused on one of the tines of a quartz tuning fork by a concave mirror. The pulse repetition frequency of the QCL is electronically matched with the resonance frequency of the QTF corresponding to its vibration that generates a piezoelectrical voltage. A computer captures the vibration signal from the QTF and records it as a function of the illumination wavelength. The periodic light beam focused on the tine surface of QTF generates periodic pressure changes due to optical heating and sets the resonant vibration. The amplitude of the tine vibrational motion is proportional to the intensity of light focused on its surface. The scattered light from the target suffers a decrease in intensity at certain wavelengths duet to selective absorption by the trace molecules. The amplitude of vibration of the QTF tine is decreased at those wavelengths where the target surface has selective absorption. A computer captures the piezoelectric signal generated by the vibration of the QTF and records it as a function of the illuminating wavelength. The photoacoustic signal appears as a dip at the wavelength of selective absorption by the trace material and constitutes the chemical signature of the molecules enabling the remote detection of the explosives or other hazardous chemicals.

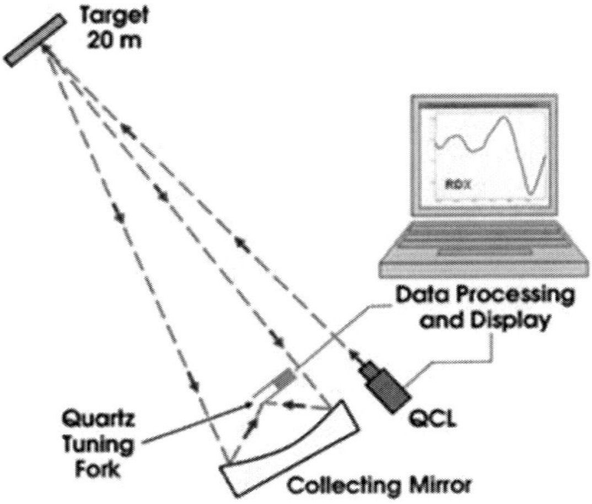

Figure 19. Schematic presentation of a standoff QTF detection system for photoacoustic signals.

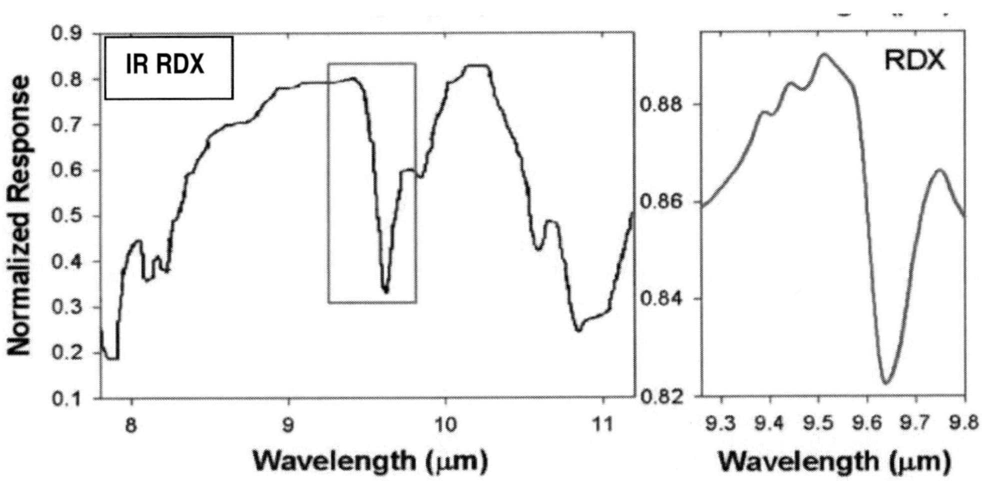

Figure 20. Standoff photoacoudtic spectrum of RDX at a distance of 20 meters from the QTF detector (right). The IR spectrum of RDX in a wider wavelength region with the range of tunable QCL enclosed in a rectangle (left) [38].

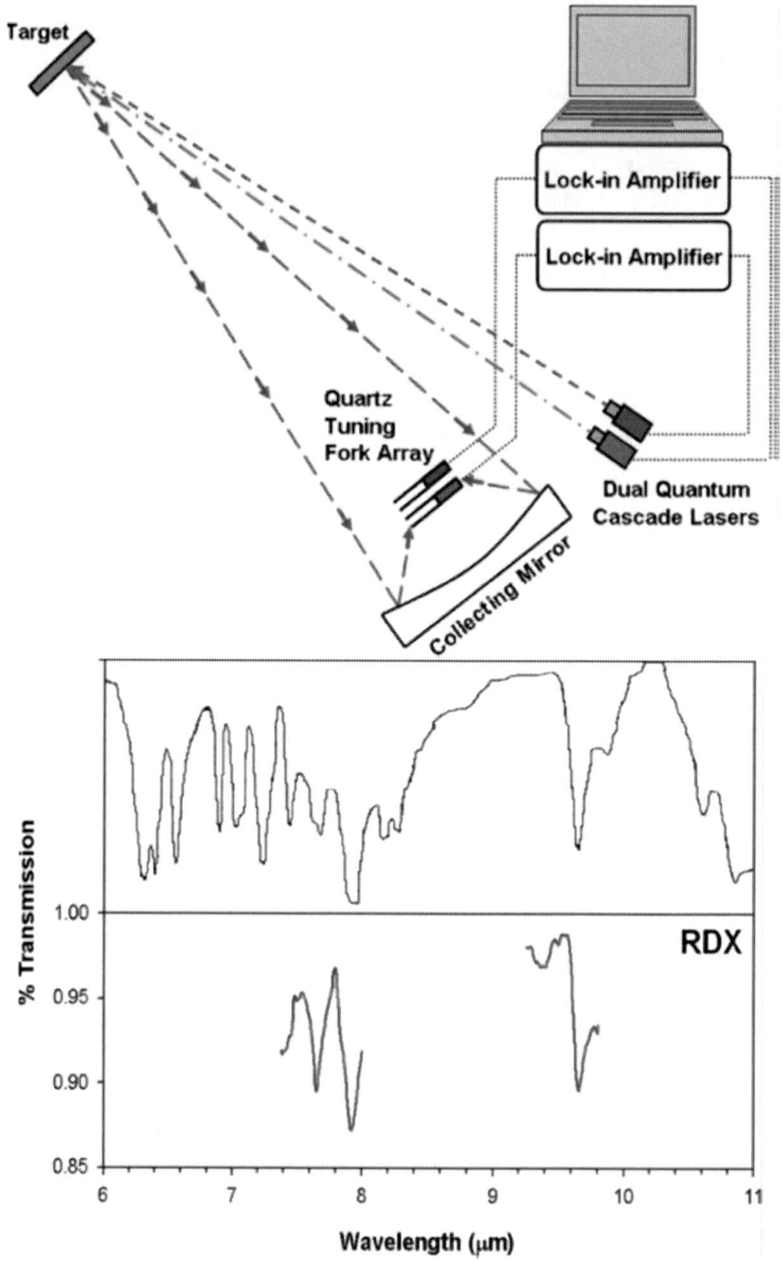

Figure 21. Experimental system for duel QTF detection system (top) the PA spectra of RDX simultaneously recorded in two different regions (bottom) with the conventional IR spectrum of RDX (middle) [39].

Neste *et al.* [38] have demonstrated the standoff detection of RDX shown in Figure 20. The scattered light from the target (its surface containing RDX residue) was collected using a mirror and focused on a Citizen quartz crystal tuning fork (QTF). The QCL light was varied in steps of 0.01nm in its tunable range from 9.25 to 9.80 μm to record the spectrum shown on the right of Figure 20. The device is claimed to have a detection limit of 100 ng of hazardous chemical per square centimeter of target at a standoff distance of 20 meters. The QCL source is a low-power IR laser and is thus eye-safe, unlike other standoff techniques that use visible lasers and because of the QTF detector it is easily operable in field conditions. According to a remark attributed to Larry Senesac, the technique derived from a chance discovery that occurred when the group tried to detect the photoacoustic signal from the target at a distance in open atmosphere.

In a subsequent improvement of their QTF detection system Neste *et al.* [39] used a duel system to extend the spectral coverage of the standoff detection as shown on the top of Figure 21. In addition to the tunable QCL source for 9.25 to 9.80 μm range shown in Figure 19, they employed an additional QCL with tuning range from 7.38 to 8.0 μm. The two QCLs could be scanned through their full tuning range in 0.8 second or could be program controlled to any longer interval with each of them tuned in increments of 0.01nm. The QTF detectors were separated from each other by a gap of about 0.5 mm. In this setup the resonant frequency difference between the two QTF was 143 Hz. This was based on the prior knowledge that the resonant frequency difference between two QTFs in air must be at least 10Hz to avoid their interference between their individual signals.

The baseline PA spectrum is recorded with the two QTF detectors from the bare surface of the target. The target is then coated with varying amounts of RDX residue prior to illumination with the QCLs. In these experiments a quartz crystal microbalance (QCM) was used as the target which allowed a quantitative measure of the material deposited on the surface. The concentrations ranged from about 100 ng to 4μg per square centimeter. The PA spectra shown in Figure 21 are the ratio of the target scan spectrum and the baseline spectrum. In the present case the RDX residue was always in small patches on the surface but not covering the entire surface. This made it possible to scan the surface and ratio the scans so that the difference exhibits as the spectrum of the residue on the surface. The spectrum in the middle of Figure 21 is conventional IR spectrum of bulk RDX and the PA spectra at the bottom clearly exhibit the spectral signatures of RDX in the 8 and 10 μm regions covered by the two QCLs.

4.3. QCL- Based Photoacoustic Spectroscopy of Trace Vapors

Chemical warfare agents (CWA) like nerve agent Sarin and explosives like Triacetone Triperoxide (TAPT) need to be detected in the form of trace vapors to ensure public safety in crowded surroundings. TAPT is easy to synthesize using commonly available chemicals, acetone and hydrogen peroxide which makes it an easy explosive choice for the terrorists. In recent years, the availability of tunable room temperature quantum cascade lasers (QCL), in the wavelength region 5 μm to 12 μm, as excitation source for photoacoustic spectroscopy, has led to the development of very sensitive gas sensors [40-47].

The QCL-PAS system developed by Patel and co-workers [40-42] has been used to record the photoacoustic spectra of TAPT in vapor phase at various temperatures and also of acetone vapor as shown in Figure 22. It is possible to use this system for TAPT detection in real environments with ultra low probability of false alarm (PFA). It is found that the lowest temperature at which TAPT can be detected is about -47°C setting its lowest detection limit of about 1 ppb. To protect the synthesis of TAPT inside an aircraft by terrorists it is necessary to also detect at least one of the precursors and with this object in view the PA spectrum of acetone shown in Figure 22 corresponds to a detection limit of 1.4 ppm.

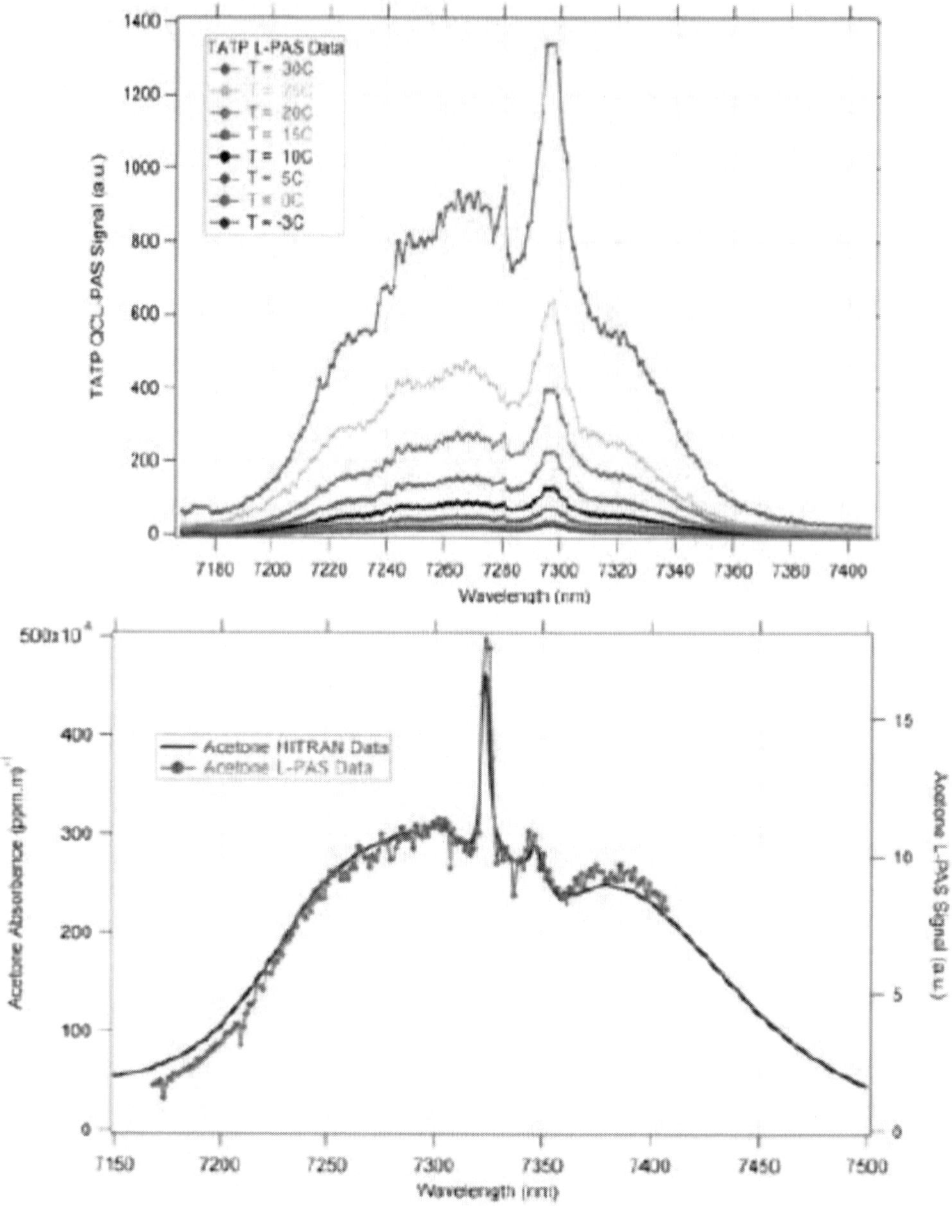

Figure 22. QCL-PA spectra of TAPT at different temperatures (top) and that for 1.4 ppm acetone along with HITRAN acetone absorption simulation shown in blue color (bottom) [42].

Holthoff *et al.* [46] have developed a microelectromechanical system (MEMS)-scale photoacoustic sensor for the detection of trace gases. The MEMS-scale differential photoacoustic cell is shown in Figure 23. The differential detection system has two resonator tubes, both fitted with a microphone, but radiation is directed through one to generate the PA signal. The two microphones are identical and the response of the reference microphone signal is subtracted from that of the photoacoustic microphone signal to allow the removal of noise (such as external vibrations) that are present in both resonant chambers.

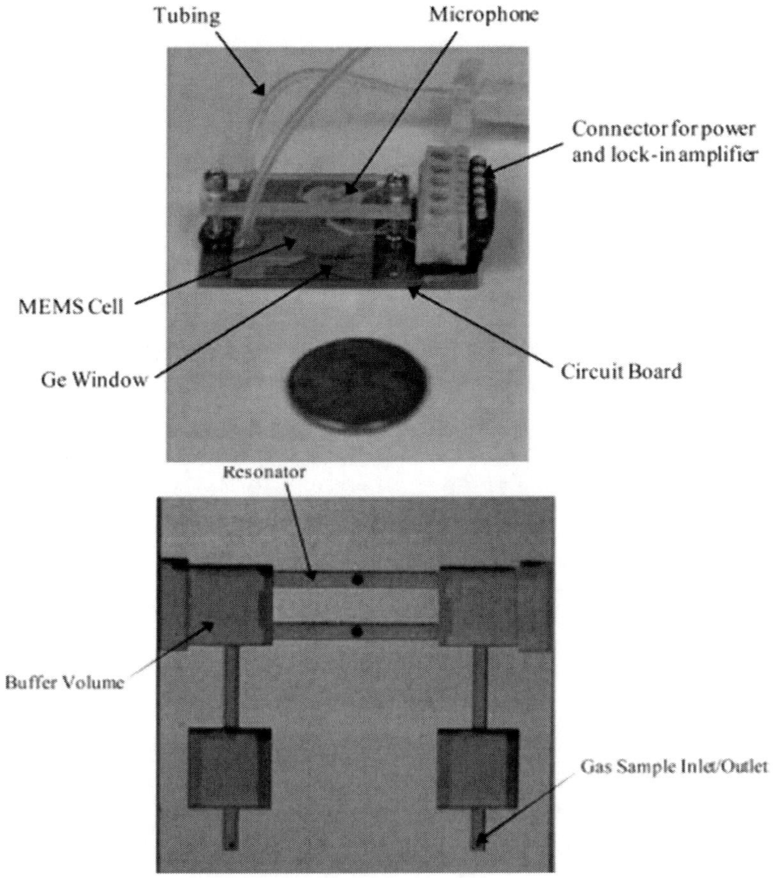

Figure 23. Photograph of the complete MEMS-scale PA cell while collecting data (top) and that of its internal structure (bottom) [46].

The PA cell consists of two 8.5 mm long open resonators having square cross sections. The resonator is flanked on both sides by a buffer volume to provide noise suppression. Each of the buffer volumes is connected by a tube to gas input and output acoustic filters (see Figure 23 bottom). The PA cell has two germanium windows which are attached to the buffer volumes on either side of the resonator. The MEMS-scale PA cell is mounted on a circuit board to allow for easy wiring of the microphones to a power supply and a lock-in-amplifier. The pulsed QCL source was continuously tunable from 1015 cm^{-1} to 1240 cm^{-1} (9.85μm to 8.06μm) and had a resolution of 1 cm^{-1}. A plano-convex lens of BaF_2 was used for focusing. The mode of signal detection in the differential PA system with two resonator tubes is schematically represented in Figure 24.

The PA spectra of acetic acid, acetone, 1,4-dioxane and vinyl acetate in the spectral region from 1050 cm^{-1} to 1240 cm^{-1} are shown in Figure 25. It is interesting to note that the spectra contain absorption features representative of the vibrational modes present in the selected molecules which may be said to provide the chemical signature of these vapors for their detection. Holthoff et.al found that their limits of detection for the four molecules were well below their airborne exposure limits recommended by the National Institute for Occupational Safety and Health (NIOSH). This shows that MEMS-scale PA detection system is very useful for field applications.

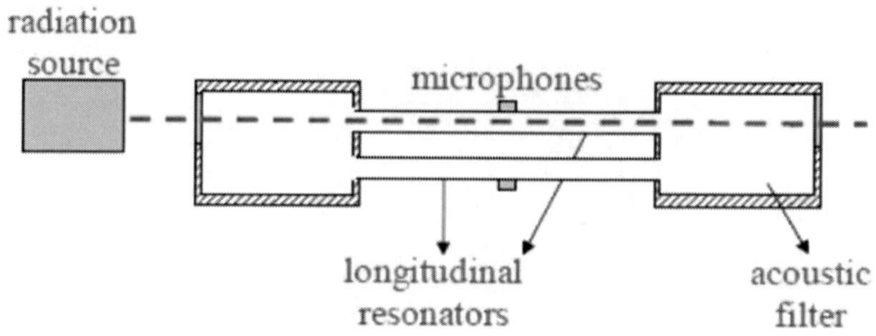

Figure 24. Schematic detection of PA signals by the two microphones in the differential system containing two resonator tubes.

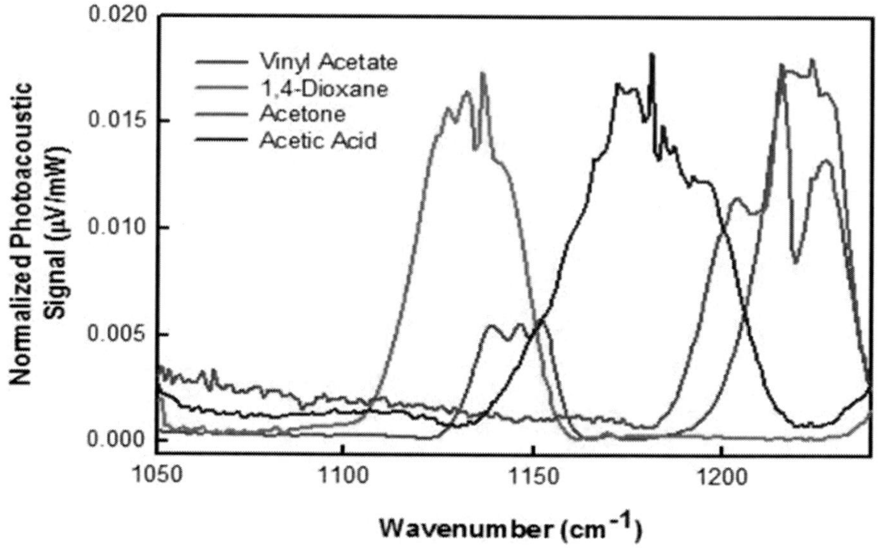

Figure 25. QCL-PA spectra of vapors of acetic acid, acetone, 1,4-dioxane and vinyl acetate recorded by MEMS-scale photoacoustic sensor.

There have been projects in the past to develop a differential absorption lidar (DIAL) and photoacoustic spectrometer for sensing hazardous chemical agents with mid-IR tunable lasers [48]. These systems can be assembled in trucks for remote detection of pollutant gases and warfare agents in the open atmosphere. A laser based PA sensor has been designed and fabricated for trace level detection of NO_2 in the atmosphere with great improvement in the signal to noise ratio of detection [49]. New research reports are regularly appearing on development of sensors for real-time detection of gaseous pollutants and gaseous nerve agents [50, 51]. At the end a futuristic configuration of a Differential Remote Photoacoustic Spectroscopy (DIRPAS) sensor is shown in Figure 26. The sensor is based on eye-safe mid-IR laser excitation with an acoustic microphone/parabolic reflector detection scheme and is in the process of development by optics and infrared sensor team of Pacific Northwest National Laboratory (PNNL) under the US Department of Energy. It is anticipated that the DIRPAS sensor would be capable of detecting fingerprint quantities of RDX on vehicle exterior surfaces at distances out to 5 meters while operating at an outdoor vehicle check point.

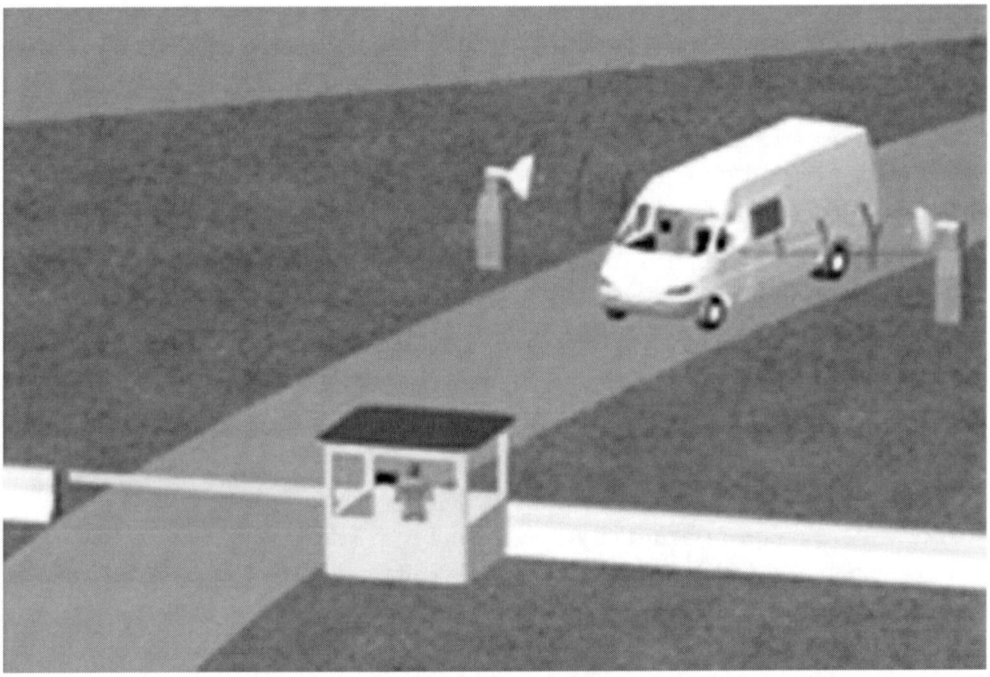

Figure 26. A possible configuration for DIRPAS to detect vehicle borne improvised explosive devices.

CONCLUSION

Optical absorption followed by its conversion to heat in the sample is the origin of generating both photoacoustic (PA) and photothermal (PT) signals. We have briefly described the basic processes within the sample or outside it that lead to a PA or a PT signal. This is followed by descriptions of various methods of detecting the PA and PT signals when a wavelength tunable laser is used as source of optical excitation. The detection techniques have been illustrated with laboratory based experimental systems. It has been shown that PT detection systems are more effective in measuring traces of solid samples whereas PA detection is better suited for gases and vapors. Developments, of practical PT and PA detection systems for use in the real world outside a laboratory, have been discussed with examples of trace detection of explosives and other chemical agents. The emergence of quantum cascade lasers (QCL) has played a very significant role in enhancing the performance capabilities of PA and PT sensors. Rapid developments, in QCL technology, have taken this very important source of mid-IR radiation from a lab curiosity to a rugged field-deployable source for PA/PT spectroscopy. There has been a constant decrease in the size of PA/PT sensors over the years coupled with an order of magnitude enhancement in their sensitivity. It is hoped that in times to come the use of PA/PT sensors for chemical detection will find many new applications in the fields of public safety, environmental science and health science.

ACKNOWLEDGEMENTS

I am thankful to Dr. Y. Dwivedi who provided some much needed literature for the preparation of this article. I met with an accident while writing the article and it would not have been in its present form without moral and material support of Rudra, Vineeta, Sudheer, Sangeeta and Tanya to whom I affectionately dedicate it.

REFERENCES

[1] A.G. Bell, *Am. J. Sci.* 20 (1880) 305.
[2] A.G. Bell, *Phil. Mag.* 11 (1881) 510.
[3] L.B. Kreuzer, *J. Appl. Phys.* 43 (1971) 2934.
[4] L.B. Kreuzer, C.K.N. Patel, *Science* 173 (1971) 45.
[5] A.A. Kosterev, Yu.A. Bakhirkin, R.F. Curl and F.K. Tittel, *Opt. Lett.* 27 (2002) 1902.
[6] D. Weidmann, A.A. Kosterev, F. K. Tittel, N. Ryan and D. McDonald, *Opt. Lett.* 29 (2004) 1837.
[7] A.A. Kosterev and F.K. Tittel, *Appl. Opt.* 43 (2004) 6213.
[8] A.C. Boccara, D. Fournier, W. Jackson and N.M. Amer, *Opt. Lett.* 5 (1980) 377.
[9] W.B. Jackson, N.M. Amer, A.C. Boccara, D. Fournier, *Appl. Opt.* 20 (1981) 1333.
[10] R.L. Swofford, M.E. Long and A.C. Albrecht, *J. Chem. Phys.* 65 (1976) 179.
[11] R.C. Sharma and S.N. Thakur, *Pramana* 56 (2001) 87.
[12] D.S. Moore, *Rev. Sci. Instrum.* 75 (2004) 2499.
[13] S.D. Huang, L. Kolaitis and D.M. Lubman, *Appl. Spectrosc.* 41 (1987) 1371.
[14] H. Li, A. Harris, B. Xu, P.J. Wrzesinski, V.V. Lozovoy and M. Dantus, *Opt. Exp.* 16 (2008) 5499.
[15] C.C. Phifer, R.L. Schmitt, L.R. Thorne, P. Hargis Jr. and J.E. Parmeter, *SANDIA REPORT* (2006) SAND2006-6697.
[16] D. Wu, J.P. Singh, F.Y. Yueh and D.L. Monts, *Appl. Opt.* 35 (1996) 3998.
[17] G.M. Boudreaux, T.S. Miller, A.J. Kunefke, J.P. Singh, F.Y. Yueh and D.L. Monts, *Appl. Opt.* 38 (1999) 1411.
[18] J.P. Singh, F.Y. Yueh, H. Zhang and K.P. Camey, *Appl. Spectrosc.* 2 (1999) 59.
[19] J.L. Gottfried, F.C. Lucia, C.A. Munson and A.W. Miziolek, *Anal. Bioanal. Chem.* 395 (2009) 283.
[20] A. A. Bol'shakov, J. H. Yoo, C. Liu, J. R. Plumer, and R. E. Russo, *Appl. Opt.* 49 (2010), C132.
[21] C. Mullen, D. Huestis, M. Coggiola and H. Oser, *Int. J. Mass Spectry.* 252 (2006) 69.
[22] C.M. Wynn, S. Palmacci, R.R. Kunz, K. Clow and M. Rothschild, *Appl. Opt.* 37 (2008) 5767.
[23] R.L. Prasad and S.N. Thakur, *J. Ind. Chem. Soc.* 80 (2003) 341.
[24] R.L. Prasad, R. Prasad, G.C. Bhar and S.N. Thakur, *Spectrochim. Acta.* A58 (2002) 3094.
[25] R.L. Prasad, S.N. Thakur and G.C. Bhar, *Pramana* 59 (2002) 487.
[26] K. Narayanan and S.N. Thakur, *Appl. Opt.* 29 (1990) 2478.
[27] A.A. Kosterev and F.K. Tittel, *IEEE J. Quant. Electron.* 38 (2002) 582.

[28] X. Chen, D. Guo, F-Sen Choi, C-Chia Wang, S. Trivedi and J. Fan, *Proc. Of SPIE* 8631 (2013) 2H-1.
[29] C.W. Van Neste, M.E. Morales-Rodriguez, L.R. Senesac, S.M. Mahajan and T. Thundat, *Sensors & Actuators* B 150 (2010) 402.
[30] S. Kim, D. Lee, X. Liu, C.V. Neste, S. Jeon and T. Thundat, *Sci. Rep.* 3 (2013) 1111.
[31] L.A. Skvortsov and E.M. Maksimov, *Quant. Eletron.* 40 (2010) 565.
[32] D. Weidmann, A. A. Kosterev, F. K. Tittel, N. Ryan and D. McDonald, *Opt. Lett.* 29 (2004) 1837.
[33] A. A. Kosterev and F.K. Tittel, *Rev. Sci. Instrum.* 76 (2005) 043105.
[34] T. H. Risby and S. F. Solga, *Appl. Phys.* B 85 (2006) 421.
[35] M.R. McCurdy, Y. Bakhirkin, G. Wysocki, R. Lewicki and J. Tittel, *Breath Res.* 1 (2007) 014001.
[36] A.A. Kosterev, G. Wysocki, Y. Bakhirkin, S. So, R. Lewicki, M. Fraser, F.K. Tittel and R.F. Curl, *Appl. Phys.* B90 (2007) 165.
[37] N. Petra, J. Zweck, A.A. Kosterev, S.E. Minkoff and D. Thomazy, *Appl. Phys.* B94 (2009) 673.
[38] C.W. Van Neste, L. Senesac and T, Thundat, *Appl. Phys. Lett.* 92 (2008) 324102.
[39] C.W. Van Neste, L.R. Senesac and T. Thundat, *Anal. Chem.* 81 (2009) 1952.
[40] I.G. Dunayevskiy, A. Tsekoun, M. Prasanna, R. Go, M. Pushkarsky and C.K.N. Patel, *Appl Opt.* 46 (2007) 6397.
[41] A. Mukherjee, M. Prasanna, M. Lane, R. Go, I. Dunayevskiy, A. Tsekoun and C.K.N. Patel, *Appl. Opt.* 47 (2008) 4884.
[42] C.K.N. Patel, *Eur. Phys. J. Special Topics* 153 (2008) 1.
[43] C. Bauer, A.K. Sharma, U. Willer, J. Burgmeier, B. Braunschweig, W. Schade, S. Blaser, L. Hvozdara, A. Muller and G. Holl, *Appl. Phys.* B92 (2008) 327.
[44] S.L. Firebaugh, F. Roignant and E.A. Terray, Modeling the Response of Photoacoustic Gas Sensors, Cosmol Inc. (2009).
[45] E.L. Holthoff, D.A. Heaps and P.M. Pellegrino, *Sensors Journal IEEE* 10 (2010) 572.
[46] E. Holthoff, J. Bender, P. Pellegrino and A. Fisher, *Sensors (Basel)* 10 (2010) 1986.
[47] R.C. Sausa and J.B. Cabalo, *Appl. Spectrosc.* 66 (2012) 993.
[48] V.V. Vaicikauskas, V. Kabelka, Z. Kuprionis, V. Svedas, M. Kaucikas and E.K. *Maldutis, Proc. SPIE* 5613 (2004).
[49] M.A. Gondal and M.A. Dastageer, *J. Environ. Sci. & Health A45* (2010) 1406.
[50] E.L. Holthoff and P.M. Pellegrino, *Proc. SPIE* 8358 (2012).
[51] K.P. Gurton, M. Felton and R. Tober, *Opt. Lett.* 37 (2012) 3474.

In: Spectroscopic Techniques for Security, Forensic ... ISBN: 978-1-63117-404-9
Editors: Y. Dwivedi, S. B. Rai and J. P. Singh

Chapter 5

MONITORING OF ENERGETIC MATERIALS USING LIBS

Shikha Rai[1], A. K. Pathak[1,2], A. K. Rai[1,*], B. K. Ghosh[1] and S. N. Thakur[3]

[1]Department of Physics, University of Allahabad, India
[2]Department of Physics, Ewing Christian College, India
[3]Department of Physics, Banaras Hindu University, India

ABSTRACT

In the article we present the review work on energetic materials using Laser Induced Breakdown Spectroscopic (LIBS) technique. To address the requirements of rapid, *in situ* and standoff detection of minor and major mineral compositions in by-product samples, the potential of LIBS technique has been emphasized in the present study. LIBS spectra of some energetic samples have been recorded in the range of 200-900 nm with detection of elements like carbon (C), hydrogen (H), nitrogen (N) and oxygen (O). Identification of nitro compounds on the basis of intensity ratio of elements and presence of molecular bands are discussed in this article.

1. INTRODUCTION

Increasing use of energetic materials in terrorist activities, poses serious threat to public safety in several areas like security at airports, railway stations, law courts, power installations etc in addition to health safety and environmental remediation. In view of the present scenario, we require an analytical method that works for remote and rapid detection/identification of energetic materials. The explosive materials are pure substances or mixtures that chemically react and rapidly liberate large amounts of heat. The three major classes of energetic materials are nitro aromatics, nitramines and nitrate esters, which undergo rapid decomposition to produce heat, light, sound and sudden increase in entropy due to

* E-mail: awadheshkrai@rediffmail.com.

expansion of hot gases. These explosives are mostly in solid form or formulations whose formation enthalpy or heat of formation is very high. They undergo burning to deflagration to detonation when subjected to mechanical, electrostatic or thermal stimulus. So, before studying these sensitive materials by any analytical technique, it is essential to go into the nature of target samples along with merits and demerits of the analytical techniques to be applied.

A number of techniques are being used, for bulk and trace detection of materials. The methods of x- ray analysis, neutron activation and terahertz imaging are currently being used or undergoing refinements, for detection of energetic materials. One of the most vital needs of the security agencies is the availability of a technique for stand-off detection. Neither X-ray imaging nor neutron activation is practical for remote detection, due to drastically reduced sensitivity at increasing distances and the generation of potentially harmful ionizing radiation [1]. Although terahertz imaging is a promising technique that employs non-ionizing radiation but absorption of water vapor and other species in the atmosphere potentially limit its application for stand-off detection [2]. The common trace explosive detection techniques, such as ion mobility spectrometry and gas chromatography rely on vapor detection, but common explosives have very low vapor pressure. In addition the explosives are concealed by packaging, to decrease the vapor concentration up to three orders of magnitude [3]. Optical techniques such as cavity ring down spectroscopy (CRDS), Raman spectroscopy, Photoacoustic spectroscopy, and Photo-fragmentation in conjunction with, resonance-enhanced multi photon ionization (PF-REMPI) or laser-induced fluorescence (PF-LIF) have been applied to trace explosive detection [4]. Of these techniques, only Raman has been demonstrated on solid nitro compounds at stand-off distances, although long integration times (i.e., multiple laser shots) were required to improve the signal-to-noise ratio [5-6]. The other vital features required, of the technique for explosive detection, are its capability of throughput behavior, multi-elemental analysis, user-friendly, cost, size, mobility, applicability to sample of all aggregation states (solid, liquid and vapor), and real-time measurements. Laser-induced breakdown spectroscopy (LIBS) meet all these requirements.

The ability of LIBS to provide rapid multi-element micro-analysis of bulk samples (solid, liquid, gas) with no sample preparation has been widely demonstrated [7-15] and is, its greatest advantage. LIBS sensors can be made rugged and field portable, all components i.e., laser, detector, computer, etc. can be miniaturized and, it also offers the flexibility of standoff or point detection. In the present work we discuss the use of LIBS for the investigations of different classes of energetic materials particularly nitro aromatic compounds (organic) and inorganic compounds. Pioneer workers have developed LIBS systems for applications like kinetic modeling of LIBS plumes [16–18], detection and discrimination, of geological materials [19–28], chemical & biological warfare agents [29–31] and a number of reviews [32- 35] have been published in the recent years. LIBS has been used in investigations relating to biomedicals [36-39], archeology [40-45], environment [46-49] and forensics [50], polymers [51–54], and thermoplasts [55–57].

1.1. LIBS for Energetic Materials

One of the attractive features of LIBS is its requirement of minute quantities of material, and it has been used to study many groups of energetic materials [58-78]. Elemental composition of the sample can be determined by LIBS because of the characteristic atomic emission lines in the spectrum of the plasma-plume created by laser-target interaction. The atomic emission lines of carbon, hydrogen, oxygen and nitrogen (from ultraviolet to near infrared regions) are commonly used to identify nitro compounds (organic compounds). Anzano et al. [79] have utilized LIBS to develop correlation techniques for sorting of a variety of organic materials (plastics), including polystyrene and high-density polyethylene. The slight differences in the ratios of carbon and hydrogen intensities, have led to great success (90% to 99%) in this method of identification. Portnov et al. [80] have used LIBS to investigate the spectral signatures of nitro aromatic and polycyclic aromatic hydrocarbon samples. They observed the atomic emission lines of C, H, N and O and attributed bands to molecular fragments due to the CN violet system and the C_2 Swan system. These fragments were successfully used to show differences between the compounds studied. Ferioli and Buckley [81] have used LIBS to study the hydrocarbon mixtures (C_3H_8, CH_4 and CO_2 in air). The strength of the C, N and O atomic emission lines was investigated in relation to the concentration of carbon and hydrogen in the samples. In addition to the absolute intensities of the atomic emission lines, the peak intensity ratios of these lines have been used to analyze the samples. Tran et al. [58] analyzed the atomic intensity ratios of several organic compounds in order to determine the empirical formula of a compound, based on the ratios from several elements. Calibration curves were made on the basis of C/H, C/O and C/N atomic emission ratios, for a variety of compounds spread over a wide range of stoichiometry. Four compounds with known stoichiometries were tested against the calibration curves. In the study of nitro aromatic and polycyclic aromatic hydrocarbon, the intensity ratios of C_2 and CN bands, and those between O and N atomic line of different samples were shown to correlate with the molecular formula [80].

Recent efforts at U. S. Army Research Laboratory (ARL) have been focused on detection of explosive residues. The ability of LIBS to detect trace amounts of materials with a single laser shot is especially important for residue detection. The success of LIBS for identifying organic compounds, based on atomic emission intensity ratios, helped the researchers of ARL to investigate the explosive compounds using LIBS technique. Carbon, hydrogen, oxygen, and nitrogen are found in most military explosives. A common characteristic of most explosives is their high nitrogen and oxygen content relative to the amount of carbon and hydrogen [82]. By tracking the amounts of oxygen and nitrogen in a sample relative to the other elements, it is possible to determine whether a compound is energetic or non-energetic. Although explosive residues can be identified, based on the elemental ratios of carbon, hydrogen, nitrogen and oxygen, the interference of atmospheric oxygen and nitrogen with the laser-induced plasma complicates the discrimination of residue. Minimizing the oxygen and nitrogen contribution from the atmosphere is necessary to track the true value of oxygen and nitrogen relative to carbon and hydrogen in a LIBS spectrum. In a laboratory arrangement, argon can be used to displace the ambient atmosphere from the surface of the sample. Use of argon gas at atmospheric pressures also results in an increase in target emission signal compared to air [83].

From the survey of the work on energetic materials using different analytical techniques we concluded that LIBS is suitable technique for the present study. Although, it has been used on several energetic materials, less attention has been paid on compounds like 4-nitroaniline, 4-nitrotoluene, ammonium nitrate and polyethylene glycol. The unique parameter, which we have taken for the identification of nitro compounds, in the present work is based on normalized LIBS intensity ratio, which is obtained by dividing the peak intensity of the atomic lines of elements with their transition probability (A) and their statistical weight (g). The choice of this parameter for the identification, is based on the fact that the intensity of emission line depends mainly on these two factors (i.e., g and A).

2. LIBS Experimental Setup

A schematic of the experimental setup for laser induced breakdown spectroscopy is shown in Figure 1 [8]. The plasma-plume is generated by multiphoton ionization (MPI) and/or tunnel ionization (TI) [84] during laser-target interaction. In the present work we used nanosecond laser pulse with irradiance in the range of 10^7 -10^{12} W/cm^2 and such laser irradiance is high enough to cause a plasma plume in single laser shot. The processes involved in plasma formation as a result of laser-matter-interaction are shown in Figure 2.

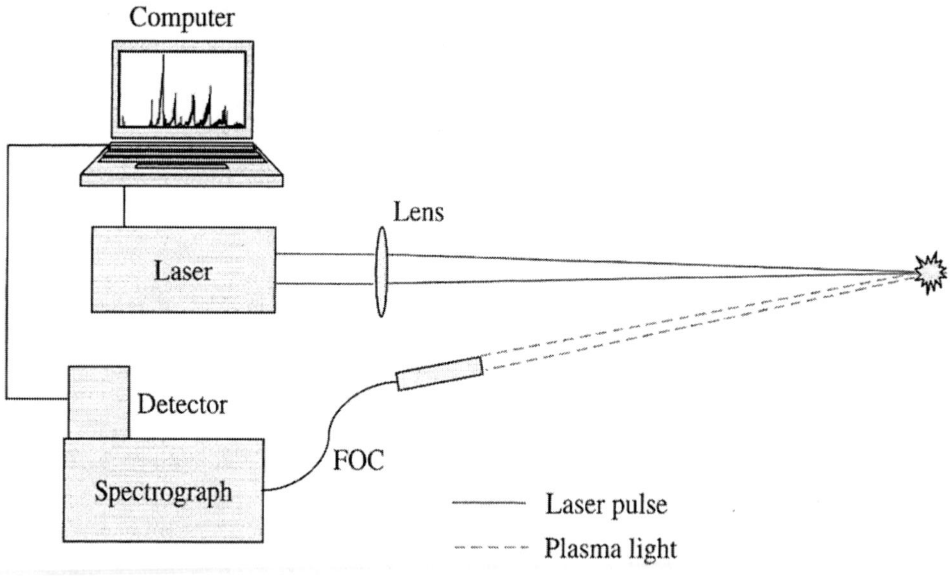

Figure 1. Schematic diagram of experimental setup for Laser induced Breakdown Spectroscopy.

When a nanosecond (ns) laser pulse is focused on the target surface, the optical-interaction causes material to leave the sample surface in the form of electrons, ions, atoms, molecules, clusters, and particles, etc. generating the plasma-plume (Figure 2). The rate of the energy dissipation, through vaporization, is relatively slow compared to energy delivery from the laser pulse. The initial part of the laser pulse causes the underlying layer of the material to reach its critical temperature and pressure before the surface layer vaporizes, and forces the surface to spread out into shock waves. The later part of laser pulse, further heats the expanded plasma plume and results in hot plasma. The plasma quickly cools at the end of the laser pulse, and emits characteristic radiation from atoms and molecules present in the plasma plume.

The removal of sample material from the target as described in the previous section is called 'Laser ablation'. Main issues in laser-ablation are the irradiance and laser pulse duration to initiate vaporization. In order to obtain optimum LIBS signal, it is essential that sample-mass retains its composition after ablation into the plasma. At lower irradiances (i.e., $<10^8$ W/cm^2) and longer pulse duration (microsecond), the thermal processes are dominant causing differential vaporization. At higher irradiances ($\geq 10^8$ W/cm^2) and nanosecond pulse duration the surface temperature is instantly increased and further vaporization is inhibited until the substrate reaches the critical temperature. This condition is more conducive to uniform heating, causing an explosive release of ejected material whose composition is close to that of the target. The phenomena of laser-target interactions have been reviewed by several authors [85–86]. When the laser pulse energy deposited into the sample is much larger than the latent heat of vaporization of all of the constituents, then all atomic species get completely vaporized and removed.

At the beginning, the emission from the plasma is dominated by the continuum radiation due to free-free (Bremsstrahlung emission) and free-bound (recombination radiation) transitions. Free-free transitions are due to the interaction of electrons with positively charged ions whereas in free-bound transition, a free electron (kinetic energy E_k) is captured into an ionic or atomic energy level (having ionization energy E_I) and as a result a photon of energy h $\nu = E_k + E_I$ is emitted. The emission from plasma is captured by the data acquisition system (assembly of fiber optic and spectrograph equipped with charge coupled detector) in the form of spectral signature of target material. This spectrum provides the information about the composition of target material, as shown in Figure 3.

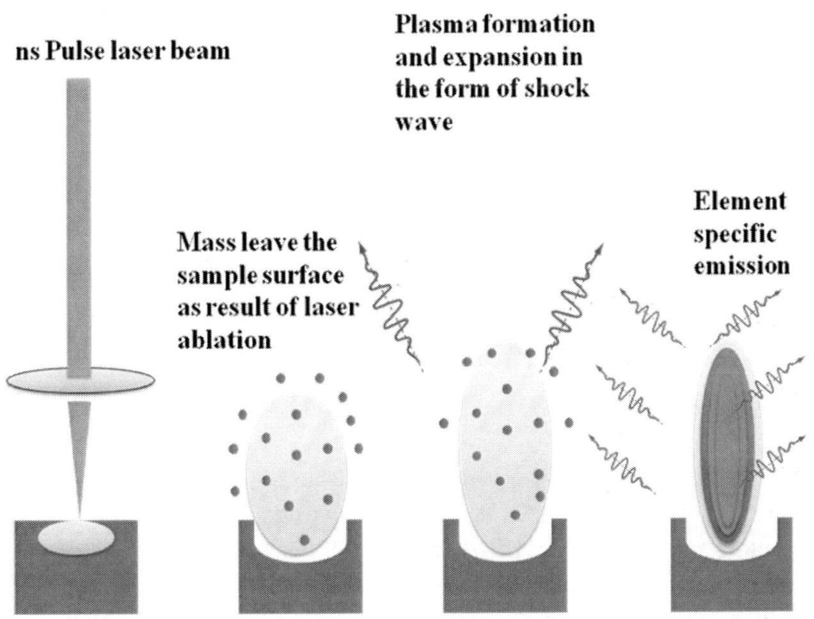

Figure 2. Depiction of the process involved in Laser induced plasma.

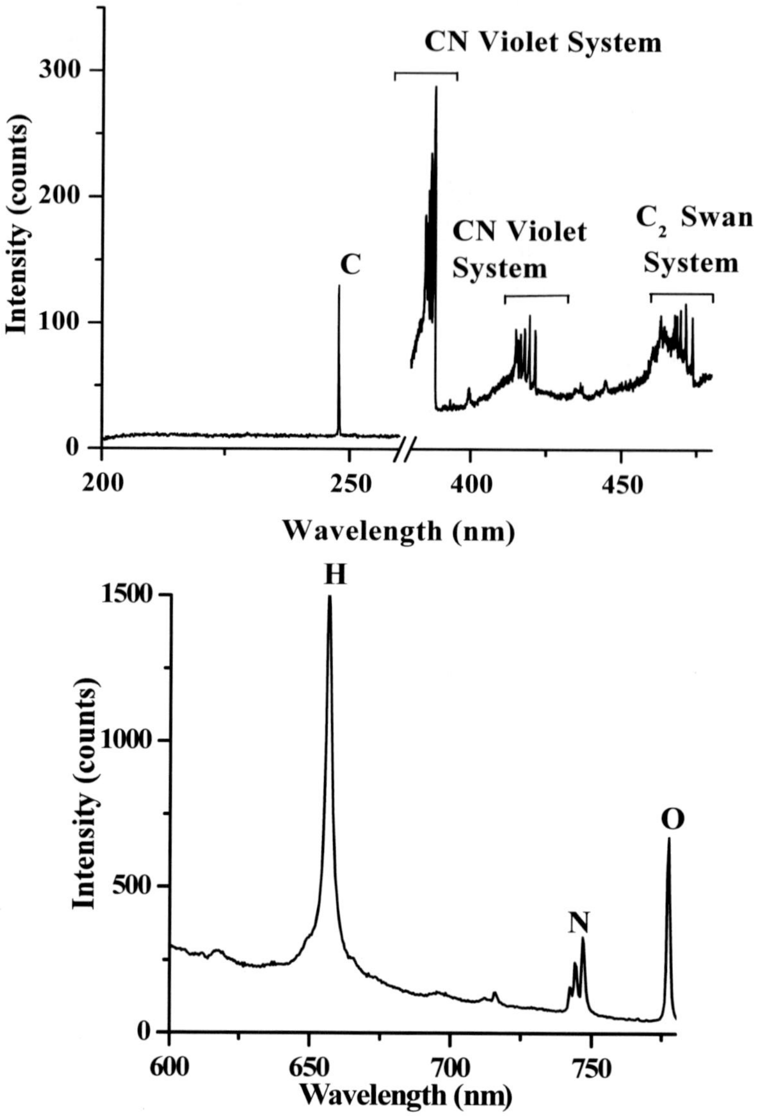

Figure 3. Typical LIBS spectra of organic compound recorded in our laboratory.

3. Comparison of LIBS to Conventional Atomic Emission Techniques and Limitations

In addition to LIBS, there are several other emission spectroscopic techniques such as: Atomic absorption spectroscopy (AAS), Inductively coupled plasma - atomic emission spectrometry - mass spectrometry (ICP-AES and ICP-MS) and X-ray fluorescence (XRF). The samples, generally in liquid or solid form, are vaporized in a flame or graphite furnace to produce free atoms. The main disadvantage of AAS- techniques is that they analyze one element at a time [87] and in this way they are time consuming as compared to LIBS. ICP can be combined with atomic emission spectrometry (AES) or with mass spectrometry (MS).

Both are used to determine the concentration of only trace element in acid solution that needs rigorous sample preparation. XRF can be used in the laboratory as well as in the field. Similar to LIBS, in XRF little or no sample preparation is needed and portable units are also available, but its main drawback is the inability to detect lighter elements with atomic number below 12.

The description of the plasma state and the evaluation of its essential physical parameters are strictly connected to the concept of thermodynamic equilibrium. For plasma to be in complete thermodynamic equilibrium (LTE), all processes are balanced and characterized by a single temperature. Only when these conditions are fulfilled, the observed spectral line intensities can be connected with the relative composition or concentrations of constituent elements of the sample.

3.1. Criteria of Optically Thin Plasma

In the case of very high plasma density, the plasma itself absorbs its own emission. The absorption causes a distortion of the spectral line profile, resulting in a broadened line. For optically thin plasma, the measured intensity I_{ki} should satisfy the following relation:

$$\frac{I_{ki}}{A_{ki} g_k} = \frac{hcN(T)}{4\pi\lambda U\,(T)}\, e^{-\frac{E_k}{K_b T}} \qquad (1)$$

where λ is the wavelength of spectral line; A_{ki} is the transition probability; g_k is the statistical weight for the upper level; E_k is the excited level energy; T is the temperature; K_b is the Boltzmann constants and U (T) is the partition function, and N(T) is the total density of neutral atoms or ions.

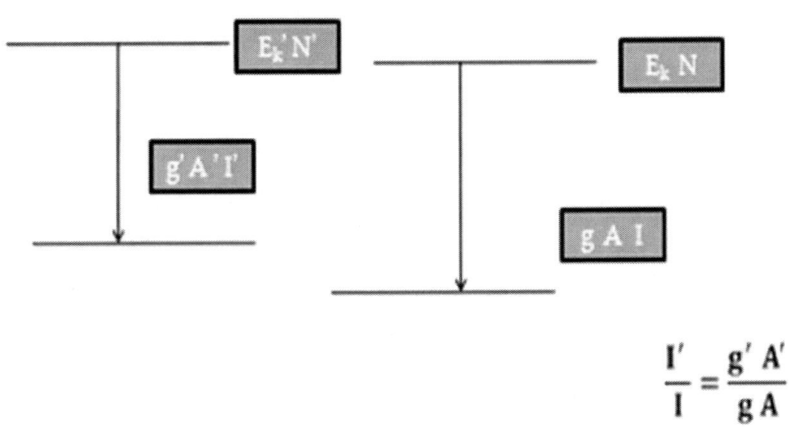

Figure 4. Pictorial representation of optically thin plasma.

So if we take two different emission lines of same species with upper energy levels E'_k and E_k as shown in Figure 4, then intensity ratio of these lines would be close to each other if and only if product of their statistical weight for the upper level and transition probability are close to each other.

3.2. McWhirter Criteria

It states that calculated electron number density should be lower than the value measured experimentally from Stark broadening. i.e.,

$$N_e \geq 1.6 \times 10^{12} \; [T]^{1/2} \; [\Delta E]^3 \tag{2}$$

where T is plasma temperature (K) and ΔE is the largest energy transition in (eV) and N_e is electron density. Equation (2) reveals that in order to verify the McWhirter criteria, the plasma temperature and electron density should be known For LTE at temperature T the emission intensity of a spectral line can be related to the total density N (T) of neutral atoms or ions by Boltzmann's law [88] as shown in Equation (1). On rearranging and taking natural log of Equation (1), finally we have

$$\log_e \frac{I_{ki}}{g_k \; A_{ki}} = -\frac{E_k}{K_b T} + \log_e \frac{C\,F}{U(T)} \tag{3}$$

where I_{ki} is the intensity of emitted spectral line, F is an experimental factor and C is species concentration.

The electron densities are calculated using the relation;

$$N_e \sim \frac{\Delta\lambda_{true}}{2w} \times 10^{16} \tag{4}$$

where Δλ is FWHM and *w* is electron impact parameter.

In a laser-induced plasma source the line-width is due to Stark broadening, which results from collisions of charged species whereas Pressure and Doppler broadenings are quite small [88-89].

4. Results and Discussion

Most of the energetic materials used by military as propellants and explosives are organic compounds containing nitro (-NO_2) groups. These nitro compounds are prototypical molecules for decomposition of energetic molecules and one of their important characteristics is to rapidly penetrate in the skin. The nitro compounds can cause the formation of methemoglobin on acute exposures. Burning, deflagration and detonation of energetic materials are ultimately related to their chemical properties. In present work we discuss the results of our investigations on organic compounds- 4-nitroaniline and 4-nitrotoluene.

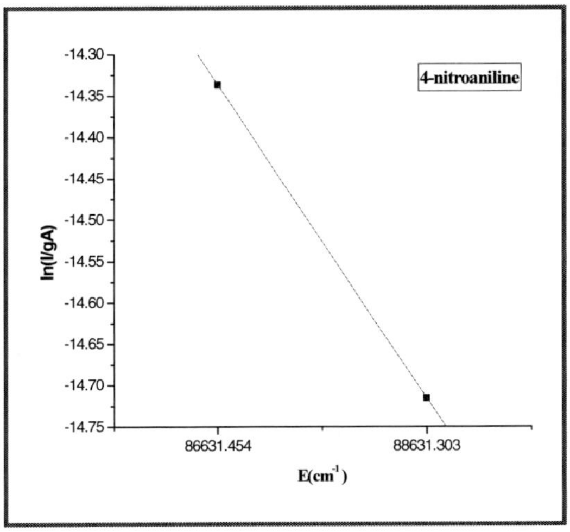

(a)

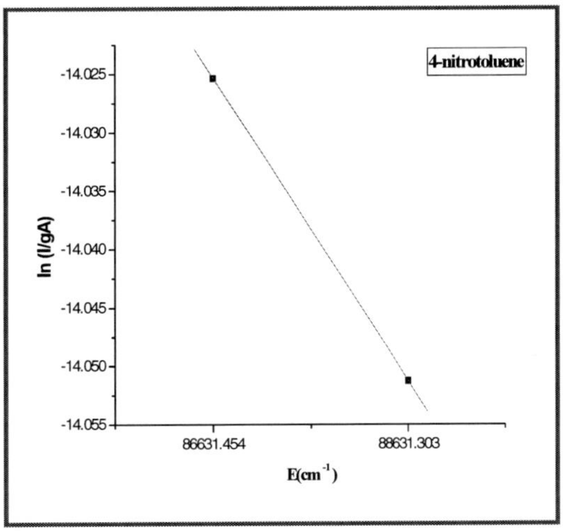

(b)

Figure 5. Boltzmann plot of O (I) for (a) 4-nitroaniline (b) 4-nitrotoluene.

In this work we chose non-resonant lines of O (I) at 777.1 nm and 844.6 nm whose upper energy levels are close to each other and the lower states do not involve the ground state of the atom. To measure the intensity ratio of these two lines, we have recorded the LIBS spectra of 4-nitroaniline and 4-nitrotoluene. The observed intensity ratio of these lines is consistent with the ratio of the products $g \times A$ of statistical weight (g) of their upper energy levels and their transition probability (A), indicating that the plasma is optically thin.

The multiplet structure is not resolved in the observed atomic lines at 777.1 nm and 844.6 nm and each of them is composed of three overlapped components. In order to obtain the separate intensity (I) of each component we have distributed the observed intensity into three different parts according to the relative intensities of the components as given in the NIST Table [90]. The graph between ln I/gA versus E is plotted for the strongest components of 777.1 nm and 844.6 nm and are shown in Figure 5.

Using these graphs, the equilibrium plasma-temperatures are found to be 2120 K and 3050 K for 4-nitroaniline and 4-nitrotoluene respectively.

Lorentzian line shape functions, fitted to the experimental data points of the emission intensities for 777.1 nm lines, are shown in Figure 6 for 4-nitroaniline & 4-nitrotoluene respectively. The observed line-widths, FWHM ($\Delta\lambda_{\text{observed}}$), have been corrected by subtracting the instrumental width ($\Delta\lambda_{\text{instrument}}$) whose value is 0.05 nm for our spectrometer. The corrected values of $\Delta\lambda$ for 777.1 nm line are 1.434 nm and 1.330 nm for 4-nitroaniline & 4-nitrotoluene respectively.

The impact parameter of O at 777.1 nm line for the plasma temperature range 2000 K to 5000 K has been obtained from reference [88] as 1.99×10^{-3} and 2.48×10^{-3} nm for 4-nitroaniline & 4-nitrotoluene respectively, and in the light of Equation 4, the values of N_e for the two samples are found to be 3.60×10^{18} cm^{-3} and 2.68×10^{18} cm^{-3} respectively.

We have also calculated the lower limit for the electron-density for a plasma exhibiting LTE using the McWhirter criterion [91-92], as already shown above in Equation (2). By substituting the value of T = 2120 K and 3050 K (computed from Boltzmann Plots) and ΔE = 1.59 eV from NIST atomic spectroscopic data Table [90] in Equation (2), the values of electron densities come out to be 2.98×10^{14} cm^{-3} and 3.58×10^{14} cm^{-3} for 4-nitroaniline and 4-

nitrotoluene respectively. Thus it is found that the electron densities obtained from equation (4) are greater than their respective lower limits given by Equation (2) and the condition for LTE is fulfilled for the laser-induced plasma in our experiments.

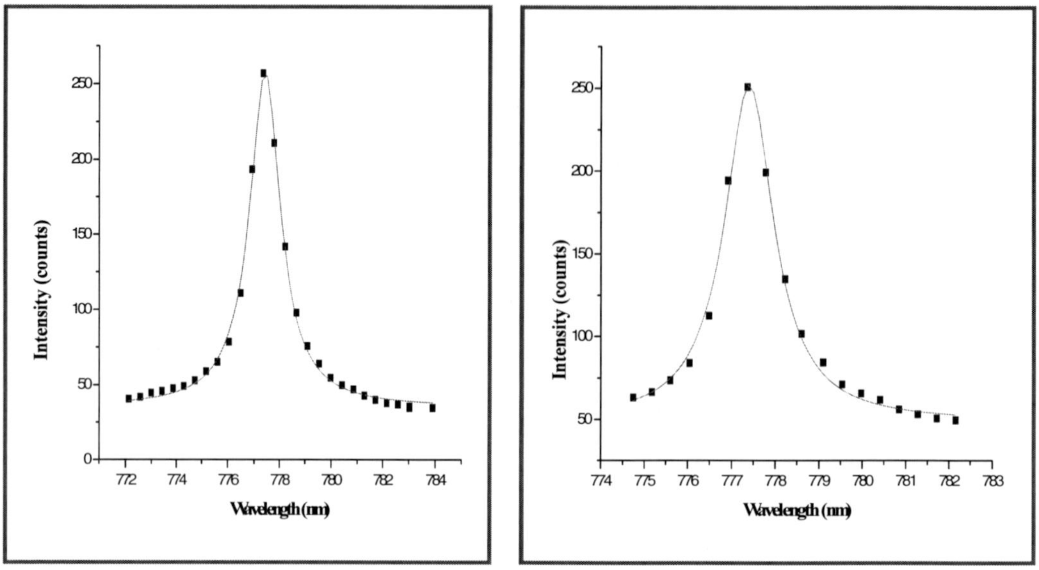

(a) (b)

Figure 6. Lorentzian profile of O (I) at 777.1 nm of (a) 4-nitroaniline and (b) 4-nitrotoluene.

We have also calculated the lower limit for the electron-density for a plasma exhibiting LTE using the McWhirter criterion [91-92], as already shown above in Equation (2). By substituting the value of T = 2120 K and 3050 K (computed from Boltzmann Plots) and ΔE = 1.59 eV from NIST atomic spectroscopic data Table [90] in Equation (2), the values of electron densities come out to be 2.98 $\times 10^{14}$ cm^{-3} and 3.58 x 10^{14} cm^{-3} for 4-nitroaniline and 4-nitrotoluene respectively. Thus it is found that the electron densities obtained from equation (4) are greater than their respective lower limits given by Equation (2) and the condition for LTE is fulfilled for the laser-induced plasma in our experiments.

4.1. Stoichiometric Identification of Nitro Compounds

The use of LIBS as an experimental tool for identification of compounds on an operational level needs extensive developments in the determination of stoichiometry of the compound using intensity ratios of spectral lines of one or more pairs of elements. We have determined the stoichiometry of the compound from the ratio of the intensity of the spectral lines of the two elements i.e., O/N, which has been found to yield good results [93-95] for distinction between different organic compounds. The measured intensities of atomic lines of O at 777.1 nm and N at 744.2 nm were determined from the area under line profile to give (O/N) ratios of 13.97 and 10.28 for 4-nitroaniline and 4-nitrotoluene respectively. The ratio of O/N obtained from the chemical formula of 4-nitroaniline ($C_6H_6N_2O_2$) is, however, 1 and that of 4-nitrotoluene ($C_7H_7NO_2$) is 2.

The mismatch between the O/N ratio obtained from the LIBS spectrum and that from the chemical formula may arise due to two factors. The first reason is the superimposition of three components of atomic line of O (I) (777.19 nm, 777.41 nm, and 777.53 nm) which were not resolved due to lack of resolution (0.75 nm) in our spectra. The second reason is the specific values of transition probability and statistical weight for upper levels of each of the observed lines. To account for these shortcomings we distribute the observed intensity (I) into three distinct intensities by fractional ratio for each of the three overlapped components on the basis of their counts reported in NIST data Table [90] and then taken the value of the ratio I/gA for each of them. This ratio (I/gA) for each component line of O relative to that of 744.2 nm line of N for average spectra of 4-nitroaniline and 4-nitrotoluene are shown in Table 1.

Table 1. Measured intensity ratio of O (I) at 777.1 nm with respect to N (I) at 744.2 nm for 4-nitroaniline and 4-nitrotoluene

Nitro compound	Stoichiometry (O/N) ratio	Measured (O/N) ratio
4-nitroaniline $C_6H_6N_2O_2$ Mw =138.12 g mol^{-1}	1/1	~1
4-nitrotoluene $C_7H_7NO_2$ Mw = 137.13 g mol^{-1}	2/1	>1.5

The present study demonstrates that LIBS is capable of characterizing organic compounds and also has potential to identify the samples on the basis of the intensity ratio of their characteristic spectral lines.

4.2. Interference of Air Atmosphere on the LIBS Signal of Nitro Compounds

Ambient air contains around (80%) nitrogen and (20%) oxygen and their influence on the laser-induced plasma complicates the goal of discrimination of residues of different nitro compounds. Therefore, it is essential to minimize the contribution of atmospheric nitrogen and oxygen in order to track the actual spectral lines of nitrogen and oxygen in LIBS spectra from chemical composition of the sample. To displace atmospheric nitrogen and oxygen from the sample surface we have purged it with argon gas. The overall LIBS signal (intensity of spectral lines as well as background) increases in presence of argon but there is decease in the intensity of lines of N and O. Since argon gas is heavier, and has lower thermal conductivity relative to air, the plasma-plume gets confined near the sample surface in presence of argon as compared to air. The observed spectral signatures of lighter elements- carbon (C), hydrogen (H), nitrogen (N) and oxygen (O) for the present investigation- are shown in Table 2.

We have calculated certain sets of LIBS intensity ratios of constituent elements C, H, N and O for the residue of nitro compounds as shown in Figure 7. It is found that ratios of O/C, O/H, N/ C and N/ H get diminished while O/ N ratio gets enhanced for the LIBS spectra, recorded in argon atmosphere. These ratios basically exhibit, the effects of reduced interference of air and confinement of plasma in the presence of argon. The decrement in the intensity ratios of O/C, O/H, N/ C and N/ H reflect the reduction in the number density of N and of O in going from air to argon, and the effect of plasma confinement which increases the signals for C and H. The increment in the intensity ratio O/N is also governed by the same factors, but in this case reduction in signals of N and O due to absence of air dominates over increment in these signals due to plasma confinement. Our results are also supported by the results of earlier workers [75].

Table 2. Spectral wavelengths of organic elements identified in energetic materials

Elements	Wavelength (nm)
Carbon	229.6 nm, 247.8 nm
Hydrogen	656.2 nm
Nitrogen	744.2 nm, 746.8 nm
Oxygen	777.2 nm, 844.2 nm

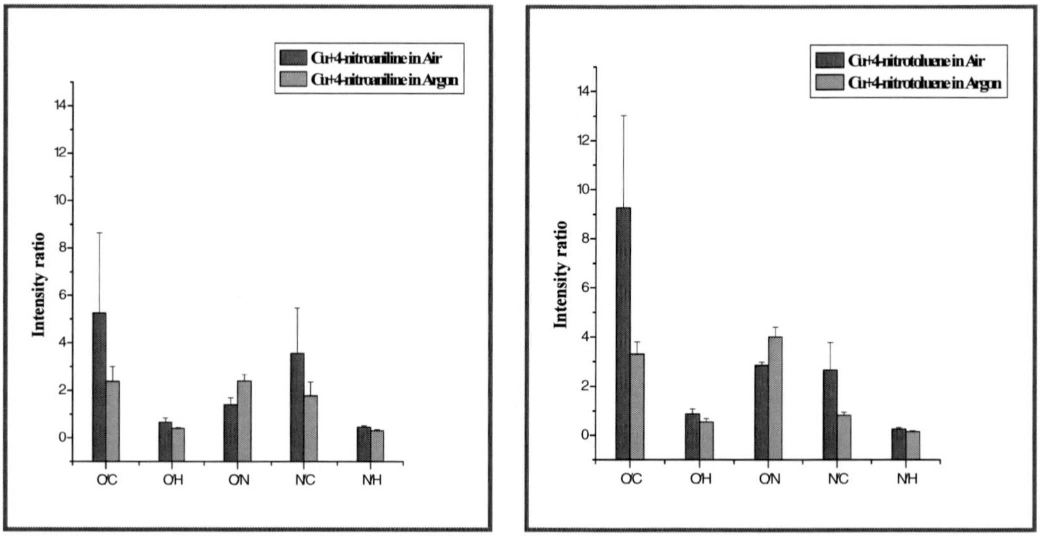

(a) (b)

Figure 7. Bar diagram of intensity ratio of different spectral lines present in the LIBS spectra of (a) 4-nitroaniline and (b) 4-nitrotoluene in Cu matrix recorded in Air and Argon atmosphere.

4.3. Intensity Ratio of C_2 to CN Bands in Air and Argon Atmosphere

Correlation of alteration in the intensities of molecular bands appearing in the LIBS spectra of both nitro compounds is useful in their identification and discrimination. Therefore, we have evaluated the intensity ratio of the C_2 (473.7 nm) band to the CN (388.3 nm) band for each nitro compounds both in the presence of air and of argon. These two bands have been chosen for calculation of the intensity ratio because of their very high intensity. It is clear from Figure 8(a) that the LIBS signal for C_2 increases in argon while it decreases for CN, as reflected in the normalized intensity ratio of C_2/CN. The intensity ratio of C_2/CN increases in the presence of argon, but its values are substantially different for the two compounds, being larger for 4-nitrotoluene in comparison to 4-nitroaniline (Figure 8(a)). The existence of the C_2 band system in the LIBS spectra of both compounds is directly related to the sample composition. We have normalized intensity of the C_2 band by dividing the band signal with the corresponding background signal and results are shown in Figure 8(b).

Correlation of the intensities of the atomic spectral lines of C, H and N with molecular bands (C_2 and CN), appearing in the LIBS spectra is also helpful in studying the intensity variation in atomic lines of different elements of organic compounds in presence of argon and air. Thus measurement of the intensities of atomic lines of C, H and N appearing in the LIBS spectra of nitro compounds is also essential to establish the correlation between their molecular structures. Hence we have also calculated the normalized intensity of the spectral lines of C, H and N at 247.8 nm, 656.2 nm and 744.2 nm respectively.

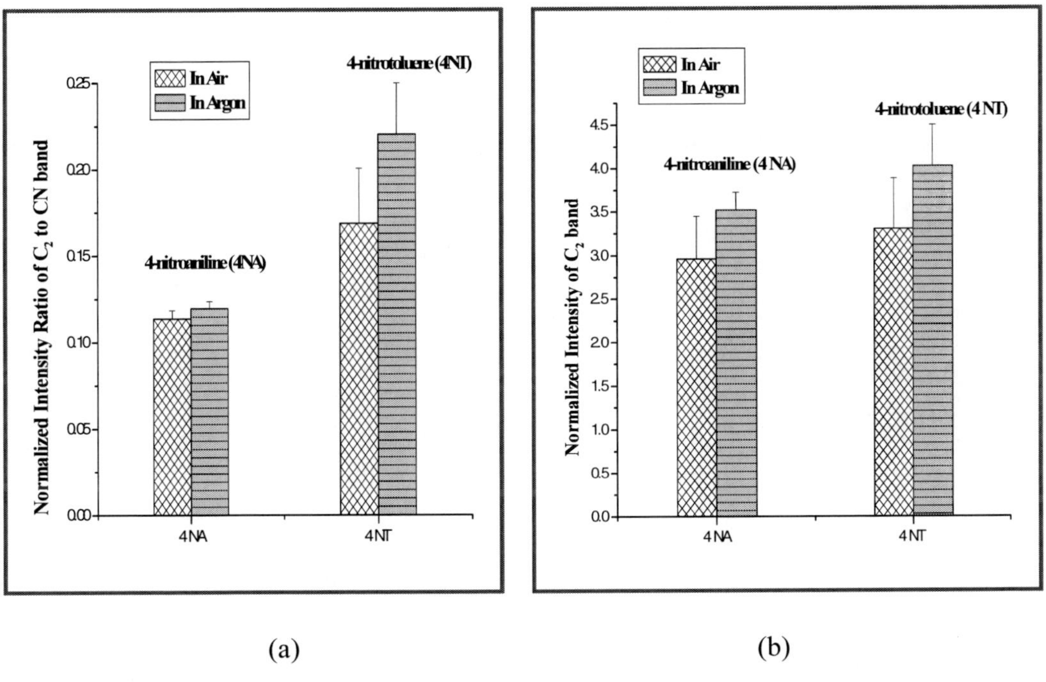

(a) (b)

Figure 8. Bar diagram of the normalized intensity ratio of two molecular bands (C_2/CN) in the LIBS spectra of nitro compounds in air and argon atmospheres (a) and that for the intensity of C_2 bands (b).

Our results show that the normalized intensities of the atomic lines of carbon and hydrogen increase while nitrogen decreases in an argon atmosphere (Figure 9). The enhancement in the normalized intensities of the spectral lines of carbon and hydrogen is due to the confinement of plasma. The reduction in intensity of the spectral lines of N is due to reduction of nitrogen content of air in laser-induced plasma. The increase in the intensity of the spectral lines of C and H and decrease in the intensity of N are higher for 4-nitrotoluene than for 4-nitroaniline. Thus the trend (increase/decrease) of atomic lines is consistent with the trend obtained from the molecular band.

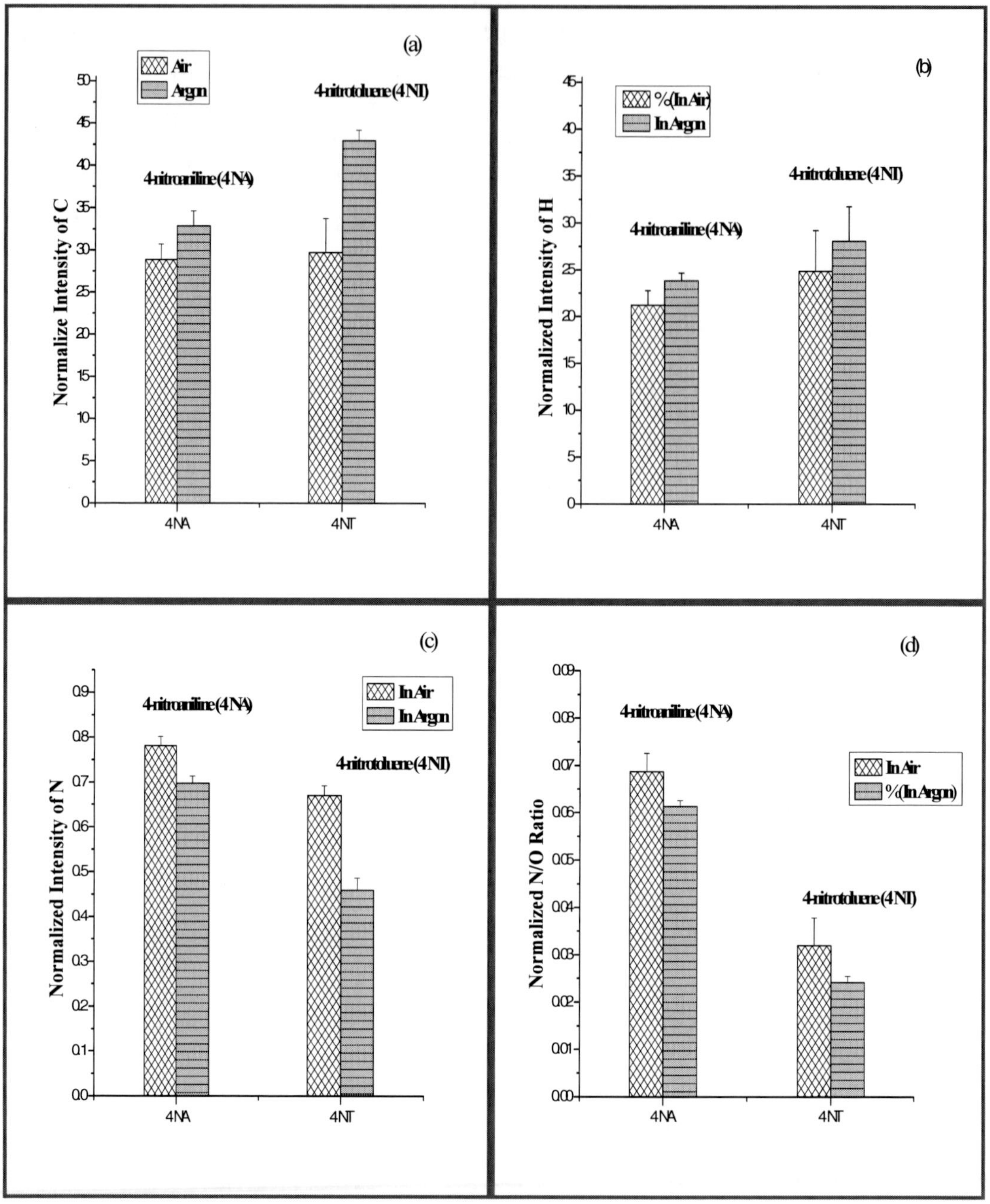

Figure 9. Bar diagrams of the normalized intensity of atomic lines of (a) Carbon (b) Hydrogen (c) Nitrogen and (d) N/O ratio from LIBS spectra of nitro compounds in air and argon atmospheres.

Acknowledgement

Most of the contents presented in the manuscript are from the D. Phil. Thesis of Dr. (Mrs.) Shikha Rai, on Spectroscopic Study of Some Selected Energetic Materials in

University of Allahabad (2011). We duly acknowledge to Springer, Asian Journal of Spectroscopy and AIP publishing for some of the contents used in chapter.

REFERENCES

[1] Parmeter, J. E. The Challenge of Standoff Explosives Detection, *38th Annual International Carnahan Conference on Security Technology,* 355 (2004).

[2] Federici J. F., Schulkin B., Huang F., Gary D., Barat R., Oliveira F., et al*., Semicond. Sci. Technol*., 20, S266 (2005).

[3] Steinfeld J. I., Wormhoudt J., *Annu. Rev. Phys. Chem*., 49, 203 (1998).

[4] Munson C. A., Gottfried J. L., de Lucia F. C., McNesby K. L., Miziolek A. W., *ARL-TR* 4279 (2007).

[5] Sharma S. K., Misra A. K., Sharma B., *Spectrochim. Acta, Part A*, 61, 2404 (2005).

[6] Carter J. C., Angel S. M., Snyder M. L., Scaffidi J., Whipple R. E., Reynolds J. G., *Appl. Spectrosc.,* 59, 769 (2005).

[7] Rusak D. A., Castle B. C., Smith B. W., Winefordner J. D., *Crit. Rev. Anal. Chem.,* 27, 257 (1997).

[8] Cremers D. A., Radziemski L. J., *Handbook of Laser-Induced Breakdown Spectroscopy,* John Wiley & Sons, Ltd., West Sussex, England (2006).

[9] Song K., Lee Y. I., Sneddon J., *Appl. Spectrosc. Rev*. 32, 183 (1997).

[10] Rusak D. A., Castle B. C., Smith B. W., Winefordner J. D., *Trends Anal. Chem*, 17, 453 (1998).

[11] Sneddon J., Lee Y. I., *Anal. Lett.,* 32, 2143 (1999).

[12] Winefordner J. D., Gornushkin I. B., Pappas D., Matveev O. I., Smith B. W., *J. Anal. At. Spectrom.* 15, 1161 (2000).

[13] Winefordner J. D., Gornushkin I. B., Correll T., Gibb E., Smith B. W., Omenetto N., *J. Anal. At. Spectrom.* 19,1061 (2004).

[14] Miziolek A., Palleschi V., Schechter I., *Laser Induced Breakdown Spectroscopy,* Cambridge University Press, Cambridge, UK (2006).

[15] Pasquini C., Cortez J., Silva L. M. C., Gonzaga F. B., *J. Braz. Chem. Soc.*, 18, 463 (2007).

[16] Babushok V. I., de Lucia F. C., Dagdigian P. J., Nusca M. J., Miziolek A. W., *Appl. Opt.,* 42, 5947 (2003).

[17] Babushok V. I., de Lucia F. C., Dagdigian P. J., Miziolek A. W., *Spectrochim. Acta, Part B*, 60, 926 (2005).

[18] Babushok V. I., de Lucia F. C., Dagdigian P. J., Gottfried J. L., Munson C. A., Nusca M. J., *Spectrochim. Acta, Part B,* 62B, 1321 (2007).

[19] Gottfried J. L., Harmon R. S., de Lucia F. C., Miziolek A. W., *Spectrochim. Acta, Part B*, 64, 1009 (2009).

[20] Harmon R. S., de Lucia F. C., Miziolek A. W., McNesby K. L., Walters R. A., French P. D., *Geochem.: Explor. Environ. Anal.,*5, 21 (2005).

[21] Harmon R. S., de Lucia F. C., McManus C. E., McMillan N. J., Jenkins T. F., Walsh M. E., *Appl. Geochem.*, 21, 730 (2006).

[22] McMillan N., McManus C., Harmon R., de Lucia F., Miziolek A., *Anal. Bioanal. Chem.*, 385, 263 (2006).
[23] McMillan N. J., Harmon R. S., de Lucia F. C., Miziolek A. M., *Spectrochim. Acta, Part B*, 62B, 1528 (2007).
[24] Harmon R. S., Remus J., McMillan N. J., McManus C., Collins L., Gottfried J. L., *Appl. Geochem.*, 24, 1125 (2008).
[25] McManus C. E., McMillan N. J., Harmon R. S., Whitmore R. C., de Lucia F., Miziolek A. W., *Appl. Opt.* 47, G72 (2008).
[26] Lazic V., Colao F., Fantoni R., Spizzicchino V., *Spectrochim. Acta B*, 60, 1014 (2005).
[27] Oujja M., Vila A., Rebollar E., Garcia J. F., Castillejo M., *Spectrochim. Acta B*, 60, 1140 (2005).
[28] Fabre C., Boiron M. C., Dubessy J., Chabiron A., Charoy B., Crespo T. M., *Geochim. Cosmochim. Acta*, 66, 1401 (2002).
[29] Munson C. A., Gottfried J. L., Snyder E. G., de Lucia F. C., Gullett B., Miziolek A. W., *Appl. Opt.*, 47, G 48 (2008).
[30] Snyder E. G., Munson C. A., Gottfried J. L., de Lucia F. C., Gullett B., Miziolek A., *Appl. Opt.*, 47, G 80 (2008).
[31] Gottfried J. L., de Lucia F. C., Munson C. A., Miziolek A. W., *Appl. Spectrosc.*, 62, 353 (2008).
[32] Babushok V. I., de Lucia F. C., Gottfried J. L. Jr., Munson C. A., Miziolek A. W, *Spectrochim. Acta, Part B*, 61, 999 (2006).
[33] Cremers D. A., Chinni R. C., *Appl. Spectrosc. Rev.*, 44, 457 (2009).
[34] Singh V. K., Rai A. K., *Lasers Med Sci*, DOI 10.1007/s10103-011-0921-2 (2011).
[35] Rai A. K., Rai V. N., Yueh F. Y., Singh J. P., *Trends Appl. Spectrosc.*, 4, 165 (2002).
[36] Onge L. S., Kwong E., Sabsabi M., Vadas E. H. B., *J. Pharm, Biomed. Anal.*, 36, 277 (2004).
[37] Samuels C., de Lucia F. C., McNesby K. L., Miziolek A. W., *Appl. Opt.*, 42, 6205 (2003).
[38] Dixon P. B., Hahn D. W., *Anal. Chem.*, 77, 631 (2005).
[39] Samek O., Telle H. H., Beddows D. C. S., *BMC Oral Health*, 1, 1 (2001).
[40] Melessanaki K., Papadakis V., Balas C., Anglos D., *Spectrochim. Acta, Part B*, 56, 2337 (2001).
[41] Melessanaki K., Mateo M., Ferrence S. C., Betancourt P. P., Anglos D., *Appl. Surf. Sci.*, 197, 156 (2002).
[42] Anzano J., Gutierrez J., Villoria M., *Anal. Lett.*, 38, 1957 (2005).
[43] Pouli P., Melessanaki K., Giakoumaki A., Argyropoulos V., Anglos D., *Spectrochim. Acta, Part B*, 60, 1163 (2005).
[44] Lopez A. J., Nicolas G., Mateo M. P., Pinon V., Tobar M. J., Ramil A., *Spectrochim. Acta, Part B* 60, 1149 (2005).
[45] Vadillo J. M., Laserna J. J., *Talanta* 43, 1149 (1996).
[46] Wainner R. T., Harmon R. S., Miziolek A. W., McNesby K. L., French P. D., *Spectrochim. Acta, Part B*, 56, 777 (2001).
[47] Gondal M. A., Hussain T., Yamani Z. H., Baig M. A., *Talanta*, 69, 1072 (2006).
[48] Pace D. M. D., D'Angelo C. A., Bertuccelli D., Bertuccelli G., *Spectrochim. Acta Part B*, 61, 929 (2006).
[49] Gondal M. A., Hussain T., *Talanta* 71, 73 (2007).

[50] Sarkar A., Aggarwal S. K. and Alamelu D., *Anal. Methods*, 2, 32 (2010).

[51] Sattmann R., Moench I., Krause H., Noll R., Couris S., Hatziapostolou A., et al., *Appl. Spectrosc.* 52, 456 (1998).

[52] Anzano J., Lasheras R. J., Bonilla B., Casas J., *Polym. Test.*, 27, 705 (2008).

[53] Siddiqui M. N., Gondal M. A., Redhwi H. H., *J. Environ. Sci. Health., Part A*, 43, 1303 (2008).

[54] Gondal M. A., Siddiqui M. N., *J. Environ. Sci. Health Part A*, 42, 1989 (2007).

[55] Fink H., Panne U., Niessner R., *Anal. Chim. Acta*, 440, 17 (2001).

[56] Fink H., Panne U., Niessner R., *Anal. Chem.*, 74, 4334 (2002).

[57] Radivojevic I., Niessner R., Haisch C., Florek S., Becker-Ross H., Panne U., *Spectrochim. Acta, Part B*, 59, 335 (2004).

[58] Tran M., Sun S., Smith B. W., Winefordner J. D., *J. Anal. At. Spectrom.* 16, 628 (2001).

[59] de Lucia F. C., Gottfried J. L., Munson C. A., Miziolek A. W., *Spectrochim. Acta Part B*, *62*, 1399 (2007).

[60] Grönlund R., Lundqvist M., Svanberg S., *Opt. Lett.*, 30, 2882 (2005).

[61] Sabasi M., Russo R., 4th International Conference on Laser Induced Plasma Spectroscopy and Applications (LIBS 2006) special issue, *Spectrochim. Acta, Part B*, 62, 1285 (2006).

[62] de Lucia F. C., Harmon R. S., McNesby K. L., Winkel R. J., Miziolek A. W., *Appl. Opt.*,42, 6148 (2003).

[63] Harmon R. S., de Lucia F. C., la Pointe A., Winkel R. J., Miziolek A. W., *Anal. Bioanal. Chem.*, 385, 1140 (2006).

[64] Munson C. A., de Lucia F. C., Piehler T., McNesby K. L., Miziolek A. W., *Spectrochim. Acta Part B*, 60, 1217 (2005).

[65] Samuels A. C., de Lucia F. C., McNesby K. L., and Miziolek A. W., *Appl. Opt.*, 42, 6205 (2003).

[66] Rehse S. J., Diedrich J., and Palchaudhuri S., *Spectrochim. Acta Part B*, 62, 1169 (2007).

[67] Diedrich J., Rehse S. J., and Palchaudhuri S., *J. Appl. Phys.* 102, 014702 (2007).

[68] Hybl J. D., Lithgow G. A., and Buckley S. G., *Appl. Spectrosc.* 57, 1207 (2003).

[69] Snyder E. G., Gullett B., Ryan S., Oudejans L., Touati A., *Appl. Spectrosc.*, 60, 860 (2006).

[70] Bousquet B., Sirven J. B., and Canioni L., *Spectrochim. Acta Part B*, 62, 1582 (2007).

[71] de Lucia F. C., Samuels A. C., Harmon R. S., Walters R. A., McNesby K. L., la Pointe A., et al., *IEEE Sens. J.*, 5, 681 (2005).

[72] Lazic V., Palucci A., Jovicevic S., Poggi C., and Buono E., *Spectrochim. Acta Part B*, 64, 1028 (2009).

[73] Moreno C. L., Palanco S., Laserna J. J., de Lucia F., Miziolek A. W., Rose J., *J. Anal. At. Spectrom.*, 21, 55 (2006).

[74] de Lucia F. C., Gottfried J. L. Jr., Munson C. A., Miziolek A. W., *Appl. Opt.*, 47, G112 (2008).

[75] Gottfried J. L., de Lucia F. C., Munson Jr. C. A., Miziolek A. W., *J. Anal. At. Spectrom.*, 23, 205 (2008).

[76] de Lucia F. C., J. L. Gottfried, A. W. Miziolek, *Opt. Exp.* 17, 419 (2009).

[77] Gottfried J. L, de Lucia F. C., Miziolek A. W., *J. Anal. At. Spectrom.* 24, 288 (2009).

[78] de Lucia F. C., Gottfried J. L., *Propellants Explos. Pyrotech., 35,* 268 *(2010)*
[79] Anzano J. M., Gornushkin I. B., Smith B. W., Winefordner J. D., *Polym. Eng. Sci.*, 40, 2423 (2000).
[80] Portnov A., Rosenwaks S., Bar I., *Appl. Opt.*, 42, 2835 (2003).
[81] Ferioli F., Buckley S. G., *Combust. Flame*, 144, 435 (2006).
[82] *Existing and Potential Standoff Explosives Detection Techniques,* Washington, National Research Council. http://www.nap.edu /catalog /10998.html (2004).
[83] Lee Y. L., Thiem T. L., Kim G. H., Teng Y. -Y., Sneddon J., *Appl. Spectrosc.*, 46, 1597 (1992).
[84] Chin S. L., *Adv. in Multipho. Proce. Spectrosc.* 16, 249 (2004).
[85] Morgan C. G., *Rep. Prog. Phys.*, 38, 621 (1975).
[86] Ready J. F., *Effect of High Power Laser Radiation,* Academic Press, New York (1971).
[87] Fairman B., Hinds M. W., Nelms S. M., Pennyd D. M., Goodalle P., *J. Anal. At. Spectrom.* 13, 233R (1998).
[88] Griem H. R., *Plasma Spectroscopy,* McGraw-Hill, New York (1964).
[89] Rai D. K., Thakur S. N., *Atomic Spectra Structure and Modern Spectroscopy,* Varanasi (2005).
[90] Sansonetti J. E., Martin W. C., Handbook of Basic Atomic Spectroscopic Data, *J. Phys. Chem. Ref. Data*, 34, 1559 (2005).
[91] McWhirter R. W. P., *Plasma Diagnostic Techniques,* Academic Press, New York (1965).
[92] Shaikh N. M., Rashid B., Hafeez S., Jamil Y., Baig M. A., *J. Phys. D: Appl. Phys.*, 39, 1384 (2006).
[93] de Lucia F. C., Harmon R. S., McNesby K. L., Winkel R. J., Miziolek A. W., *Appl. Opt.*, 42, 6148 (2003).
[94] Tran M., Sun Q., Smith B. W., Winefordner J. D., *Appl. Spectrosc.,* 55, 739 (2001).
[95] Kaski S., Hakkanen H., Korppi-Tommola J., *J. Anal. At. Spectrom.* 19, 474 (2004).

In: Spectroscopic Techniques for Security, Forensic ... ISBN: 978-1-63117-404-9
Editors: Y. Dwivedi, S. B. Rai and J. P. Singh

Chapter 6

APPLIED MICROWAVE ROTATIONAL SPECTROSCOPY

K. P. Rajappan Nair **and** ***Jens-Uwe Grabow***[†]
Institut für Physikalische Chemie und Elektrochemie, Lehrgebiet A,
Gottfried Wilhelm Leibniz Universität Hannover, Hannover, Germany

1. INTRODUCTION

The history of scientific developments often show a regular pattern. First the general ideas are expressed and tested in laboratories which might be called fundamental science and then their results are utilized in Applied Science or what we call technology. But the development of microwave spectroscopy went into the opposite direction. The researchrelated to radar techniques during World War II had given rise to the development of microwave sources in different regions. These microwave sources have since then been used in the development of microwave spectroscopy. The first historical paper in microwave spectroscopy was that of Cleeton and Williams [1] in the year 1934 who measured the ammonia spectrum in the region 1 to 4 cm, observing the inversion line at 1.25 cm. Since then, a number of papers, textbooks [2-31] and book chapters have been published, with the most recent review provided by Grabow and Caminati [26, 27]. A number of more specific works focusing on the quantitative nature of microwave spectroscopy have also been published [2, 12, 13, 14].

Though microwave spectroscopy was originally used by many workers purely as a research tool for structural determinations of gas phase molecular species, its applications have been increased enormously and the frequency range has been largely widened. Now this technique is used from the centimeter region to submillimeter region, giving rise to detailed information on molecular structure and molecular interactions, the accuracy of which are unparalleled and these accurately determined data eventually lead to many new theoretical developments in the understanding of molecular properties. The microwave region of the spectrum extends roughly over a frequency range of 300 MHz to 3 THz or in wavelength units of about 1 m to 0.1 mm. The upper portion of the microwave region is often referred to

* E-mail: kprnair@gmail.com, kprnair@yahoo.co.uk.
† E-mail: jens-uwe.grabow@pci.uni-hannover.de.

as millimeter region (30 GHz to 300 GHz), submillimeter or far infrared region (300 GHz to 3 THz), the latter lately often also referred to as Teraherz region. The microwave region became widely known from its usage in radar techniques, speeding traps, household microwave ovens, wireless and satellite communication, and its application in radio astronomy. However, microwaves are increasingly popular to uncover the structure and properties of molecules as the microwaves can interact with molecules that possess a permanent dipole moment by rotational excitation in this region. The resulting absorption (or re-emission) of the microwave radiation provides a means to detect and quantify molecules. The appealing fact is its ability to measure rotational transitions with extremely high resolution. Unlike in an atom, particular interest in the spectroscopy of molecules deals with three types of quantized energy levels and the transitions between them. The particle motions which are associated with a molecule and of interest to the present discussions give rise to three types of energy contributions:

1. Rotational
2. Vibrational
3. Electronic

The molecules can undergo transitions between rotational states, between vibrational states and between electronic states. The transitions between rotational levels generally lie in the microwave region, between vibrational levels in the infrared region and between electronic energy levels in the visible or ultraviolet regions. Such transitions can be either spontaneous or can be induced by the application of appropriate external electromagnetic radiations at the resonant frequencies. The rotational transitions occur in the microwave region and are observed typically in the frequency range from $0.3 \cdot 10^9$ to $3.0 \cdot 10^{12}$ Hertz. The spontaneous transition rate as an atom or molecule goes from an excited state to a lower state varies approximately to the cube of the frequency of the transition. Thus, if we consider the microwave transitions which is the prime theme of discussion here, they occur much less rapidly than transitions at visible or ultraviolet regions do. Consequently most of the studies in the radiofrequency or microwave regions are done by forcing a sample of molecules to absorb radiations instead of waiting to emit spontaneously. The only exception is in the study of astrophysical objects by radio astronomy where these astronomical sources are large enough so that spontaneous emission can be observed and collected by large antennas and then amplified and detected by cooled low noise electronic detectors.

Initial interests in microwave spectroscopy as a potential analytical tool focused on the quantitative detection of chemical species by direct absorption [10, 32]. The use of microwave spectroscopy as a chemical tool has essentially developed from two different approaches. At the beginning of its application towards the structure and properties of molecules, studies were conducted on direct induced absorption between rotational levels. Though a commercial microwave spectrometer was available in the market for some time, by Hewlett Packard and Cambridge Scientific Instruments, they were not suited for all problems and hence met with limited success [32]. Advanced spectrometers and techniques had to be developed to suit individual needs which differ from lab to lab. Subtle and new experimental techniques have improved the resolution of the spectrometers and hence the data produced on molecular structure and interactions have remarkable durability. Subsequent advancements in microwave technology and techniques have spurred the progress of the field. Though usually microwave spectra could only be observed on polar molecules as the rotational transitions occur due to the interaction of electromagnetic radiation with the electric dipole moment of the molecule, detection of microwave spectra on nonpolar molecules such as CH_4, originating from the centrifugal effects, though very weak, had been achieved with high sensitive methods. Starting with the video spectrometers, passing through the generally used Stark modulation spectrometers, many developments have been achieved in these techniques. The observation of weak hydrogen bonded complexes and inert gas complexes were made possible. Spectra of a number of short lived diatomic molecules have been carried out with great precision and internuclear distances could be determined with an accuracy better than the sixth decimal place of an Angstrom [33-43]. The recent developments are based on the free induction decay of the microwave radiation. The important step in the development of instrumentation had been obtained by the introduction of Fourier Transform techniques (FTMW). FTMW investigations on molecular jets are being carried out in many laboratories now. New theoretical developments and various computer programs are developed in order to interpret the complex molecular spectra in the microwave region. The unrivalled precision of the systematic characterization of molecules has led to significant improvements in molecular theories.

As early as the 1960's, astrochemists realized the potential of microwave spectroscopy for the identification and mapping of compounds in the interstellar medium. Laboratory based results are used to guide searches for the presence of the same compounds in the interstellar medium since the rotational spectra would be identical with the exception of the relative velocity shifts of the cloud movement with respect to the Earth based systems. This has been an active field of study for many years and is still growing today. About 180 compounds have been identified and studied in the interstellar medium by this method [44-71].

Dicke and co-workers [72] initiated a most significant contribution to the technique of microwave spectroscopy by the introduction of time-domain techniques which is commonly referred to as FT-MW spectroscopy. These methods were established by utilizing a hollow waveguide static gas sample cell in the mid-1970s by Flygare and co-workers [73]. It took another few years of development for FT-MW to become mature enough to compete with continuous wave (CW) MW absorption techniques after these early pulsed-excitation experiments. This is similar to the success of Fourier transform nuclear magnetic resonance (FT-NMR) over CW-NMR. These improvements occurred in the 1980s when FT-MW spectroscopy rapidly evolved by including a pulsed molecular jet and a Fabry-Perot type resonator [74-76]. The use of the low temperature molecular expansion, $\approx$ 1 K, enabled the study of hundreds of previously undetected molecular species, dimers, trimers, van der Waals complexes and hydrogen bonded species. Moreover, the high sensitivity of this technique facilitated measurements on trace level species. For instance, with this technique, most ^{13}C isotopomers can be studied in natural abundance without requiring laborious synthetic enrichment experiments (for example Figure 1).

A number of recent advancements and developments have further increased the sensitivity and general capabilities of Fourier Transform Microwave Spectroscopy. The signal to noise ratio is now closer to the theoretical limits with the relatively recent development of the coaxial beam resonator arrangement (COBRA), developed at Kiel and Hannover [77, 78]. To maximize the overlap between the molecular expansion and the microwave radiation, [77] multiple excitation of a single sample pulse [79], adaptation of the cyclically ordered phase sequence (CYCLOPS) detection system from NMR spectroscopy [78], improved coupling between resonator and spectrometer [80], and cryogenic cooling of the cavity mirrors [80] have been developed. The development of programmable and broad banded coaxial components, e.g., synthesized sources, solid-state low-noise and power amplifiers, fast switches and mixers, and the use of single-sideband generation instead of a second oscillator for super heterodyne detection [81], now allow the construction of automated, multi-octave FT-MW spectrometers [82].

A renewed interest in efforts to demonstrate the quantitative capabilities of microwave spectroscopy began in the mid 1990s. Microwave fingerprints along with laser techniques have been developed by many groups for trace detection of chemical species in environmental applications and industrial process control.

Quantitative measurements have been developed by a number of groups including Alder and Baker, [83-87] Thirup et al. [88], Zhu et al. [89-92], Andressen and Drezler [93], and Suenram et al. [94-101]. The newer efforts focused on developing analytical methods using FTMW spectroscopy. Currently, most of the FTMW measurements on Chemical Warfare Agents (CWA) and their derivatives reported in the open literature have been performed by Suenram et al. [94-101]. There continue to be on-going efforts to demonstrate the analytical capabilities of FTMW as well as in the further development of the instrumentation.

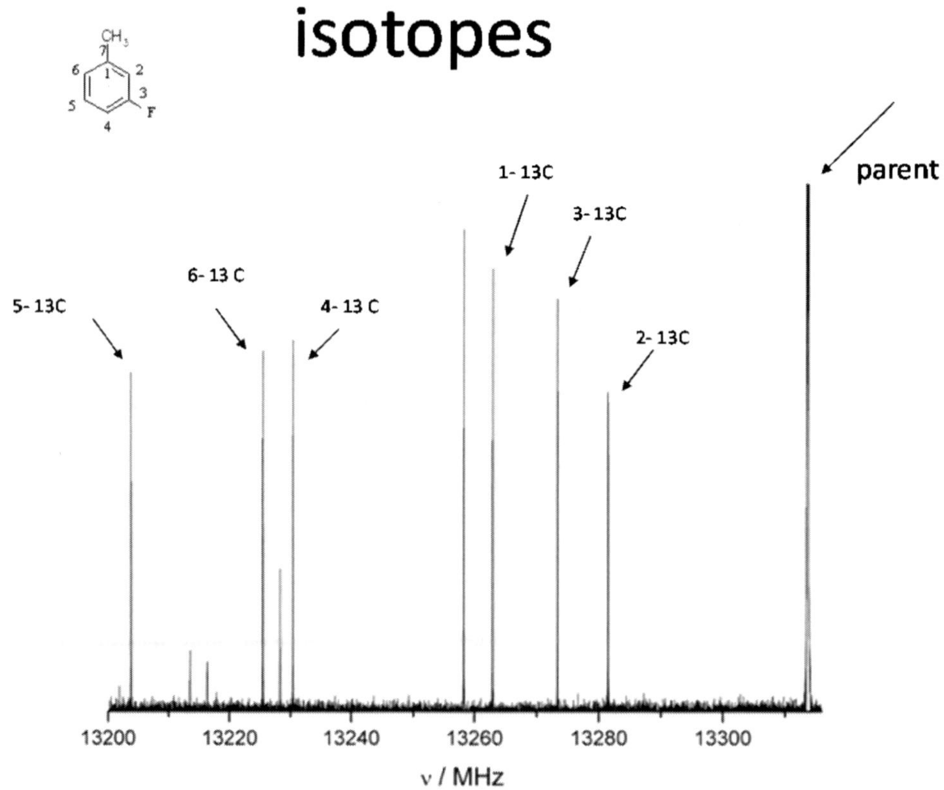

Figure 1. ^{13}C isotopic transitions in metafluorotoluene recorded at the Hannover molecular jetFT-MW Spectrometer (Nair K. P. R. and Grabow J. U. unpublished).

2. ROTATIONAL SPECTROSCOPY

Rotational spectroscopy is concerned with the measurement of transitions between the quantized rotational energy levels of a molecule in the gas phase. In condensed phase, the rotations and vibrations of a molecule are restricted because the atoms are thickly packed and hence the spectroscopic studies are generally carried out in the gas phase. Only the rotational spectrum of polar molecules can be measured and studied by microwave spectroscopy as the molecule requires a permanent dipole moment to interact with the electromagnetic radiation. Nonpolar molecules can be studied by Raman or infrared spectroscopy, though not of high precision as obtained in microwave spectroscopy. The rotational spectra in the microwave region are also termed as pure rotational spectra in order to distinguish it from the rotation-vibration spectra, originating from the transition between the rotational levels in one vibrational state to the rotational levels in another vibrational state. Rotation-Vibration spectra are observed generally in the infrared region. It should be remembered that rotational, vibrational and electronic transitions can occur simultaneously in a molecule, but we can distinguish them by studying the spectra in different regions. The electronic spectra under higher resolutions contain the rotational transitions from the vibrational levels of a particular electronic state to the rotational levels of vibrational level of another electronic state.

In studies of rotational or microwave spectroscopy, the molecules are commonly classified according to the symmetry and mass distribution. We distinguish:

1. Spherical top molecules, e.g.: CH_4, CCl_4, SiH_4 where $I_A = I_B = I_C$
2. Linear molecules, e.g.: HCl, AgCl, OCS, HCN where $I_A = 0$, $I_B = I_C$
3. Prolate Symmetric top molecules, e.g.: CH_3Cl, CH_3CCH where $I_A < I_B = I_C$
4. Oblate Symmetric top molecules, e.g.: NH_3 or PH_3, C_6H_6 where $I_A = I_B < I_C$
5. Asymmetric top molecules, e.g.: H_2CO, H_2O where $I_A \neq I_B \neq I_C$

I_A , I_B and I_C are moments of inertia in three perpendicular body fixed axes which depend on the mass distribution in the molecule. Of these, as mentioned earlier, the spherical tops normally do not give rise to a rotational spectrum as they do not possess an electric dipole moment to interact with the applied microwave radiations. Remember that they also have rotational energy levels as in other cases, however rotational transitions are only observable due to centrifugal distortion if no center of inversion is present. Larger molecules often belong to the fifth category viz. asymmetric top. Though analytical expressions can be derived for linear and symmetric top molecules (with spherical tops treated as a special case of symmetric top) by solving the Schrödinger equations with or without approximations, such expressions are difficult to derive in the case of asymmetric top molecules other than in some low angular momentum energy levels, and the observed transitions are generally treated using numerical diagonalization methods. The rotational energies are derived by Schrödinger equations, often adding extra terms to the rigid rotor expressions to account for the nonrigidity of the molecule and for the other finer details like rotation vibration interaction, fine and hyperfine structure, Coriolis coupling, internal rotation etc. The observed spectra are fitted by theoretical expressions in order to obtain the molecular parameters. The obtained moments of inertia are used to derive bond lengths and bond angles of the molecule. Many of the earlier spectrometers have used Stark effect modulation techniques in order to increase the signal to noise ratio which in turn could also be used to measure the Stark effects and to determine electric dipole moments components.

The previous classification of the molecule is based on the moment of inertia of the molecules in the body centered principal axis system. Once the moment of inertia tensor can be calculated by knowing the atomic mass and the distance in three mutually perpendicular axes x, y, z, the off-diagonal elements of the tensor can be eliminated by suitably transforming them into body centered principal axes system. The resulting moments of inertia are named as I_A, I_B and I_C about the three mutually perpendicular axes a, b and c. The general convention is to consider $I_A \leq I_B \leq I_C$ with the axis A corresponding to the smallest moment of inertia.

3. Microwave Spectroscopy and Molecular Structure

3.1. Microwave Spectra

The microwave region is the part of the electromagnetic spectrum with wavelengths ranging from about one tenth of a mm to one meter. Hence microwave spectroscopy is sometimes regarded as an extension of radiofrequency spectroscopy or as an extension of infrared spectroscopy. However, there are ample reasons to consider it as an independent subject. The instrumentation for generating, transmitting and detecting microwave radiations are considerably different from those of radiofrequency or of optical regions. Hence microwave spectroscopy can find a place in an electronic laboratory, or in a spectroscopy laboratory or even in a physical chemistry laboratory.

The microwave spectra of molecules correspond to changes in the rotational motions often with other fine or hyperfine structure effects superimposed. As stated earlier, the first experiment in microwave spectroscopy was conducted on the inversion frequency of the ammonia molecule. The structure of the ammonia molecule is a simple pyramid with the nitrogen atom symmetrically placed above the plane of the three hydrogen atoms. It is possible for the nitrogen atom to occupy any side of the hydrogen plane. In fact at room temperature the nitrogen atom vibrates rapidly between these two stable configurations, exhibiting an inversion frequency of about 24,000 MHz corresponding to a wavelength of 1.25 cm. As the energy levels are very close together, the populations of molecules in these two levels are almost equal. However, one can achieve a population inversion by using suitable magnetic fields and shift the levels apart because of the opposite symmetry properties of the levels. Thus it was possible to collect molecules, which are in the upper state, thereby having the essential requirement for maser action.

Microwave spectroscopy has been widely used to study the molecular structure and molecular interactions. A large amount of information on inter-atomic distances and angles, electric dipole moments, nuclear quadrupole coupling constants, barrier to internal rotation and other properties of molecules can be obtained from the analysis of the microwave spectra.

3.1.1. Rotational Energies

A number of text books are available on fundamentals of spectroscopy [2-31]. Rotational energy, along with all other forms of molecular energy, is quantized which implies that the molecules cannot have any arbitrary amount of energy, but their energy is limited to certain definite values depending on the size and shape of the molecule concerned. The permitted energy values, here the rotational energies, can in principle be calculated by quantum mechanics solving the Schrödinger equation for rotating molecule.

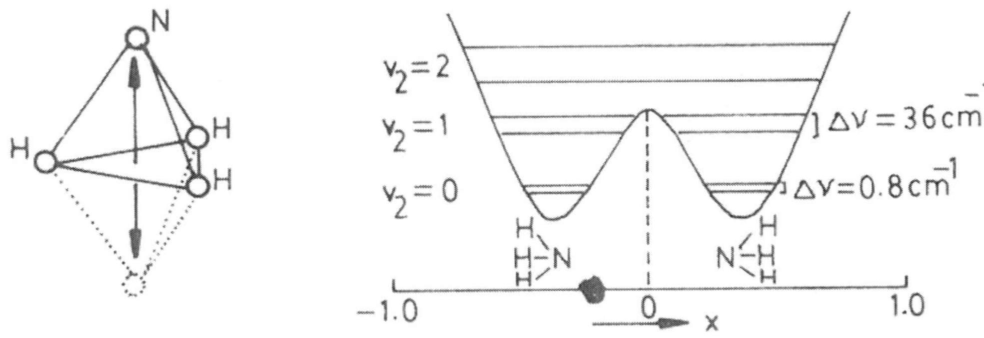

Figure 2. The ammonia molecule (left) and the potential curve of the NH_3 inversion (right).

4. Quantum Mechanical Derivation of Rotational Energy

The general method of solving quantum mechanical problem is

1) writing down the classical Hamiltonian,
2) transforming it to the corresponding quantum mechanical operator, and
3) finding the solution to the eigenvalue equation.

$$\hat{H}\psi_i = E\psi_i$$

In order to solve the Schrödinger equation for a rigid rotor we have to solve the Hamiltonian H for a rigid body of reduced mass μ and V=0. There is no potential energy associated with the rotation as long as the rotation is rigid and no external fields are present.

$$H = \frac{1}{2\mu}\left(P_x^2 + P_y^2 + P_z^2\right) = -\frac{h^2}{2\mu}\nabla^2$$

By solving the Schrödinger equation the allowed rotational energy levels are obtained. The allowed energies for a rigid linear molecule are expressed by the equation

$$E = \left(\frac{h^2}{8\pi^2 I}\right)J(J+1) = hBJ(J+1)$$

J = 0, 1, 2, 3, …

Here $I = \sum_i m_i r_i^2$ is the moment of inertia of the molecule, $B = \frac{h}{8\pi^2 I}$ is the rotational constant and J is the rotational (total angular momentum) quantum number. For a diatomic molecule $I=\mu r^2$ where μ is known as the reduced mass equal to $m_1 m_2/(m_1+m_2)$ and m_1 and m_2 are the masses of the constituent atoms; r is the inter nuclear distance. The rotational transitions among various rotational levels follow certain rules known as selection rules. Figure 3 shows the progression of rotational energy levels with the total angular momentum quantum numbers up to J = 4.

4.1. Rotational Spectra

The first requirement for the intra-molecular motion to give rise to radiation of light is that it must accompany a change in the space-fixed dipole moment components. For a rigid rotor this can only be possible if it possess a permanent dipole moment which lies in a direction perpendicular to the axis of rotation. The rotating dipole can be thought of interacting with the electric field of the radiation and allowing energy to be exchanged between the molecule and the radiation. During rotation, the component of the dipole moment in a fixed direction changes periodically with a frequency which is the rotational frequency. Quantum-mechanically, the emission or absorption of radiation takes place when the molecule jumps from one energy state to another state which follows certain selection rules.

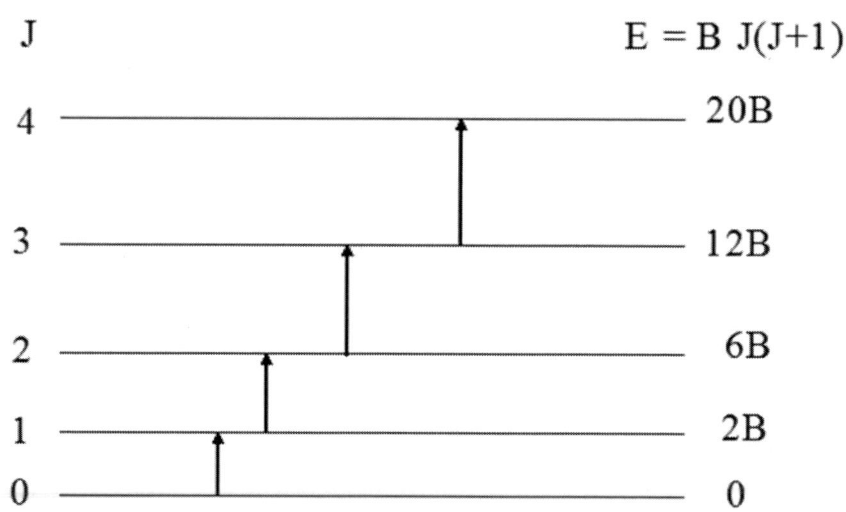

Figure 3. Rigid rotor energy levels with angular momentum quantum numbers up to J = 4.

For a rotational transition of a linear molecule to occur (in absorption or emission) the selection rule is

$\Delta J = \pm 1$

This selection rule can be obtained by evaluating the matrix elements of the dipole moment operator

$$\left\langle J'M'\middle|\mu\middle|JM\right\rangle \neq 0$$

We can write the rotational frequency of a diatomic molecule as

$$\nu = \frac{1}{h}[E(J') - E(J")]$$

and following the selection rule ΔJ = +1 as we have always the choice J' = J" + 1

$$\nu = \frac{1}{h}[E(J+1) - E(J)]$$

= B (J+1) (J+2) – BJ (J+1)

= 2B (J+1), J = 0, 1, 2, …

Thus the rotational spectrum of a diatomic molecule consists of equidistant lines with separation 2B. However, if we consider the interaction of vibration on rotation and the non-rigidity of the molecule, the solution of the Schrödinger equation gives the energy eigenvalues as depicted in Figure 4.

$$\frac{E_{v,J}}{h} = \omega_e(v+\frac{1}{2}) - \omega_e x_e(v+\frac{1}{2})^2 + B_v J(J+1) - D_v J^2(J+1)^2 + H_v J^3(J+1)^3 + .$$

v represents the vibrational quantum number, ω_e is the vibrational frequency and $\omega_e x_e$ is the anharmonicity constant.

Thus the frequency of rotational transitions becomes

$$\nu = 2B_v(J+1) - 4D_v(J+1)^3 + H_v(J+1)^3\left[(J+2)^3 - J^3\right] + ..$$

In the case of polyatomic molecules, the molecules in the gas phase are free to rotate relative to a set of orthogonal axes of fixed orientation in space centered on the centre of mass of the molecule. Rotation about these unique axes a, b and c is associated with a rotational constant and consequently quantized energy levels depending on the rotational constants which is inversely proportional to the moments of inertia bout the axes. For diatomic and linear molecules there is only one rotational constant to consider and hence one rotational quantum number. For symmetric tops there are two rotational constants and two quantum number numbers usually defined as J and K. For asymmetric top molecules there are three rotational constants.

In spherical tops all moments of inertia are equal i.e., $I_A = I_B = I_C$. Linear molecules have $I_A \neq I_B = I_C$, approaching the limiting case $I_A = 0$. In Symmetric tops we have $I_A = I_B < I_C$ (oblate symmetric top - disc shaped) or $I_A < I_B = I_C$ (prolate symmetric top - cigar shaped), whereas in asymmetric top all moments of inertia are different and generally $I_A < I_B < I_C$.

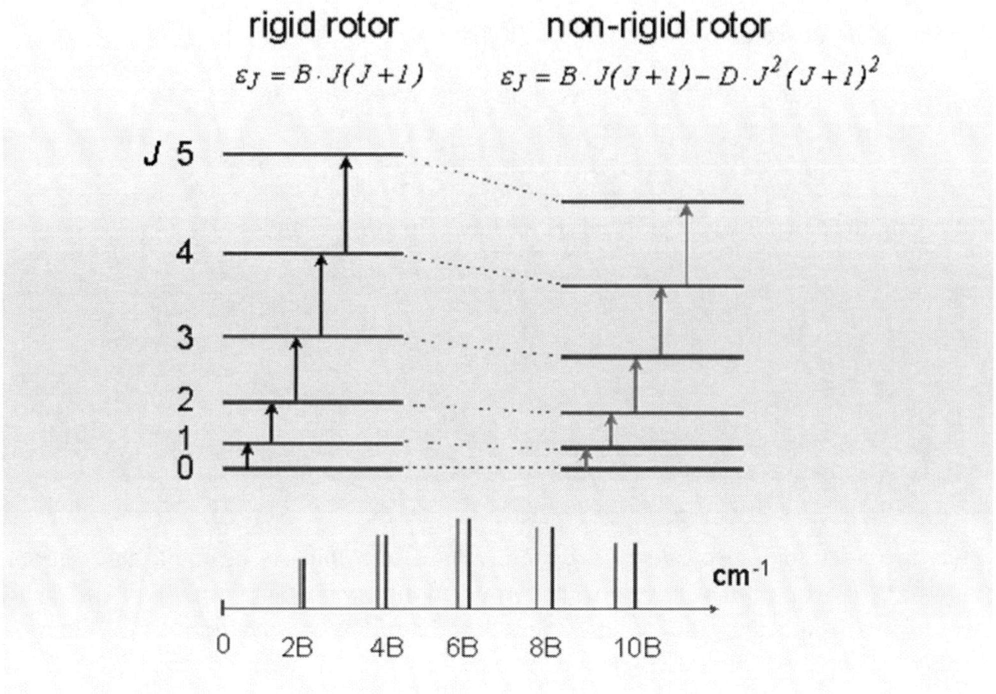

Figure 4. Distorted rotor energy levels with angular momentum quantum numbers up to J = 5.

Table 1. The rotational constants of the TlF diatomic molecule [a)]

V	$^{205}Tl^{19}F$	$^{203}Tl^{19}F$
0	13334.684(5)	13345.818(5)
1	13244.894(7)	13255.916(7)
2	13155.481(10)	13166.391(10)
3	13066.453(40)	13077.273(40)
4	12977.839(40)	12988.475(60)
5	12889.483(40)	12899.984(80)
6	12801.645(60)	-
7	12714.100(80)	-

[102]J. Hoeft and K. P. R. Nair, Z. Phys. D. - Atoms, Molecules and Clusters 5, 345 (1987).

In the case of a diatomic molecule there is only one independent moment of inertia (two are equal and one is zero). Generally, isotopic substitutions can give rise to the complete determination of molecular structure from the rotational constants of the parent molecule and of the isotopologues. When the data on isotopic molecules are not available, the rotational constants of the parent molecule can be fitted by a constrained set of structural parameters with an assumed structure to begin with. Though microwave spectra could be measured in vibrational excited states (of the ground electronic state), the intensity of the spectra in the vibrationally excited states tend to be weak at room temperature. While the lowest state is appreciably populated, the population in the excited vibrational states decreases quickly according to the Boltzmann distribution law. However, the rotational spectra of a number of vibrational states have been measured in some molecules [102]. As the inter-nuclear distances in the excited vibrational states are larger, the moments of inertia in the excited vibrational states are larger and hence the rotational constants decrease. E.g., the rotational constants obtained from measurements up to v = 7 for TlF are collected in Table 1.

The microwave spectrum of spherical top molecules cannot be observed as they do not possess a permanent dipole moment. Rotational Raman spectra also cannot be obtained if the polarizability is isotropic. However, rotational constants can be obtained for such molecules from the study of their ro-vibrational spectra. Also, in spherical top molecules that do not possess a center of inversion as in the case of CH_4, a small dipole moment can be obtained from centrifugal distortion.

Since its early development, microwave spectroscopy has emerged as a power tool for studying a variety of fundamental molecular properties. Although best known for its ability to provide the most precise values of bond lengths and bond angles, microwave spectroscopy is also one of the best means to obtain information on other molecular parameters such as nuclear quadrupole interactions (See Figure 5), nuclear spins, nuclear masses, molecular magnetic moments, low lying vibrational modes, barrier to internal rotation, and rotational isomerism in molecules. A compilation is given in Table 2. A typical spectrum [103] showing the nuclear quadrupole hyperfine structure observed in the rotational spectrum of p-chlorotoluene is shown in Figure 5.

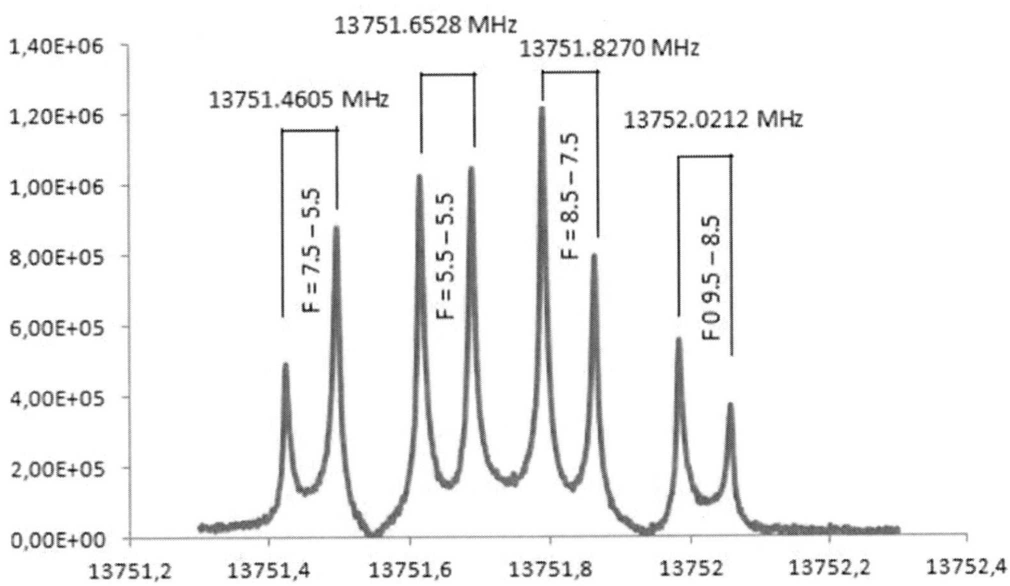

Figure 5. ^{35}Cl Nuclear quadrupole hyperfine structure of the $J = 8_{1,8} - 7_{1,7}$ rotational transition of parachlorotluene for the torsional A-species. The spectrum taken with the supersonic-jet FT-MW spectrometer in Hannover (K. P. R. Nair and J.-W. Grabow, unpublished).

4.2. Linear Molecules

Besides diatomic molecules, many polyatomic molecules like OCSand HCN are linear too. Of special astrophysical interest are carbon-chain type linear molecules HC_xNC. Eleven such carbon chain molecules were studied in the year 2000 by Mc Carthy et al. [104]. The energy level structures of HC_4NC and HC_6NC H—C≡C—C≡C—C≡C—N≡C are shown in Figure 6.

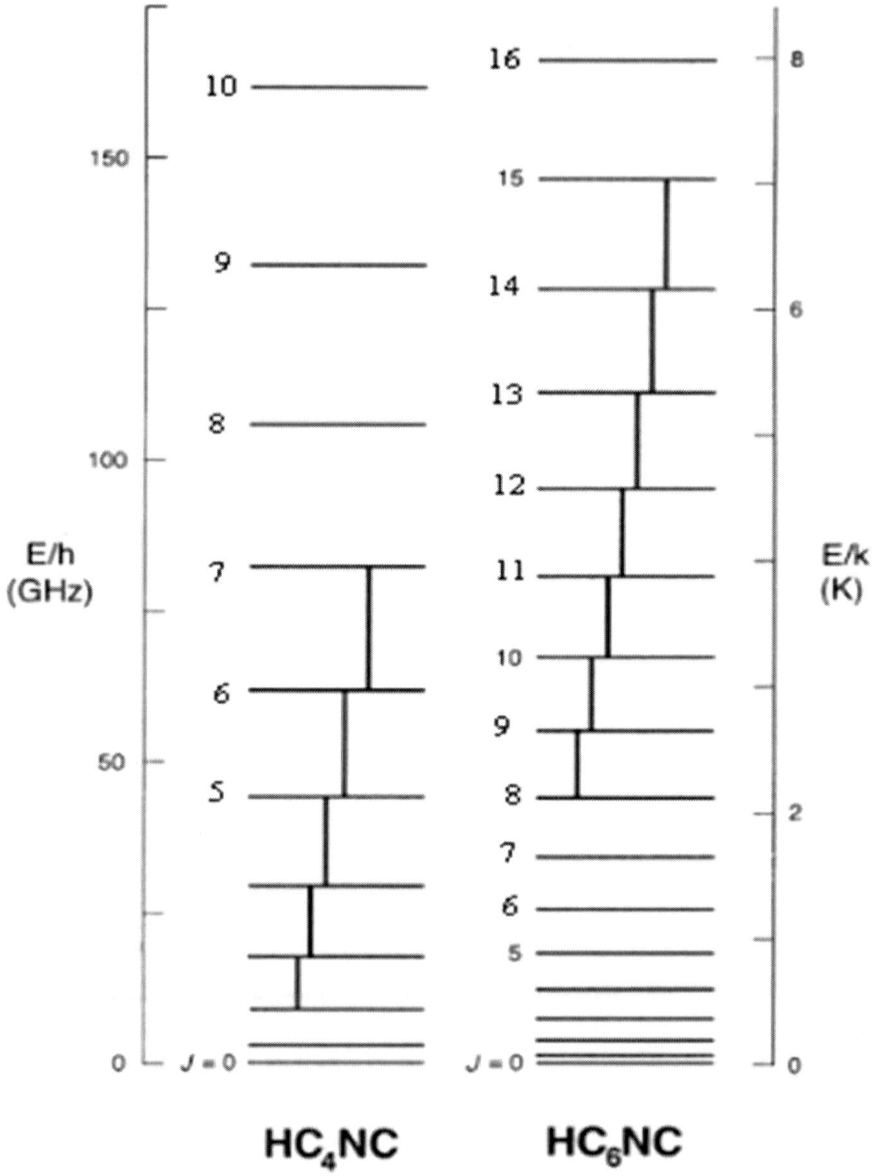

Figure 6. Energy levels of the linear molecules HC_4NC and HC_6NC.

Table 2. Application of Microwave Spectroscopy as a physical tool

Molecular Structure	Bond lengths and bond angles
Molecular interactions	Fine structure, hyperfine structure, internal rotation, low lying vibrations
Stark effect and Zeeman effect	Electric dipole moments and magnetic moments
Structural flexibility	Different conformers, torsional barriers
Double resonance, multi-resonance, multi-photon experiments	Study of complicated and difficult to access energy levels
New Maser and Laser sources	Studying the energy levels for possible laser/maser action.
Radioastronomy	Detection of molecules in space
Analytical spectroscopy	Quantitative monitoring of the environment and detection of chemical agents

Also the electric and magnetic properties could be obtained from Stark and Zeeman studies.

4.3. Symmetric Tops

In a prolate top, the c axis and b axis moments of inertia are equal, that is, $I_c = I_b > I_a$.

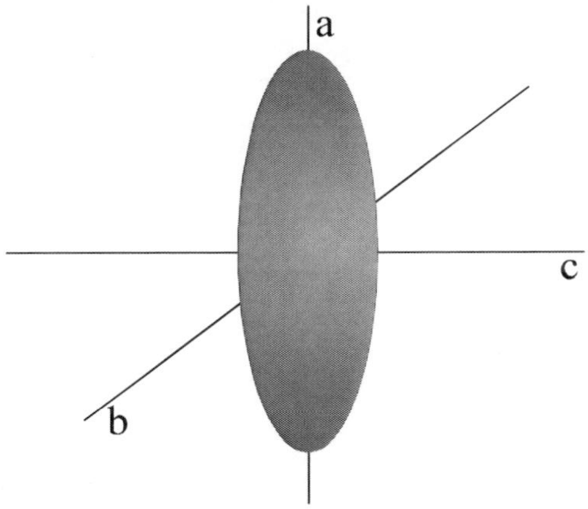

Figure 7. Cigar-shaped or football-shaped prolate top molecule.

Prolate tops are "cigar-shaped" or "football-shaped" as given in Figure 7. The Hamiltonian for a prolate top is

$$\hat{H} = \frac{\hat{J}_a^2}{2I_a} + \frac{\hat{J}_b^2}{2I_b} + \frac{\hat{J}_c^2}{2I_c} = \frac{\hat{J}_a^2}{2I_a} + \frac{\hat{J}_b^2 + \hat{J}_c^2}{2I_b} = \frac{\hat{J}_a^2}{2I_a}\left[-\frac{\hat{J}_a^2}{2I_b} + \frac{\hat{J}_a^2}{2I_b}\right] + \frac{\hat{J}_b^2 + \hat{J}_c^2}{2I_b}$$

$$= \frac{1}{2}\left(\frac{\hat{J}_a^2}{I_a} - \frac{\hat{J}_a^2}{I_b}\right) + \frac{\hat{J}_a^2 + \hat{J}_b^2 + \hat{J}_c^2}{2I_b} = \frac{\hat{J}_a^2}{2}\left(\frac{1}{I_a} - \frac{1}{I_b}\right) + \frac{\hat{J}^2}{2I_b}$$

$$\hat{J}_a^2\psi = \hbar^2 K^2 \psi$$

where K is quantum number describing the projection of the total angular momentum along the symmetry axis, in this case the "a" axis. Therefore, the energy for a prolate top is

$$E = \frac{\hbar^2}{2I_b} J(J+1) + \frac{\hbar^2}{2}\left(\frac{1}{I_a} - \frac{1}{I_b}\right) K^2 = hBJ(J+1) + h(A-B)K^2$$

Rotations of symmetric tops are thus characterized by the J, K, and M quantum numbers:

- The wave functions of symmetrical tops, $|\mathrm{J\,K\,M}\rangle$, are linear combinations of spherical harmonic functions.
- We will not consider the explicit form of wave function because we know how the $|\mathrm{J\,K\,M}\rangle$ wave function behaves with the angular momentum operators.

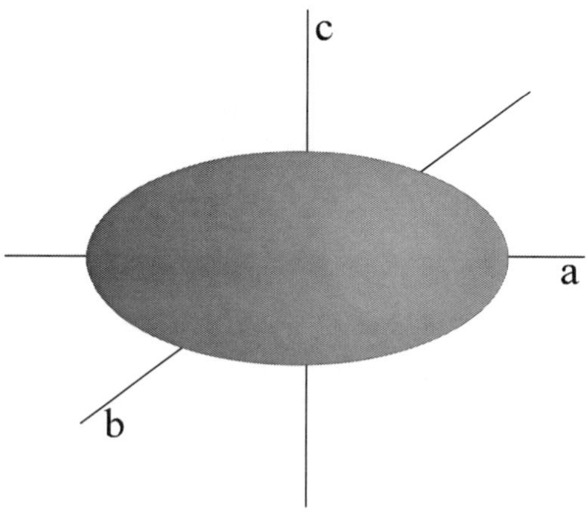

Figure 8. Disk-shaped oblate top molecule.

4.4. Oblate Top

In an oblate top, the a axis and b axis moments of inertia are equal, that is, $\mathrm{I_c > I_b = I_a}$. Oblate tops are "disc shaped" as seen in Figure 8. The Hamiltonian for an oblate top is

$$\hat{H} = \frac{\hat{J}_a^2}{2I_a} + \frac{\hat{J}_b^2}{2I_b} + \frac{\hat{J}_c^2}{2I_c} = \frac{\hat{J}_a^2 + \hat{J}_b^2}{2I_b} + \frac{\hat{J}_c^2}{2I_c} = \frac{\hat{J}_a^2 + \hat{J}_b^2}{2I_b}\left[+\frac{\hat{J}_c^2}{2I_b} - \frac{\hat{J}_a^2}{2I_b}\right] + \frac{\hat{J}_c^2}{2I_c}$$

$$= \frac{\hat{J}_a^2 + \hat{J}_b^2 + \hat{J}_c^2}{2I_b} + \frac{1}{2}\left(\frac{J_c^2}{I_c} - \frac{J_c^2}{I_b}\right) = \frac{\hat{J}^2}{2I_b} + \frac{J_c^2}{2}\left(\frac{1}{I_c} - \frac{1}{I_b}\right)$$

$$\hat{J}_c^2\psi = \hbar^2 K^2 \psi$$

where K is quantum number describing the projection of the total angular momentum along "c" axis, which is the symmetry axis for this case. Therefore, the energy for an oblate top is

$$E = \frac{\hbar^2}{2I_b}J(J+1) + \frac{\hbar^2}{2}\left(\frac{1}{I_a} - \frac{1}{I_b}\right)K^2 = hBJ(J+1) + h(C-B)K^2$$

Selection rules are $\Delta J = \pm 1$, $\Delta K = 0$. $\Delta M = 0, \pm 1$. Because of the selection rule $\Delta K = 0$, the axial rotational constant, A or C, is not generally determinable from microwave spectra.

The energy level diagrams of prolate and oblate symmetric tops are given in Figure 9.

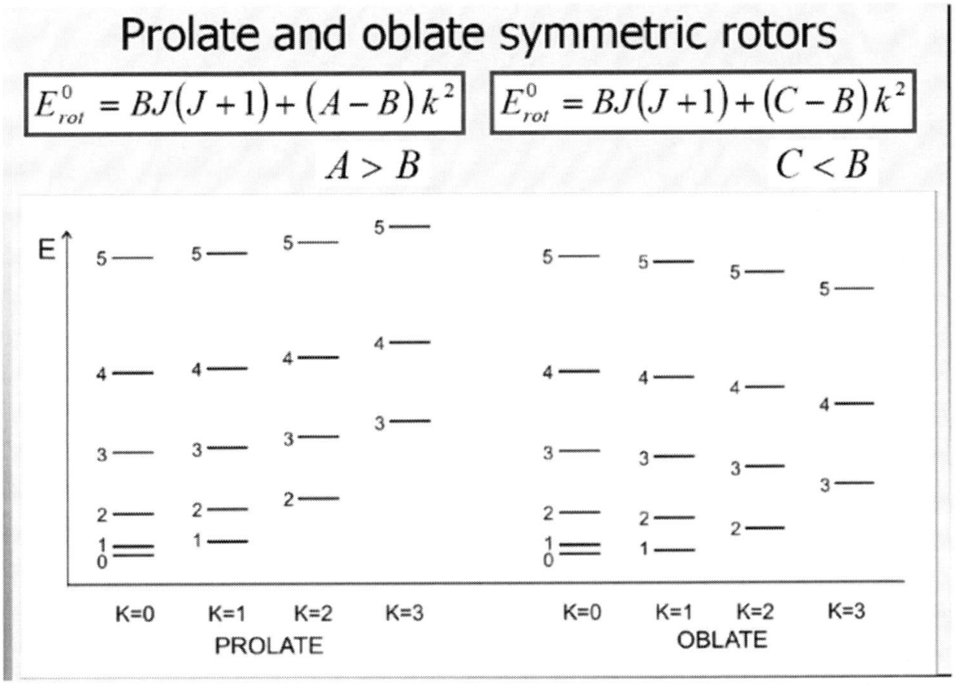

Figure 9. The energy level diagrams of prolate and oblate symmetric tops.

Asymmetric top molecules are characterized by the following properties:

- No analytic expression for energy levels,
- Labeling of the energy levels: JK_aK_c, where K_aK_c are pseudo-quantum numbers,
- Selection rules : $\Delta J = \pm 1$; $\Delta M = 0, \pm 1$
  - a-type ($\mu_a \neq 0$) transitions: $\Delta K_a = 0, \pm 2, \ldots$; $\Delta K_c = \pm 1, \pm 3, \ldots$
  - b-type ($\mu_b \neq 0$) transitions: $\Delta K_a = \pm 1, \pm 3, \ldots$; $\Delta K_c = \pm 1, \pm 3, \ldots$
  - c-type ($\mu_c \neq 0$) transitions: $\Delta K_a = \pm 1, \pm 3, \ldots$; $\Delta K_c = 0, \pm 2, \ldots$

$$\langle J,k,m | \mathbf{H}^0_{rot} | J,k,m \rangle = \frac{\hbar^2}{4}\left\{\left(\frac{1}{I^0_{xx}} + \frac{1}{I^0_{yy}}\right)\left[J(J+1) - k^2\right] + \frac{2k^2}{I^0_{zz}}\right\}$$

$$\langle J,k,m | \mathbf{H}^0_{rot} | J,k \pm 2,m \rangle = \frac{\hbar^2}{8}\left(\frac{1}{I^0_{xx}} - \frac{1}{I^0_{yy}}\right)\left[J(J+1) - k(k \pm 1)\right]^{1/2}\left[J(J+1) - (k \pm 1)(k \pm 2)\right]^{1/2}$$

The energy level structure of an asymmetric top molecule is shown in Figures 10 and 11 which correlates with the prolate and oblate symmetric top energy levels as limiting cases. Explicit expressions for a few low J levels are given in Table 3. Levels within a J can't cross. Level with K > 0 start degenerate with the degeneracy being lifted as soon as the limiting symmetric top case becomes asymmetric.

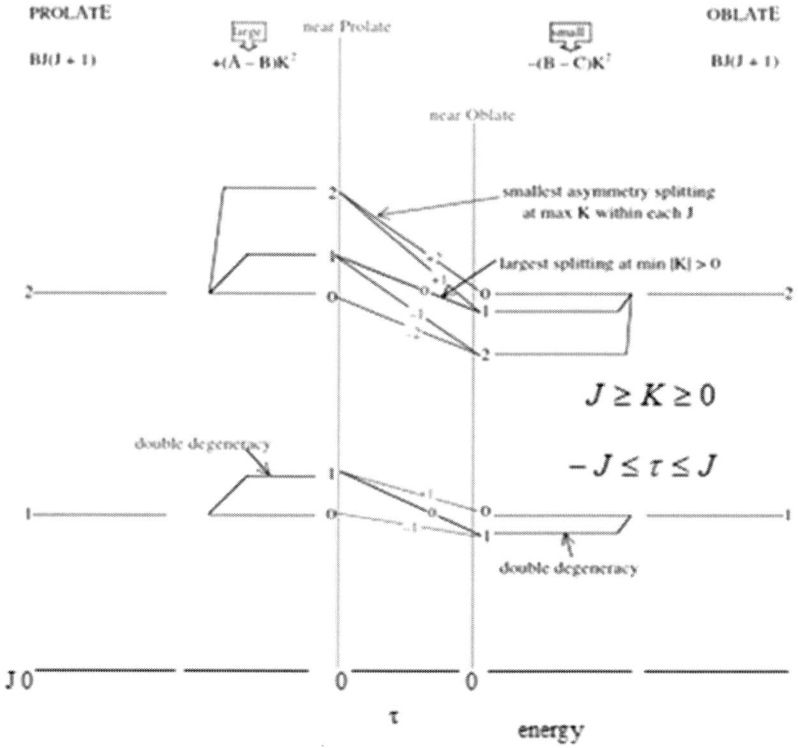

Figure 10. The energy level structure of an asymmetric top molecule which correlate with the prolate and oblate symmetric top energy levels as limiting cases.

The correlation diagram is only qualitative and does not accurately represent the true level spacing. From the moments of inertia one can define an asymmetry parameter κ as

$$\kappa = \frac{2B - A - C}{A - C}$$

which varies from -1 for a prolate symmetric top to 1 for an oblate symmetric top.

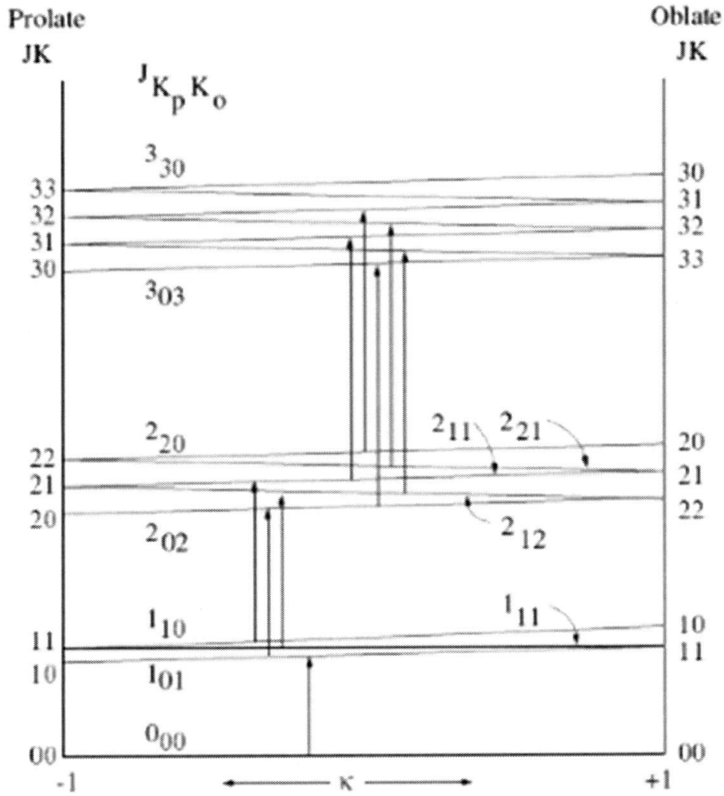

Figure 11. Energy level diagram for J = 0,1,2,3 of an asymmetric rotor.

Table 3. Analytical energy expressions for low J asymmetric rotor states

$J_{K_pK_o}$	Rotational Energy
0_{00}	0
1_{01}	$B+C$
1_{11}	$A+C$
1_{10}	$A+B$
2_{02}	$2A+2B+2C - 2\sqrt{(B-C)^2+(A-C)(A-B)}$
2_{12}	$A+B+4C$
2_{11}	$A+4B+C$
2_{21}	$4A+B+C$
2_{20}	$2A+2B+2C + 2\sqrt{(B-C)^2+(A-C)(A-B)}$
3_{03}	$2A+5B+5C - 2\sqrt{4(B-C)^2+(A-B)(A-C)}$
3_{13}	$5A+2B+5C - 2\sqrt{4(A-C)^2-(A-B)(B-C)}$
3_{12}	$5A+5B+2C - 2\sqrt{4(A-B)^2+(A-C)(B-C)}$
3_{22}	$4A+4B+4C$
3_{21}	$2A+5B+5C + 2\sqrt{4(B-C)^2-(A-B)(A-C)}$
3_{31}	$5A+2B+5C + 2\sqrt{4(A-C)^2-(A-B)(B-C)}$
3_{30}	$5A+5B+2C + 2\sqrt{4(A-B)^2+(A-C)(B-C)}$

5. Microwave Spectrometers

The earlier forms of microwave spectrometers were of simple video type absorption detection. Then, to increase sensitivity, the Stark modulation spectrometers and source modulation spectrometers emerged. Most research laboratories now exploit pulsed excitation Fourier Transform techniques.

5.1. Stark Modulation Spectrometers

The commonly used microwave absorption spectrometers were based on Stark Effect modulation technique. Still a number of laboratories use these types of spectrometers for molecular structural studies. The disadvantages are line-broadening due to the inhomogeneous distribution of the Stark field in a waveguide used as sample cell and thus can only be used in the low frequency regions and hence for low-lying molecular rotational transitions. In Stark modulation, as shown in Figure 12, a square wave periodic potential is used on the Stark electrode [8, 10, 65].

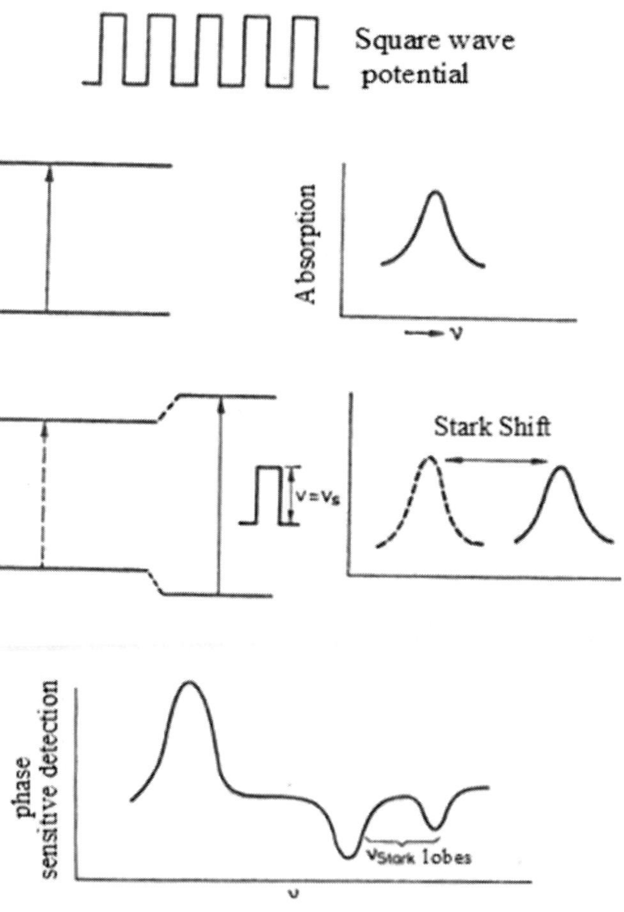

Figure 12. Stark Modulation, use of a square wave periodic potential on the Stark electrode.

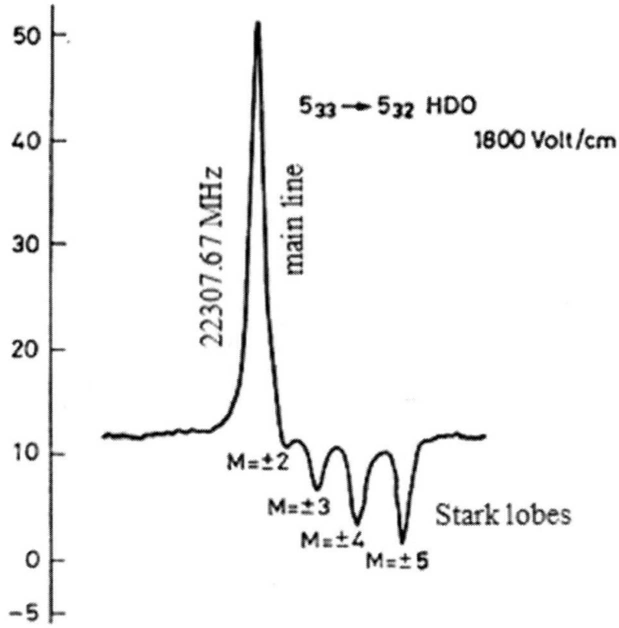

Figure 13. Stark spectrum of the HDO molecule J=$5_{3,3}$ - $5_{3,2}$ rotational transition.

The electric potential changes from zero to the desired potential at rates of say 100 kHz. Thus at certain intervals the molecules are in a Stark field and at other time no field exists. Thus the energy levels are intermittently changed by the intermittent electric field. That means, when we tune the microwave frequency to a transition, the transition is obtained at intervals. The output signal can be fed into a lock-in-amplifier and other signal-conditioning electronics to detect the line (Figure 12).

A typical Stark spectrum obtained for HDO is shown in Figure 13 [105]. The spectrum was obtained with 1800 Volt/cm in a waveguide cell. The M = ± 1 is not seen in this case because it needs higher electric fields to become resolved. The various levels of developments in experimental techniques in microwave spectroscopy are given in Table 4.

Table 4. Various levels of developments in experimental techniques in microwave spectroscopy

1.	Video spectrometers
2.	Stark effect spectrometers
3.	Source modulation spectrometers
4.	Power modulation spectrometers
5.	Source modulation spectrometers
6.	High temperature spectrometers
7.	Waveguide Fourier transform spectrometers
8.	Absorption spectrometers on free jets
9.	Pulsed molecular jet expansion Fourier transform spectrometers

6. Fourier Transform Microwave (FTMW) Spectroscopy

The developments in FT-NMR had prompted researchers to develop a similar experimental frame work in microwave spectroscopy as well. The behavior of the system is described by quasi-optical Bloch equations. Experimentally, a short pulse of (typically 0-3 microsecond duration) microwave radiation is introduced in resonance with a rotational transition of the molecule. The molecules which absorb the microwave energy from the pulse are induced to rotate coherently in phase with the incident microwave radiation. Deactivation of the polarization pulse results in microwave emission which accompanies deactivation of coherence of the molecular ensemble. Free induction decay occurs and the measurement in the time domain is transformed into frequency domain by Fourier transformation methods. The first ideas on this technique were put forward by Dicke and Romer in 1950s [72] and the first waveguide FTMW spectrometer was constructed by Eggers and Flygare [73] in 1975.

It was in 1980s that Balle, Campbell, Keenan and Flygare [75, 76] demonstrated the FTMW microwave technique within a free space cell comprising of an evacuated chamber containing a Fabry-Perot cavity. In this technique a sample is cooled by supersonic expansion to a few Kelvin and in less than a millisecond time the sample is probed. Supersonic cooling of molecules brings more of them in the ground vibrational state and the use of coaxial expansion with respect to the Fabry-Perot resonator brings more resolution and thereby increasing the accuracy of measurements. The technique also allows to study molecular complexes especially hydrogen bonded species as there is insufficient energy for these weakly bound molecules to undergo fragmentation or chemical reaction at such low temperatures. William Klemperer has used this technique extensively to study many weakly-bound molecular complexes.

The Balle-Flygare technique had undergone many innovations with the advent of digitizers and related electronic technology. Pate and coworkers at the University of Virginia improved the technique by using a high speed (4GS/s) arbitrary wave form generator to provide a microwave polarization "chirp" that sweeps up to 12 GHz in frequency in less than a microsecond and by using a high speed oscilloscope to digitize and Fourier transform the molecular free induction decay. Modified versions are constructed in various laboratories.

A new FT-MW spectrometer (Figure 14) on molecular jets was constructed by Grabow et al. at the University of Kiel and at Hannover. [77, 78].

The primary difference in the techniques of the earlier direct absorption method and the FT-MW technique are in the method used to probe the molecules and the manner of treating the associated noise. The direct absorption technique uses a CW source to probe a gas, typically modulating the source (source modulation) or resonance frequency (Stark effect modulation) to reduce the $1/f$ noise. This in turn creates side bands in the microwave source frequency allowing an intensity change to be detected rather than measurement of the source intensity. Using this method, the source noise is reduced to a levels low enough to become insignificant, rather the detector noise now becomes the most critical noise source [106]. In the 3 to 30 GHz region, the modulation of molecular absorption frequencies by means of Stark effect in conjunction with phase sensitive detection is generally used, whereas in the millimeter region source modulation techniques are used because introducing Stark electrodes in the comparatively smaller dimensioned microwave cells is difficult. Additionally a larger number of Stark components of higher rotational transitions are encountered in the millimeter region.

In FT-MW spectroscopy the presence of a molecular electromagnetic field is detected in the absence of the probe radiation by a pulsed microwave source. After the probe radiation decays in the cavity, the weak microwave impulse-response of the molecular sample is immediately amplified to a level that further processing has no impact on the signal-to-noise (SNR) ratio. The analog signal is processed by heterodyne detection and the frequency of the signal is then converted to one that can be easily digitized. The down-converted signal can then be repeatedly recorded and phase-coherently averaged thereby increasing the signal-to-noise ratio. A Fourier transform is performed on the data to obtain the frequency resolved amplitude information. The time-domain signal may also be fit directly to a model of the so-called free-induction-decay (FID) and the amplitude and frequency determined from the model parameter. In the case of transient FID signals, the amplifier noise becomes critical [106]. However, broadband (2 GHz to 26.5 GHz) microwave amplifiers are available in this region with very low noise figures, typically around 2 dB. Amplifiers even lower in noise can be obtained for octave bands.

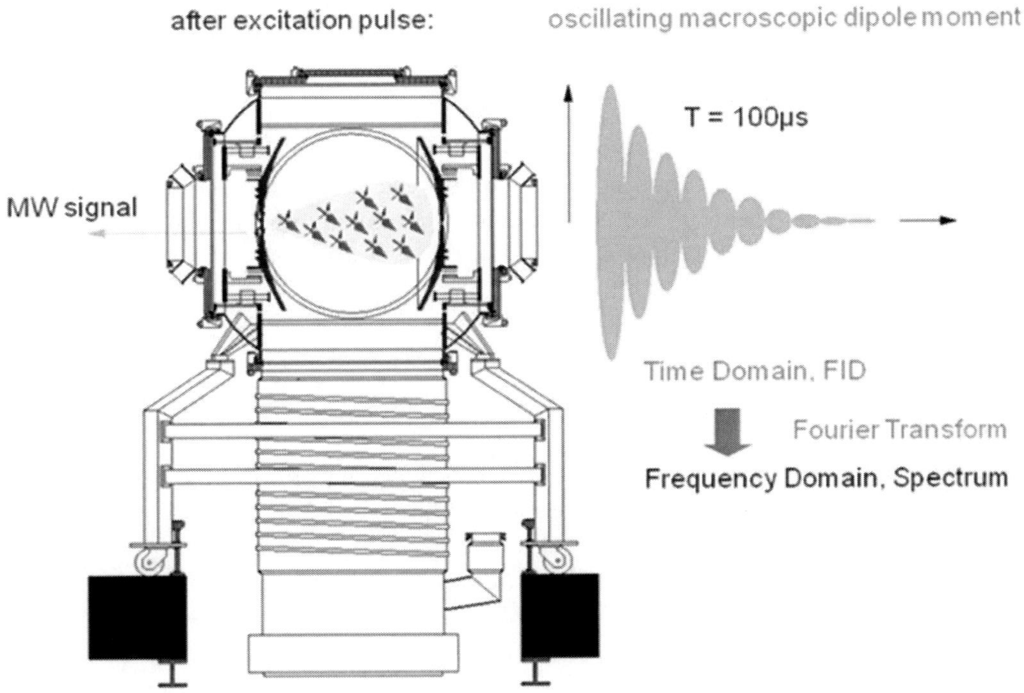

Figure 14. Schematic diagram of a supersonic-jet Fourier transform microwave spectrometer.

Even though the present technique gives high resolution and improved signal to noise ratio, there also exists many similarities between this technique and earlier microwave spectroscopic techniques. For example, these techniques can employ multi-pass resonator cavities with high values of Q to overcome the lower molecular absorption coefficients lying in this frequency region. More over the sample pressure can be reduced in order to decrease the line broadening or experiments can be performed on molecular beams. A supersonic molecular expansion is a technique which significantly cools samples to low rotational temperatures, thereby localizing the rotational population to a few relatively low energy levels, enhancing the sensitivity of the technique. Finally, there are common factors that affect the intensity of absorption in both techniques.

The most advantage of microwave spectroscopy in comparison with other alternative methods is its ability to measure rotational transitions with extremely high resolution and the determination of molecular constants with high precision. This ability is a result of the narrow banded radiation sources and low pressure measurements that have been developed since the establishment of microwave spectroscopy. Currently, many continuous wave (CW) microwave sources are commercially available with resolutions of 1 Hz and tunable over a wide frequency range. High resolution techniques, such as Fourier transform microwave spectroscopy with a pulsed microwave source, considerably enhance the ability of spectral methods to separate and detect multiple analytes in a complex mixture by differences in their transition frequency, a characteristic referred to as specificity. Since the number of chemical warfare agents (CWAs) that can be deployed in possible chemical attacks is in the thousands, techniques that exhibit high specificity are very attractive in detecting these agents. The complex environments in which the CWAs can be deployed may also host numerous interferers that can either obscure CWA detection or create false positives with instruments that have poor specificity.

As highly stable solid state microwave sources are commercially available now, it gives another advantage of microwave spectroscopy. These sources, when coupled with frequency standards, reduce the frequency drift in the source to virtually zero, relative to the line widths of the microwave transitions. This, in turn, provides the possibility of long term source stability that few techniques can rival. Furthermore, the turnkey operation of some of these sources offers a low maintenance level which is invaluable to point sensors deployed in the field.

7. APPLICATIONS

The first accidental application of microwave spectroscopy is the discovery of lasers. The forerunner of laser was the maser. The theoretical frame work for the maser principles was first formulated by Nikolay Basov and Alexander Prokhorov from Lebedev Institute of Physics; the results were published in 1954. Independently, Charles H. Townes, James P. Gordon, and H. J. Zeiger built [107, 108] the first ammonia maser at Columbia University in 1953. This device used stimulated emission in a stream of energized ammonia molecules to produce an amplification of microwaves at a frequency of about 24.0 gigahertz. Townes and Schawlow later proposed the principle of an optical maser or laser [109] and Theodore H. Maiman built the first working model of a laser in 1960 [110]. The population inversion in the close-lying two inversion levels of the ammonia molecule was attained by physically removing the lower state by a non-uniform magnetic field which can be termed as state selection method. The details of the ammonia maser can be obtained in many text books. Later the maser techniques were used to obtain optical masers or the so called lasers which has wide applications in all fields of science and technology including medical sciences.

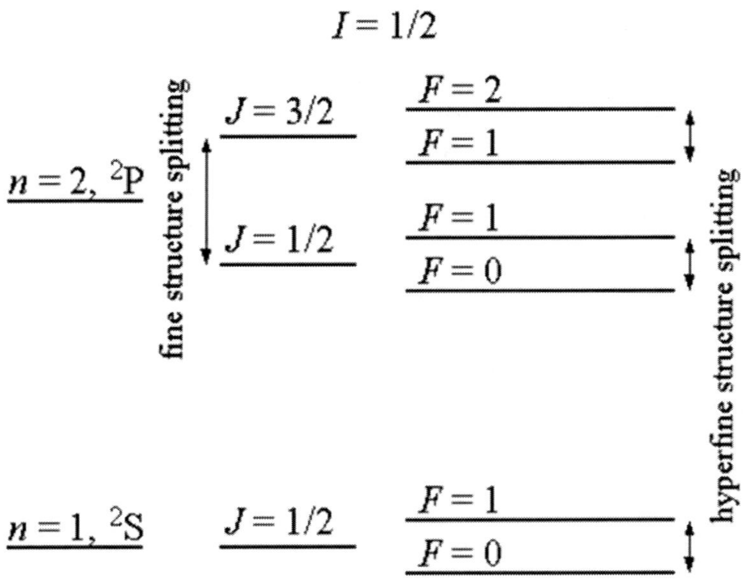

Figure 15. Energy level splitting of hydrogen atom.

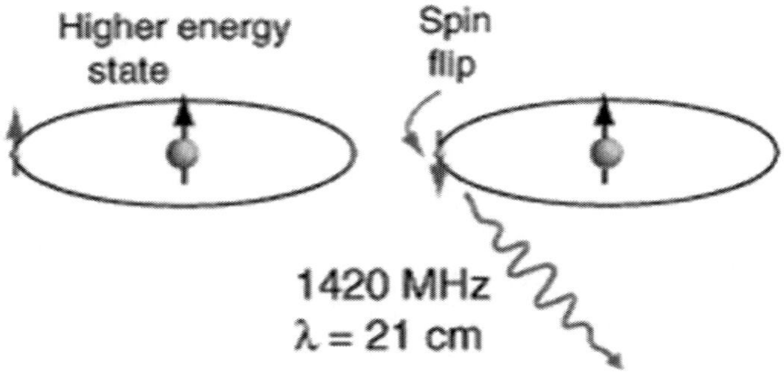

Figure 16. Spinning electron and proton in atomic hydrogen.

7.1. Astrophysical Applications

One of the most important applications of microwave spectroscopy is in the study of astrophysical molecules. A field that has already developed under the name radio astronomy. Microwaves can be monitored to probe the interstellar media and other heavenly bodies. There are now ground based telescopes and airborne telescopes. Mars and Saturn missions all have carried spectrometers to unravel the space. A list of molecules observed from the heavenly bodies are collected in the Cologne Molecular Spectra database compiled by Müller and coworkers and are available in internet [45]. There is also a website of NIST maintained by Frank Lovas [111].

The 21 cm line of hydrogen atom is of special interest to astrophysics: The hydrogen line, 21 centimeter line or HI line refers to the spectral line created by changes in the energy state of neutral hydrogen and occurs at a frequency of 1420.40575 MHz, equivalent to a vacuum wavelength of around 21cm. This line falls within the microwave region and is used extensively in astronomy, since it can penetrate dust clouds that are opaque to visible wavelengths. The radiation comprising the hydrogen line comes from the transition between the two hyperfine levels of the hydrogen 1s ground state [112], see Figures 15-17. The first observation of the hydrogen atom hyperfine transition in space was observed in 1951 by Harold Inving Ewen and Edward Mills Purcell.There are many molecules detected in space now. Up-to-date lists can be found online, e.g., at the NIST website [111].

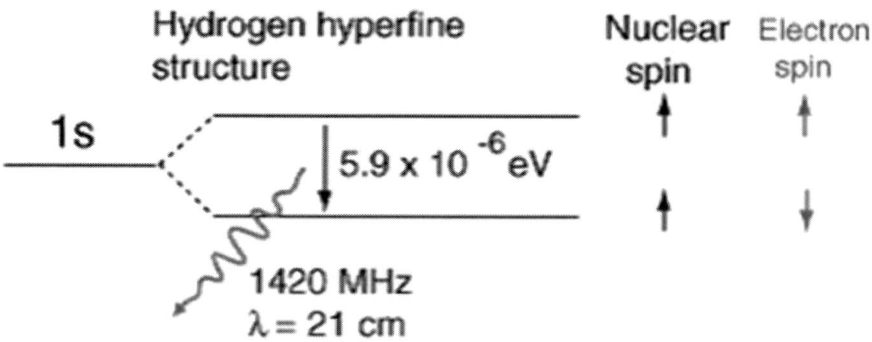

Figure 17. Hydro hyperfine structure.

7.2. Astrophysical Maser

Thus since 1965 it has been found that many molecules from space show monochromatic coherent stimulated emission at the microwave frequencies. This astrophysical maser comes from a naturally occurring source of stimulated emission. In 1965 Weaver and coworkers [113] observed an unknown line at 1665 MHz coming from a far source. The line could not be identified from any known microwave spectrum. The line was first named *Mysterium,* however later analysis showed that the line belongs to the OH radical from compact sources within the cloud. In subsequent observations, maser lines from water molecules were observed in 1969 [114]. The CH_3OH emission was observed in 1970 [115] and SiO emission in 1974 [116], which were found to originate from within molecular clouds.

A maser is characterized by stimulated emission (or seeded) of monochromatic and coherent radiation. The maser emission frequency correspond to the energy difference between two quantum energy levels of the species in the gain medium which have been pumped into a non-thermal population distribution. The main difference between a naturally occurring maser and a terrestrial laboratory maser is that the former lack the resonant cavity as in the case of terrestrial laboratory masers. The emission from an astrophysical maser is due to a single pass through the gain medium. And hence, generally, they lack spatial coherence and mode purity as expected in the case of laboratory based masers.

It is generally difficult to observe an astrophysical maser as the radiation from such sources is quite weak and sometimes may escape detection due to limited sensitivity. Moreover they may be obscured due to the sometimes overwhelming spectral absorption from unpumped molecules of the maser species in the surrounding space. Furthermore, there should be velocity coherence in the line of observation, since the line frequencies are shifted by Doppler effect. In the case of laboratory masers the polarization is largely achieved by selectively oscillating the different modes whereas in the case of astrophysical masers this depends on the presence of a polarization-state dependent pump or the presence of a magnetic field in the gain medium.

However, the observation of masers in interstellar medium or in astrophysical objects gives rise to much valuable information on the physical conditions such as temperature, number density, magnetic fields in these media. Velocity of the particle and many other parameters could be determined by these observations. It should be remembered that the redshift found in atoms and molecules observed from space has revealed an expanding universe which may point to stellar evolution and is the backbone of many theories pertaining to the origin of the universe.

The first astrophysical maser was discovered in 1968 around highly-evolved Late-type stars (named OH/IR stars) and was due to OH emission [117], then H2O emission by S. H. Knowles, Townes and coworkers in 1969 [118] and SiO emission in 1974 by Buhl et al. [119]. Snyder and Buhl in 1974 [120] also discovered masers in external galaxies. In 1973, Whiteoak and Gardner observed maser emission in comet halos from our own solar system [121]. Extragalactic sources are also found to give maser lines with extraordinary luminocity which are termed as Megamasers since 1982. Yet another phenomenon [122] was observed in stellar and interstellar objects which show evidence of an anti-pumped (dasar) sub-thermal population in the 4830 MHz transition of formaldehyde (H2CO). Dasar is an acronym for Darkness Amplification by Stimulated Absorption of Radiation. Several species have been identified, which are listed in Table 5.

Masers in star forming regions are found moving away and hence the Doppler shift measurements on these narrow lines could give the velocity of the objects which eventually lead to three dimensional mapping of the maser environment.

Table 5. Identified hitherto species in stimulated emission from astronomical sources

Diatomic	OH. CH, SiO, SiS, ^{29}SiO, ^{30}SiO, H (atom)
Polyatomic	H_2O, H_2CO, NH_3, $^{15}NH_3$, CH_3OH, HC_3N, HCN, $H^{13}CN$

7.3. Environmental and Chemical Reagents

There were some early studies on the detection of toxic constituents in tobacco smoke by Kadaba et al. in the mid-1970s [123]. The microwave spectrometer developed for this study was a Stark cell modulated microwave spectrometer and was based on the fact that the various constituents present in a complex gaseous mixture absorb energy at different microwave frequencies. The tobacco smoke was transferred into the microwave cell which was covered with dry ice. The measurements at frequency range 22 – 38 GHz, the line due to a number of molecular species like acrolein, acetone, methyl chloride, methanol, acetaldehyde, formaldehyde, acetonitrile, HCN and so on which were identified by the appropriate rotational frequencies of the individual pure gas samples published in the literature (Townes and Schalow 1955), US Department of Commerce, National Bureau of

Standards, Microwave Spectral Tables NBS Monograph 70 #5 June 1968). These kinds of instrumentation could readily be used to detect atmospheric pollutants as well.

As mentioned earlier, more interest and efforts have been taken in mid-1990s to demonstrate the quantitative capabilities of microwave spectroscopy in detecting trace gases in molecular mixtures. In 1995 Lovas et al. [94-96,124]developed a pulsed molecular beam Fabry-Perot cavity Fourier Transform spectrometer at NIST and have shown high sensitivities for many molecular species in the low parts per million (ppm) to parts per billion (ppb) range. Minimum detectable concentration for several representative compounds was provided. These include acetaldehyde, acrolein, propionaldehyde, benzldehyde, p-tolualdehyde, methanol, SO_2, propane, methyl t-butyl ether, ethyl t-butyl ether and others.

Very reliable and most accurate knowledge of the accessible molecular rotational transitions are imperative in order to identify specifically the molecular rotational transitions which gives conformations of CWAs that are most favorable to quantitative analysis. And it is very crucial as there are a multitude of accessible rotational lines. This requires accurate knowledge of the frequency and intensity of the rotational transitions and therefore the construction of a resonator suitable for the desired frequency region. Additionally, analysis of the rotational spectra can give information pertaining to the gas phase geometry and possible various conformers of CWAs and their simulants thereby increasing our understanding of the chemical nature of these molecules and facilitating the development of theoretical models.

Hence a database was also seemed necessary for the comparison of the detected data of frequencies. The development of such a data base for microwave spectroscopic quantitative and qualitative studies for chemical warfare agents provides the fundamental data needed for current and future endeavors The efforts are being initiated for surveying and assigning rotational spectra of a number of compounds. Spectroscopic information and the rotational frequencies are obtained for some of the most common CWAs and their simulants. These measurements [125] have been carried out using the pulsed molecular beam FT-MW spectrometers. The database provides sufficient information for identifying thc rotational lines in the observed spectra of CWAs.

In preparing a database, the basic approach is as follows: First, a prediction of the possible geometric structures and the dipole moment components of the molecule are obtained using already established theoretical tools like ab initio calculations, with programs such as Gaussian [126], GAMESS [127], etc. Next the molecular structures thus obtained by theoretical calculations are used to predict the rotational transition frequencies and intensities. Frequency surveys are then conducted on the FT-MW spectrometer to look for the predicted rotational transitions. Once several rotational transitions are detected in the predicted regions, the sample conditions are optimized and all accessible rotational transitions are recorded and analyzed. The final measured rotational spectrum is analyzed and modeled and molecular parameters are extracted and then these experimental results are reported in the database.

One of the major difficulties of these gas phase CWA measurements for the database is the low volatility of these compounds at room temperature. In order to get sufficient vapor pressure to acquire their spectra, most substances were placed in a pulsed solenoid valve equipped with a liquid reservoir [128] that can be heated to temperature up to approximately 200 °C. For compounds with poor thermal stability, an alternate sample preparation technique which flash vaporizes a liquid is being developed [129]. This technique also makes it possible for gas phase analysis of a liquid mixture and will be beneficial the detection of CWAs, as many agents are found in the field as condensates.

The CWAs and simulants that have been studied by these methods to date at Edgewood Chemical and Biological Center (ECBC) and the National Institute Standards and Technology (NIST) are listed by Bousquet et al. [130] and are collected in Table 6. In many cases, the spectra are quite complex, exhibiting evidence of multiple conformers and internal molecular motions.

Table 6. Chemical weapons agents and simulants examined by Fourier-transform microwave spectroscopy[a)]

Agent type	**Compound**
Blister agents	bis-2-chloro ethylsulfide (mustard gas,HD)
Blister agent	2-(ethylthio) ethanol (HOEES)
simulants	Thiodiglycol (TDG)
	2-chloroethyl ethyl sulfide (ClEES)
	Ethyl sulfide
	divinyl sulfide
	divinyl sulfoxide
	divinyl sulfone
Nerve agents	o-ethyl dimethylamidophosphorylcyanide (tabun, GA)
	isopropyl methylphosphonofluoridate (sarin, GB)
	o-pinacolyl methylphosphonofluoridate (soman, GD)
	cyclohexyl methylphosphonofluoridate (cyclosarin, GF)
Nerve agent	methyl phosphonyldifluoride (DF)
simulants	methoxy difluorophosphinoxide
	dimethyl methylphosphonate (DMMP)
	diethyl methylphosphonate (DEMP)
	diethyl ethylphosphonate (DEEP)
	diisopropyl methylphosphonate (DIMP)
	Pinacolone
	pinacolyl alcohol
	N-methyl diethanolamine
	N-ethyl diethanolamine

a) Ref. [130].

The spectral pattern obtained in the case of dimethyl methylphosphonate (DMMP) is a unique example to show how difficult it is to analyze a complicated spectrum. DMMP is a molecule having three methyl tops that can undergo hindered internal rotation within the molecule. Internal rotations give rise to additional splittings with respect to the rotational spectrum of a (semi-)rigid molecule. Additionally there is also an interchange motion between the two methoxy groups in the molecule that splits the transitions further [131-133]. Similar behaviors of methyl torsions have been observed in the P-CH_3 groups of all G-series agents. For the G-series of agents the methyl top motion is quite hindered leaving most of the rotational transitions unsplit. The internal rotations in the G-series of compounds and DMMP have been compared to provide additional insight into the behavior of the entire class of molecules [134].

Recently, microwave spectra of many reagents, proteins and lipids have been studied by the supersonic-jet FT-MW technique. E.g. the conformational scan calculated byemploying the quantum-chemical program Gaussian 09 showed nine low-energy conformers (see Figure 18) of vitamin C; three of these conformers (see Figure 19) could be unambiguously identified by by comparing the theoretical and experimental results [135].

The possibility of the detection of impurities in a commercial sample was recently described by Walter Caminati's group at Bologna with the identification of the impurity methyl pyruvate (MEPY-CH3-CO-CO-OCH_3) in a commercial sample of methylactate (MELA-CH_3-CHOH-CO-OCH_3) which exhibits a similar structure [136].

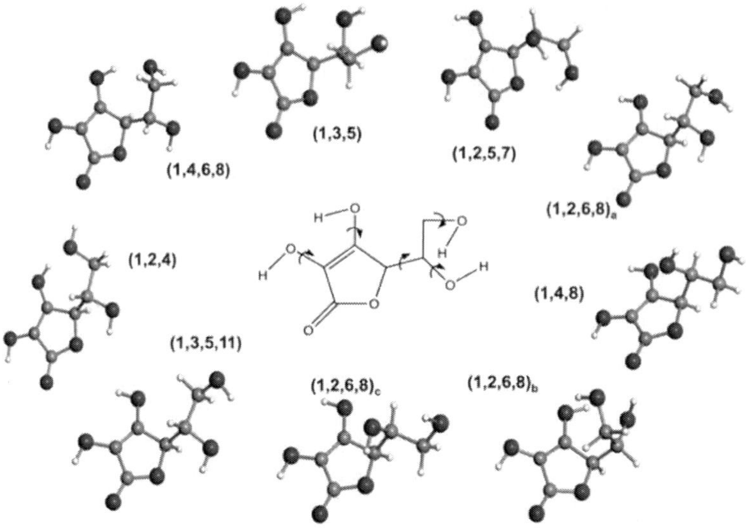

Figure 18. Quantum-chemically predicted low -energy conformers of Vitamine C (Reprinted with permission from J. Phys. Chem. Let. 4, 65-69 (2013).

Molecules having different bonding structures can be detected and unambiguously identified in a microwave spectroscopic study using a molecular jet [137]. Two such molecules, for example, are succinic acid ($C_4H_6O_4$) and succinic anhydride ($C_4H_4O_3$) whose structures are different by the presence and absence of an H_2O group. Both molecules could be identified without any blending problems and with high precision in the mixture. Succinic acid is an important platform chemical with current global production of 30000 to 50000 tons per year. This molecule and the corresponding anhydride find industrial application as ingredients to stimulate animal plant growth, additives in food pharmaceutical products, surfactants, detergents and as a feed stock chemical for the production of polymides and polysters as well as biodegradable plastics. The US department of Energy outlines succinic acid as one of the top 10 bio based products with research needs; being a metabolic in the tricarbocyclic acid cycle of several anaerobic microorganisms, succinic acid can be produced by fermentation of mixed sugars as renewable carbon source. Precise determination of their structure is important to characterize the physical and chemical properties of these molecules. The rotational spectra obtained from a supersonic-jet Fourier transform microwave spectrometer could unambiguously reveal the structure of these molecules and their presence in a sample mixture. All carbon ^{13}C isotopic species could be identified in succinic anhydride and a substitution structure (r_s structure) could be determined. Similar experiments have also been done on many isomeric halogenated toluenes at different positions of the benzene ring showing that they are easily distinguished and identified from big effects in the moments of inertia (while, in mass spectrometry, they still have the same mass signal and might be difficult to be separated by chromatography if that has to be done fast).

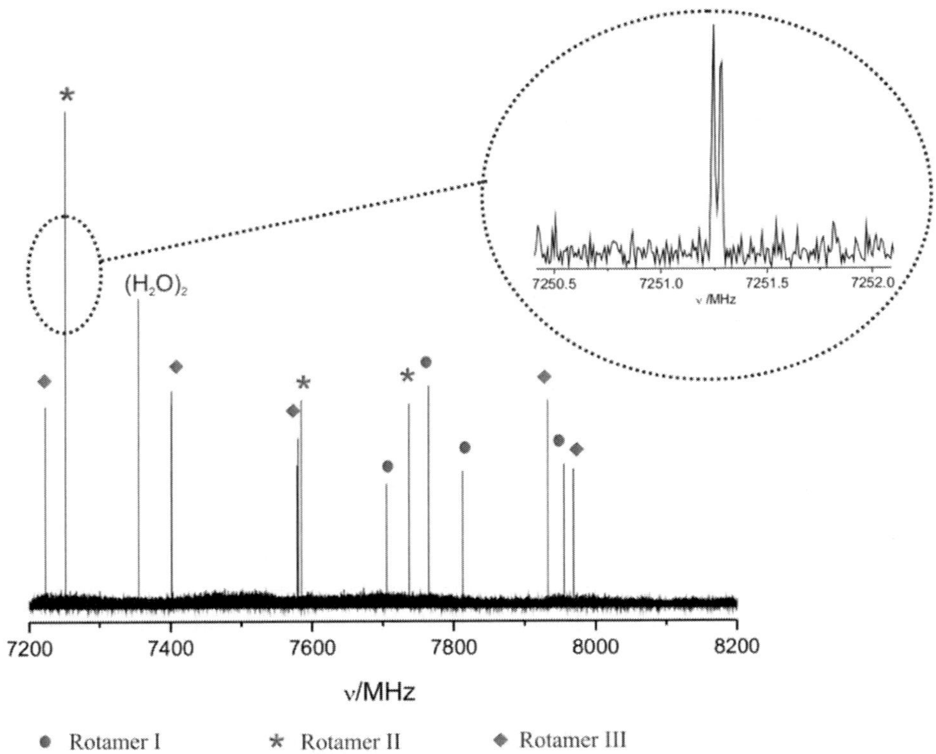

Figure 19. Spectrum of three detected Rotomers of Vitamine C (Reprinted with permission from J. Phys. Chem. Let. 4, 65-69 (2013), ACS.

8. The Developmentof a Microwave Based Trace Gas Sensor

A trace gas sensor such as for the detection of, e. g. CWA, requires many properties for it to be effective and the critical attributes include high sensitivity, high selectivity, stability and portability. Moreover, the spectrometer should support analytical measurements both in the laboratory and in the field. An ideal experiment would generally correlate to the number of molecules in a gaseous mixture. However, there are a number of factors like the absorption coefficient, size and shape and mass distribution of the molecule and, above all, the line width and resolution of the microwave spectrometer are important factors that must be considered in order to quantitatively interpret the measured absorption. The absorption of the linearly polarized microwave radiation is described by the following equation [8]:

$$\frac{dP}{P} = -\alpha_{ab} S(\omega_{ab} - \omega)dy = -\frac{{}^1Nw^0 B_{ab} \hbar\omega_{ab}}{c} S(\omega_{ab} - \omega)dy$$

$$= -\frac{2\pi\,{}^1Nw^0 \omega_{ab} \,|\, \mu_{ab} \,|^2}{c\hbar} S(\omega_{ab} - \omega)dy$$

where ${}^1N$ is the total number of molecules per unit volume, w^0 represents the fractional difference of molecules being in states a and b; B_{ab} is the Einstein coefficient of absorption, ω_{ab} is the transition frequency, $|\mu_{ab}|$ represents the transition matrix element of the dipole moment operator μ_z, c is the speed of light, h the Planck constant, and S (ω_{ab}-ω) a normalized line shape function.

The square of the transition dipole moment $|\mu_{ab}|^2$ is proportional to the probability of transition from the state a to b and is representative of the interaction between the permanent dipole moment of the molecule and the electromagnetic field. The ability to detect a transition is limited by the magnitude of the permanent dipole moment and molecules that possess a weak or having no electric dipole moment generally remain undetected. The term ${}^1Nw^0$ represents the population difference per unit volume of molecules occupying the states a and b and is described by the thermal population distribution which is governed by the Boltzmann distribution law and the degeneracy of the state. A population of probed molecules does not necessarily occupy a specific rotational 2-level system; rather, they occupy a number of states described by the Boltzmann distribution. When making rotational measurements, it is advantageous to cool the molecules to very low temperatures since it localizes the rotational population into a few low energy states thereby increasing the value of w^0 for a given transition originating from a low-energy state. For example, Figure 20 shows the rotational distribution of isopropyl methylphosphonofluoridate (sarin, GB) at several different temperatures.

The cooling of the molecules is created by introduction of a relatively high pressure gas in to a low pressure environment referred to as a supersonic expansion. Furthermore, the expansions are typically performed in a pulsed manner to reduce the pumping loads required to maintain the low pressure conditions. The rare gas provides substantial rotational cooling in the molecular rotation states and results in higher sensitivity compared to an expansion with nitrogen. The design of the vacuum chamber and sample cavity are important in the minimization of the first two conditions while clustering can be controlled to some degree by the nozzle geometry and reservoir pressure.

8.1. Line Broadening

As previously mentioned, microwave sources are available with narrow frequency and do not contribute to the observed line widths. The line broadening, associated with a supersonic expansion is dependent on the orientation of the nozzle, e.g., coaxial or perpendicular to the propagation of the electromagnetic wave. In the perpendicular position, the broadening is due to a combination of the transition time of the molecular beam through the microwave field (time of flight), i.e., transition-time broadening, and the directional velocity components of the jet expansion relative to the propagation of the wave, velocity broadening. In the coaxial position, the directional velocity of molecules from the molecular jet results in a pair of lines, a Doppler doublet. The line broadening of each of the doublet components primarily arises from the reduction in molecular density during the jet expansion.

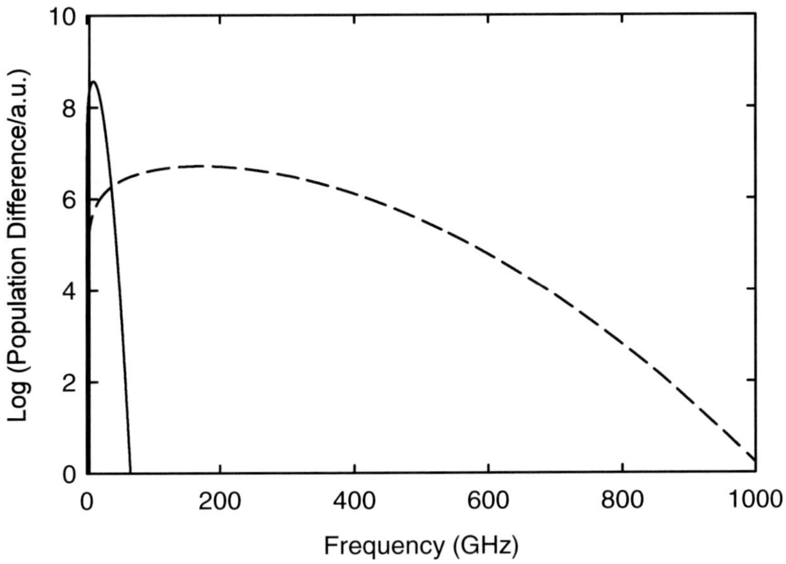

Figure 20. The fractional difference w^0 of molecules being in the two states a and b as a function of the corresponding transition frequency for isopropyl methylphosphonofluoridate (Sarin, GB) at 298 K and 1 K (Reference 130. Reprinted with permission).

8.2. Selectivity

The selectivity of FT-MW spectrometers is generally limited by the transition line widths, which are approximately within a few kilohertz. The degree of selectivity is illustrated in Figure 21, where the 104 KHz difference of two Doppler split lines of DF and GB are resolved. Thus the transition frequencies for DF and GB which are separated by 104 kHz can easily be identified. The doublets in each case arise from the fact that the measurements were obtained with the gas sample injected coaxially into the resonator cavity. The line broadening within each doublet results predominantly from the evolving distribution of the molecular density during the jet expansion. The ability to identify analytes in a complex mixture [138] is demonstrated in Figure 22. This spectrum was produced over a 2 GHz region of the microwave spectrum of a mixture containing seven CWA simulants; Thioxane, CEES, DEMP, DES, DMMP, Pinacolyl Alcohol, Pinnacolene [139-142].

8.3. Stability of Real Time Measurements

A number of modifications have been recently made to the FT-MW spectrometer to improve the long term stability of the instrument [143]. These include the development of a pressure stabilized sample delivery system and modification of the pulsed valve nozzle.

Currently, the stability has demonstrated to be approximately 1% standard deviation over a 24 hour period with sampling 2 minute intervals. At this point, the pulsed valve is believed to be the limiting factor in the signal stability. This would be sufficient to meet most field measurement requirements.

8.4. Portability

The portability of the FT-MW spectrometer is ultimately limited by the requirements of the Fabry-Perot-type resonator. The frequency range needed to detect the compounds of interest dictates that the cavity mirrors have a minimum diameter of 10 cm [144, 145]. Suenram et al. have optimized the size and sensitivity of an instrument for CWA detection.[144] A number of spectrometers have been constructed on portable laboratory carts; these instruments can be transported to a fixed location (e.g., roof of a building or subway station) or integrated into mobile platform (e.g., vehicle).

8.5. Calibration and Standards

The calibration of the FT-MW instrument has been performed with primary gravimetric standards at NIST for some atmospheric contaminants. To date the detection of CWA agents lies in the nmol/mol to mmol/mol region. An extensive study linking the CWAs to primary standards has not been performed to calibrate the FT-MW, however, the detection limits for many of the species listed in table 6 have been determined to lie in the mmol/mol to nmol/mol region.

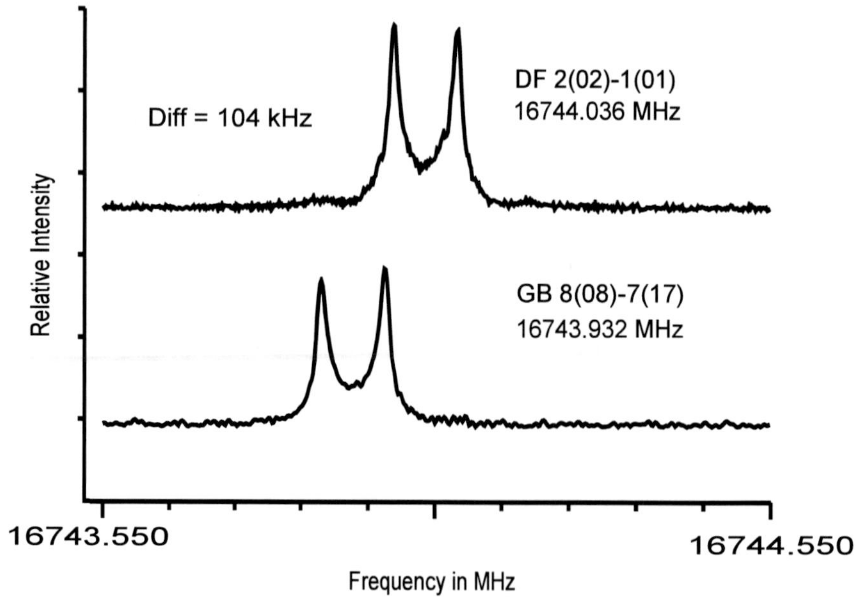

Figure 21. Spectra of CF and DF (Reference 130. Reprinted with permission).

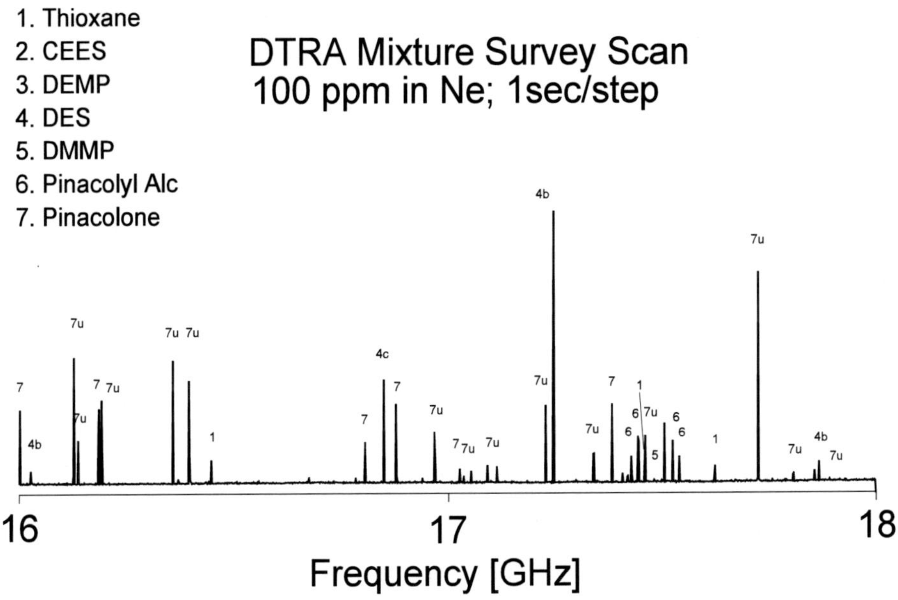

Figure 22. Survey spectrum of a complex mixture (Reference 130 Reprinted with permission).

REFERENCES

[1] Cleeton C. E. and Williams N. H., *Physical Review,* 45 (1934) 234.

[2] Gordy W., Smith W. V. and Trambarulo R. F., (1953). *Microwave Spectroscopy* Wiley, New York. (1966) Dover, New York.

[3] Delucia F. C., Krupinov A. F., and Burenin A. V., (1976). In *Molecular Spectrsocopy Modern Research*, Rao K. N., (ed.), Vol. 2, Chapter 2, Academic Press, New York.

[4] Carrington A., *Microwave Spectroscopy of Free Radicals.* Academic Press, New York (1974).

[5] Sugden T. M., and Kenney C. N., *Microwave Spectroscopy of Gases D.* van Nostrand, New York (1965).

[6] Wollrab J. E. *Rotational Spectra and Molecular Structure.* Academic Press, New York (1967).

[7] Harmony M. D., (1972). *Introduction to Molecular Energies and Spectra,* Chapter 7, Holt Rinehart and Winston, New York.

[8] Gordy W. and Cook R. L., *Microwave Molecular Spectra,* New York: John Wiley & Sons, (1970).

[9] Kroto H. W., *Molecular Rotation Spectra*, New York: Dover Publications, 1992.

[10] Townes G. W., and Schawlow A. L., *Microwave Spectroscopy,* New York: Dover Publications (1975).

[11] Suenram R. D., and Andrews A. M., *Microwave Spectroscopy, Experimental Methods in the Physical Sciences,* Vol. 29b, Academic Press, New York. (1996).

[12] Johnson D. R., and Pearson R., (1976). In Spectroscopy, D. Williams (ed.), *Methods of Experimental Physics,* Vol. 13, Part B, Chapter 4.3 Academic Press, New York.

[13] Varma R., and Hubresh L. W., (1979). *Chemical Analysis by Microwave Spectroscopy* Chemical Analysis, John Wiley and Sons, New York.

[14] Kroto H. W., *Molecular Rotation Specta,* John Wiley and Sons, London, (1975).
[15] Herzberg G., *Spectra of Diatomic Molecules,* D. van Nostrand, (1950).
[16] Mizushima M., *Theory of Rotating Dioatomic Molecules,* John Wiley and Sons, New York (1975).
[17] Carrington A., *Microwave Spectroscopy of Free Radicals,* Academic Press, London (1974).
[18] Hirota E., *High Resolution spectroscopy of Transient Molecules,* Springer Verlag, Berlin 1985.
[19] Hollas J. M., *Modern Spectroscopy,* John Wiley and Sons, Chichester, England 1996.
[20] Hollas J. M., *High Resolution Spectroscopy,* 2nd Ed. Wiley, Chichester, England 1998.
[21] Bernath P. F., *Spectra of atoms and molecules,* 2nd Edition, Oxford University Press, New York 1995.
[22] Steinfeld J. I., *Molecules and Radiation, An Introduction to Modern Molecular Spectroscopy,* 2nd Edition, Dover Publications, 1985.
[23] Demtröder W., *Laser Spectroscopy,* Springer Verlag, Berlin 1982.
[24] Aruldhas G., *Molecular Structure and Spectra,* Prentice Hall of India, New Delhi II Edition 2007.
[25] Brown John and Carrington Alan, *Rotational Spectroscopy of Diatomic Molecules,* Cambridge University Press, 2003.
[26] Grabow J. –U. and Caminati W., *Microwave Spectroscopy; Experimental Technique* in Jaan Laane Ed. Frontiers of Molecular spectroscopy, Elsevier, Amsterdam, 2009.
[27] Caminati W. and Grabow J. –U., *Microwave Spectroscopy: Molecular Systems* in Jaan Laane Ed. Frontiers of Molecular spectroscopy, Elsevier, Amsterdam, 2009.
[28] Grabow J. –U., Microwave Coherence Spectroscopy, High Resolution Technique, Challenges in Space and on Earth, Invited talk, *97th Indian Science Congress, 2010.*
[29] Nair K. P. R., Spectroscopy with Coherent Sources, from Microwaves to Lasers, Presidential Address, *84th Indian Science Congress, 1997.*
[30] Nair K. P. R., *Atoms, Molecules and Lasers,* Narosa, New Delhi and Alpha Science International, Oxford 2006, 2008.
[31] Grabow J. –U., Fourier Transform Microwave Spectroscopy: Measurement and Instrumentation in Martin Quack, Frederic Merkt, Edt. *Handbook of High Resolution Spectroscopy,* John Wiley & Sons, Chichester, 2011.
[32] Alder J. F. and Baker J. G., *Quantitative Millimeter Wave Spectrometry,* Royal Society of Chemistry, UK, 2002.
[33] Hoeft J., Lovas F. J., Tiemann E. and Toerring T., *J. Chem. Phys.* 53, 2736 (1970).
[34] Lovas F. J. and Lide D. R. Jr. *Adv. High Temp. Chem.* 3, 177 (1971).
[35] Lovas F. J. and Tiemann E., *Phys. Chem Ref. Data* 1974.
[36] Nair K. P. R., Schuetz Pahlmann H. U. and Hoeft J., *Chem. Phys. Lett.* 80 (1981) 149.
[37] Hoeft J. and Nair K. P. R., *Z Phys. D Atoms, Molecules and Clusters,* 4 (1986) 189.
[38] Hoeft J. and Nair K. P. R., *Z Phys. D Atoms, Molecules and Clusters,* 29 (1994) 203.
[39] Hoeft J.and Nair K. P. R., *Z Phys. D Atoms, Molecules and Clusters,* 8 (1988) 85.
[40] Hoeft J.and Nair K. P. R., *Chem. Phys. Lett.* 129 (1986) 538.
[41] Hoeft J.and Nair K. P. R., *Chem. Phys. Lett.* 15 (1989) 273.
[42] Hoeft J. and Nair K. P. R., *Chem. Phys. Lett.* 164 (1989) 33.
[43] Hoeft J. and Nair K. P. R., *Chem. Phys. Lett.* 215 (1993) 371.

[44] Saito S., Laboratory microwave spectroscopy of interstellar molecules *Appl. Spectosc. Review,* vol. 25 (1990) pp. 261-296.
[45] *Cologne Data Base on Astrophysical Molecules.*
[46] Zu Li-Hong and Lovas F. J., *J. Phy. Chem. Ref. Data,* 26 (1997) 17.
[47] Beaudet R. A., and Poynter R. L., *J. Phy. Chem. Ref. Data,* 7 (1) (1978) 311.
[48] Lafferty W. J., Lovas F. J., *J. Phy. Chem. Ref. Data,* 7(2) pp. 441-493 (1978].
[49] Boucher D., Burie J., Bauer A., Dubrulle A., Demaison J., *J. Phy. Chem. Ref. Data,* 9(3) pp. 659-719 (1980).
[50] Johnson D. R., Lovas F. J. and Kirchhoff W. H., *J. Phy. Chem. Ref. Data,* 1(4) (1972) 1011.
[51] Kirchhoff W. H., Johnson Donald R., Lovas Frank J., *J. Phy. Chem. Ref. Data,* 2(1) pp. 1-10 (1973).
[52] Helminger P., de Lucia F. C., and Kirchhoff W. H., *J. Phy. Chem. Ref. Data,* 2(2) pp. 215-224 (1973).
[53] de Lucia F. C., Helminger Paul, Kirchhoff William H., *J. Phy. Chem. Ref. Data,* 3(1) pp. 211-219 (1974).
[54] Maki A. G., *J. Phy. Chem. Ref. Data,* 3(1) pp. 221-244 (1974).
[55] Lovas F. J., Krupenie Paul H., *J. Phy. Chem. Ref. Data,* 3(1) pp. 245-257 (1974).
[56] Tiemann E., *J. Phy. Chem. Ref. Data,* 3(1) pp. 259-268 (1974).
[57] Winnewisser G., Hocking W. H., Gerry M. C. L., *J. Phy. Chem. Ref. Data,* 5(1) pp. 79-101 (1976).
[58] Tiemann E., *J. Phy. Chem. Ref. Data,* 5(4) pp. 1147-1156 (1976).
[59] M. C. L. Gerry, Yamada K., Winnewisser G., *J. Phy. Chem. Ref. Data,* 8(1) pp. 107-123 (1979).
[60] Bauer A., Boucher D., Burie J., Demaison J., Dubrulle A., *J. Phy. Chem. Ref. Data,* 8(2) pp. 537-558 (1979).
[61] Bauder A., *J. Phy. Chem. Ref. Data,* 8(3) pp. 583-618 (1979).
[62] Lovas F. J., Lutz H., Dreizler H., *J. Phy. Chem. Ref. Data,* 8(4) pp. 1051-1107 (1979).
[63] Willemot E., Dangoisse D., Monnanteuil N. and Bellet J., *J. Phy. Chem. Ref. Data,* 9 (1980) 59.
[64] Ozier I., Gerry M. C. L., Robiette A. G., *J. Phy. Chem. Ref. Data,* 10(4) pp. 1085-1095 (1981).
[65] Lovas Frank J., *J. Phy. Chem. Ref. Data,* 11(2) (1982) 251.
[66] Lovas F. J., *J. Phy. Chem. Ref. Data,* 14(2) (1985) 395.
[67] Kleiner I., Lovas F. J., Godefroid M., *J. Phy. Chem. Ref. Data,* 25(4) (1996) 1113.
[68] Ilyushin V. and Lovas F. J., *J. Phy. Chem. Ref. Data,* 36(3) (2007) 1141.
[69] Ilyushin V., Kleiner I. and Lovas F. J., *J. Phy. Chem. Ref. Data,* 37(1) (2008) 97.
[70] Lovas F. J., *J. Phy. Chem. Ref. Data,* 33(1) (2004) 177.
[71] Pearson J. C., Sastry K. V. L. N., Winnewisser M., Herbst Eric, de Lucia Frank C., *J. Phy. Chem. Ref. Data,* 24(1) (1995) 1.
[72] Dicke R. H. and Romer R. H., Pulse technique in microwave spectroscopy, *Rev. Sci. Instrum.* (1955) 915.
[73] Eggers J., and Flygare W. H., *Rev Sci Instrum. Rev. Sci. Instrum.* 47(1976) 448.
[74] Balle T. J. and Flygare W. H., *Rev Sci Instrum.,* 52, (1981) 33.
[75] Balle T. J., Campbell E. J., Keenan M. R., and Flygare W. J., *J Chem Phys,* 72, (1981) 922.

[76] Campbell E. J., Buxton L. W., Balle T. J., Keenan M. R. and Flygare W. J., *J Chem Phys,* 74, (1981) 829.
[77] Grabow J.-U. and Stahl W., Naturforsch Z., 45a (1990) 1043.
[78] Grabow J.-U., Stahl W. and Dreizler H. *Rev. Sci. Instrum.,* 67 (1996) 4072.
[79] Chuang C., Hawley C. J., Emilsson T. and Gutowsky H. S., *Rev. Sci. Instrum.,* 61 (1990) 1629.
[80] Grabow J.-U., Palmer E. S., McCarthy M. C. and Thaddeus P., *Unpublished.*
[81] Suenram R. D., Lovas F. J., Fraser G. T., Gillies J. Z., Gillies C. W. and Onda M., *J. Mol. Spectrosc.* 137 (1989) 127.
[82] Andresen U., Dreizler H., Grabow J.-U. and Stahl W., *Rev. Sci. Instrum.* 61 (1990) 3694.
[83] Rezgui N. D., Baker J. G. and Alder J. F., *Anal Chim Acta,* 311, 99, 1995.
[84] Rezgui N. D., Baker J. G. and Alder J. F., *Anal Chim Acta,* 312, 115, 1995.
[85] Baker J. G., Rezgui N. D. and Alder J. F., *Anal Chim Acta,* 319, 277, 1996.
[86] Baker J. G. and Alder J. F., *Anal Chim Acta,* 367, 245, 1998.
[87] Rezgui N. D., Baker J. G. and Alder J. F., *Anal Chim Acta,* 433, 269, 2001.
[88] Thirup G., Benmakroha F., Takianakos Leon, and Alder J. F., *J. Phys. E:Sci Instrum.,* 19, 823, 1986.
[89] Zhu Z., Matthews I. P. and Samuel A. H. *Sci Instrum.* 64, 103, 1993.
[90] Zhu Z., Gibson C., Samuel A. H., and Matthews I. P., *Rev Sci Inst.* 66 (10) October 1995.
[91] Zhu Z., Matthews I. P., and Samuel A. H., *Rev Sci Inst* 67 (7) July, 2496, 1996.
[92] Zhu Z., Matthews I. P. and Dickinson W. *Rev Sci Inst* 68 (7) July 1997 PP 2883.
[93] Andressen H., Dreizler H., Kretschmer U., Stahl W., and Thomsen C., *Fresenius Anal Chem,* 349, 272, 1994.
[94] Suenram R. D., Lovas F. J., and Sams R. L., *Proc. 1994 U.S. EPA/A&WMA Intern. Symp.* - Measurement of Toxic and Related Air Pollutants, Durham, NC. 1994, pp 551-561.
[95] Lovas F. J., Pereyra W., Suenram R. D., Fraser G. T., Grabow J. -U. and Hight-Walker A., *Proc. 1994 U.S. EPA/A&WMA Int. Symp.* Measurement of Toxic and Related Air Pollutants, McLean, VA. 1994 pp.58-69.
[96] Suenram R. D., Lovas F. J., Grabow J. -U., Hight Walker A. R., Lugez C. L. and Chu P. M., (1996) *Proceedings On-Site Analysis Conference,* Orlando, Florida.
[97] Hight Walker A. R., Suenram R. D., Samuels A. C., Jensen J. O., Woolard D., Wiebach W., *SPIE Conference on Air Monitoring and Detection of Chemical and Biological Agents,* Boston, MA, Nov 1998 122-127.
[98] Suenram R. D., Grabow J. –U., A. Zuban, and Leonov I., *Rev. Sci Inst.,* 70, 2127, 1999.
[99] Suenram R. D., Hight Walker A. R., Fraser G. T., Plusquellic D. F., Dabell R. S., Chu P. M., Rhoderick G. C., Samuels A. C., Jensen J. O., Ellzy M. W. and Lochner J. M. *Proceeding of the First Joint Conference on Point Detection for Chemical Agents and Related Compounds,* October 23-27, pp 515-521, 2000.
[100] da Bell R. S., Chu P. M., Fraser G. T. and Suenram R. D., *SPIE,* 4574, 216-224 (2002).
[101] Tubergen Michael J., Conrad Andrew R., Chavez Roberto E. III, Hwang Injung, Suenram Richard D., Pajski Jason J., Pate Brooks H., *J. Mol. Spectrosc.,* 251, 330 (2008).

[102] Hoeft J. and Nair K. P. R., *Z. Phys. D Atoms, Molecules and Clusters,* 5, 345-349 (1987).

[103] Nair K. P. R. and Grabow J. –U., *Unpublished.*

[104] McCarthy M. C., Chen W., Travers M. J. and Thaddeus P., *The Astrophysical Journal Supplement Series,* 129:611-623, 2000 August.

[105] Nair K. P. R., *Unpublished.*

[106] Grabow J. -U., Palmer E. S., McCarthy M. C., and Thaddeus P., *Rev. Sci. Instruments.*

[107] Basov N. G., Prokhorov A. M., Townes C. H. and Schawlow A. L., Gordon J. P., Zeigler H. J. and Townes C. H., *Phys. Rev.* 95 p-282 (1954) Microwave Spectrum of NH_3.

[108] Gordon J. P., Zeiger H. J. and Townes C. H., *Phys. Rev.* 99, 1264 (1955).

[109] Schawlow A. L. and Townes C. H., *Phys. Rev.* 112, 1940 (1958).

[110] Maiman T., *Nature* (London) 187, 493 (1960).

[111] *NIST Scientific and Technical Database,* NIST Data Gateway, www.researchgate.net.

[112] Ewen H. I., Purcell E. M., *Nature* 168 (1951) 356.

[113] Weaver H., Dieter N. H., Williams D. R. W., Lum W. T., *Nature* 208 (1965) 29-31.

[114] Cheung A. C., Rank D. M., Townes C. H., Thornton D. D., Welch W. J., Crowther J. H., *Nature* 221 (1969) 626.

[115] Ball A., Gottlieb C. A., Lilley A. E., Radford H. E., *Astrophys. J.* 162 L203-10 (1970).

[116] Buhl D., Snyder L. E., Lovas F. J., Johnson D. R., *Astrophys. J.* 192 L97-100 (1974).

[117] Wilson W. J., Darrett A. H., *Science* 161 778-9 (1968).

[118] Knowles S. H., Mayer C. H., Cheung A. E., Rank D. M. and Townes C. H. *Science* 163, (1969) 1055.

[119] Buhl D., Snyder L. E., Lovas F. J., Johnson D. R., *Astrophys. J.* 192 L97-100 (1974).

[120] Snyder L. E., Buhl D., *Astrophys. J.* 189 (1974) L31-3.

[121] Whiteoak J. B., Gardner F. F., *Astrophys. Lett.* 15 (1973) 211.

[122] Haar D. and Pelling M. A., ***Rep. Prog. Phys.*** 37 (1974) 481.

[123] Kadaba A. P. K., Bhagat P. K. and Goldberger G. N., Application of Microwave Spectroscopy for Simultaneous Detection of Toxic Constituents in Tubacco Smoke, *Bull. Environ. Contamination & Toxicology,* Springer Verlag, NY (1978).

[124] Lovas F. J., Pereyra W., Suenram R. D., Fraser G. T., Grabow J. U. and Height Walker Angela R., Using Fourier transform microwave spectroscopy to detect hazardous air pollutants in Optical Sensing for Environmental and Process Monitoring, *Proc. SPIE* 2365, 58 May 31 (1995).

[125] Marrs T. C., Maynard R. L.and Sidell F. R., *Chemical Warfare agents Toxicology and Treatment,* Wiley NY 1996.

[126] *Gaussian 03, Revision A. 1,* Frisch M. J., Trucks G. W., Schlegel H. B., Scuseria G. E., Robb M. A., Cheeseman J. R., Montgomery J. A. Jr., Vreven T., Kudin K. N., Burant J. C., Millam J. M., Iyengar S. S., Tomasi J., Barone V., Mennucci B., Cossi M., Scalmani G., Rega N., Petersson G. A., Nakatsuji H., Hada M., Ehara M., Toyota K., Fukuda R., Hasegawa J., Ishida M., Nakajima T., Honda Y., Kitao O., Nakai H., Klene M., Li X., Knox J. E., Hratchian H. P., Cross J. B., Adamo C., Jaramillo J., Gomperts R., Stratmann R. E., Yazyev O., Austin A. J., Cammi R., Pomelli C., Ochterski J. W., Ayala P. Y., Morokuma K., Voth G. A., Salvador P., Dannenberg J. J., Zakrzewski V. G., Dapprich S., Daniels A. D., Strain M. C., Farkas O., Malick D. K., Rabuck A. D., Raghavachari K., Foresman J. B., Ortiz J. V., Cui Q., Baboul A. G., Clifford S.,

Cioslowski J., Stefanov B. B., Liu G., Liashenko A., Piskorz P., Komaromi I., Martin R. L., Fox D. J., Keith T., Al-Laham M. A., Peng C. Y., Nanayakkara A., Challacombe M., Gill P. M. W., Johnson B., Chen W., Wong M. W., Gonzalez C. and Pople J. A., Gaussian, Inc., Pittsburgh PA (2003).

[127] Schmidt M. W., Baldridge K. K., Boatz J. A., Elbert S. T., Gordon M. S., Jensen J. H., Koseki S., Matsunaga N., Nguyen K. A., Su S. J., Windus T. L., Dupuis M., Montgomery J. A., *J. Comput. Chem.* 14 (1993) 1347.

[128] Suenram R. D., Golubiatnikov G. Y., Leonov I. I., Hougen J. T., Ortigoso J., Kleiner I. and Fraser G. T., *J. Mol. Spectrosc.* 208 (2001) 188-193.

[129] Sharpe W., Johnson T. J., Chu P. M., Kleinmeyer J. and Rowland B., Quantitative, Infrared Spectra of Vapor Phase Chemical Agents, *Proc of SPIE* Vol. 5085 (2003) pp 19-27.

[130] Bousquet R. R., Chu P. M., Dabell R. S., Grabow J. –U. and Suenram R. D., *IEEE Sensors Journal,* 5, 656 (1005).

[131] Suenram R. D., Lovas F. J., Plusquellic D. F., Lesarri A., Kawashima Y., Jensen J. O., and Samuels A. C., *J. Mol. Spectrosc.* 211(2002) 110.

[132] Ohashi N. and Hougen J. T., *J. Mol. Spectrosc.* 211(2002) 119.

[133] Ohashi N., Pyka J., Golubiatnikov Yu. G., Hougen J. T., Suenram R. D., Lovas F. J., Lessari A. and Kawashima Y., *J. Mol. Spectrosc.* 218(2003) 116.

[134] da Bell R. S., Suenram R. D., Lavrich R. J., Lochner J. M., Ellzy M. W., Sumpter K., Jensen J. O., and Samuels A. C., *J. Molec. Spectrosc.* 228, 230(2004).

[135] Pena Isabel, Daly A. M., Cabezas C., Mata S., Bermudez C., Nino Amaya, Lopez J. C., Grabow J. –U., and Alonso J. L., *J. Physical Chem. Lett.* 4, 65-69 (2013).

[136] Velino B., Favero L. B., Ottaviani P., Mans A. and Caminati W., *Analyst,* 138 (2013) 1959.

[137] Jahn Michaela, Nair K. P. R. and Grabow J. U., *unpublished.*

[138] Suenram R. D., Walker A. R. H., Fraser G. T., Plusquellic D. T., Dabell R. S., Chu P. M., Rhoderick G. C., Samuels A. C., Jenson J. O., Ellzy M. W. and Lochner J. M., *Rotational Signatures of Chemical Agents and Related Compounds* pp. 515, 521 (2000).

[139] Walker A. R. H., Suenram R. D., Samuaels A., Jenson J., Ellzy M. M., Lochner J. M. and Zeroka D., *J. Mol.Spectrosc.* 207(2001) 77.

[140] Suenram R. D., Dabell R. S., Walker A. R. H., Lavrich R. J., Plusquellic D. F., Ellzy M. W., Lochner J. M., Cash L., Jenson J. O., Samuels A. C., *J. Mol.Spectrosc.* 224(2004) 176.

[141] Durig J. R., Stanley A. E. and Li Y. S., *J. Mol. Structure.* 78(1982) 247.

[142] Gillies C. W., Rickus L., Gillies J. Z., Justnes H., Suenram R. D., Lovas F. J. and Diguiseppe T., *J. Mol. Structure,* 223 (1990) 273.

[143] Dabell R. S., Chu P. M., Fraser G. T. and Suenram R. D., *Evaluation of Fourier transform microwave spectroscopy as a tool for quantitative analysis, presented at SPIE Environmental and Industrial Sensing 2002.*

[144] Suenram R. D., Grabow J. U., Zuban A. and Leonov I., *Rev. Sci. Instrum.* 70(1999) 2127.

[145] Harmony M. D., Beren K. A., Angst D. M. and Ratzlaff K. L., *Rev. Sci. Instrum.* 66 (1995) 5196.

In: Spectroscopic Techniques for Security, Forensic ... ISBN: 978-1-63117-404-9
Editors: Y. Dwivedi, S. B. Rai and J. P. Singh

Chapter 7

DETECTION OF EXPLOSIVES AND HAZARDOUS CHEMICALS: VARIOUS FACETS OF RAMAN SPECTROSCOPY

***Sanchita Sil*[1,2] *and Siva Umapathy*[*1]**

[1]Department of Inorganic & Physical Chemistry, Indian Institute of Science, Bangalore, India

[2]High Energy Materials Research Laboratory, Sutarwadi, Pune, India

Today we live in an era where wars are not fought across the boundaries in a one to one combat scenario; it has taken a very different form. State and non-state actors are terrorizing the innocent civilians with bombs and improvised explosive devices (IEDs). Detection of these explosives has been a prime challenge for the scientists and researchers for several decades. Although various analytical and spectroscopic techniques are commercially available and have been used in public places like airports, railway stations, shopping areas etc., however, none of them are 100% full proof and have several limitations with respect to sensitivity, selectivity, detection time etc. Hence, there is a need to find novel techniques to overcome the limitations.

Raman spectroscopy is a powerful, non destructive tool which provides molecular information and is a potential candidate for detection of explosives. In addition, different variants of Raman spectroscopic techniques such as Resonance Raman Spectroscopy, Ultraviolet Resonance Raman Spectroscopy and Surface Enhanced Raman Spectroscopy (SERS), to name a few are available which ensures sensitive and selective detection and identification of molecules. Spatially offset Raman spectroscopy, a recently developed Raman spectroscopic method has also emerged as a potential tool for detection of hazardous chemicals concealed in non-metallic containers. SERS is a very useful analytical technique for the ultra trace detection and identification of molecules. Noble metal nanoparticles have surface plasmons in the visible range of the electromagnetic spectrum enabling them to be used as SERS substrates. These form the backbone for SERS studies and have been synthesized in different ways which have been widely reported. SERS can be a potential tool for the detection of explosives at very low concentrations. In addition, the limitation of conventional Raman spectroscopy to study deep layers of diffusely scattering media has led to the development of a new Raman

[*] Email: umapathy@ipc.iisc.ernet.in.

based technique in our laboratory called, “Multiply Amplified Raman Spectroscopy (MARS)” for the detection of hazardous chemicals. This technique allows us to probe sub-surfaces non-invasively and obtain the information of the layers within other layers. We have used MARS to study the chemical composition of deep layers in heterogeneous samples. MARS is a noninvasive technique and can be employed in detecting or screening hazardous chemicals packed in different types of non-metallic containers, envelopes etc. In this chapter, we discuss about the potential of these Raman based techniques.

1. Introduction: The Detection Challenge

Low Intensity Conflicts including terrorism has attracted attention of the research community all over the globe towards development of methodologies and technologies for the detection of explosives. Sensitive and timely detection of explosive materials are necessary to ensure safety and security of the personnel involved and the civilian population as well (Figure 1). Detection of explosives and explosive-like compounds are therefore, necessary at different complex environments e.g., mine fields, munitions storage facilities, waste water treatment facilities, transportation areas and blast sites in addition to public places like airports, railways stations, shopping malls *etc*. A wide variety of detection techniques, such as, electron capture detector (ECD), chemiluminescence (CL), thermo redox (TR) and ion mobility spectrometry (IMS) are currently available [1-2]. Although, the above mentioned techniques are used, detection by explosives-sniffing dogs has been the most prevalent practice. However, there are many challenges associated with the use of dogs which includes training of these dogs, their work hours, efficiencies to name a few [3].

Nitrated explosives, such as tri-nitro toluene (TNT), nitramines such as cyclo-trimethylene-trinitramine, (commonly known as research development explosive (RDX)) and nitrate esters, such as penta-erythritol tetranitrate (PETN), have been widely studied [4-7]. Military explosives comprise a combination of these explosives together with fuels and other excipients [8]. Low levels of traces of explosives available for detection due to low vapour pressure pose a major difficulty in explosive detection. The vapour pressure of nitroaromatic explosives are very low and ranges between 10^{-6} to 10^{-9} mbar (corresponds to a concentration in parts per billion to parts per trillion at ambient conditions). This causes a major difficulty in detection of explosives by conventional vapour phase detection methods [9]. Willer *et al.* [10] in a review of photonic sensor devices for explosive detection incorporated the vapor concentration data for different explosives. For nitroaromatics, advanced methods are required to be used to enable detection at much lower equilibrium concentrations at ambient conditions. One way to address this issue is by using a preconditioning setup, where the explosive vapour is led over a trap, which can be a cooled surface or a filter where the molecules adsorb. Subsequently, on heating the trap, the vapour molecules desorb and result in a much higher vapour concentration than present in the original material, using which nitro- aromatic explosives e.g., TNT with a vapour concentration in parts per trillion range at ambient condition is expected to be detected [11]. Nadezhdinski *et al*. [12] used tunable diode laser spectroscopy to decompose the explosive materials to smaller molecules to assess the threat. They found it easier to investigate smaller molecules compared to the explosive itself with narrow line-width tunable diode lasers. Similar photo-fragmentation schemes have been applied to investigate explosive materials, and the decay products were detected with various techniques e.g., laser induced fluorescence spectroscopy [13], resonance-enhanced multiphoton ionization [14,15], absorption spectroscopy of photofragments [16], chemiluminescence [17].

Figure 1. The challenge of detection of explosives in different scenarios.

In a laboratory setup, significantly low detection limits have been attained [18], however, for most practical purpose in defence and homeland security, where an untrained personnel is in charge of handling the detection devices attaining the desired sensitivity still remains a challenge. Additionally, the selectivity required for unique identification of an explosive in an outdoor environment, with the background of interfering elements (chemical agents with atomic compositions including sulphur, phosphorous, carbon, hydrogen, nitrogen, chlorine *etc.*) in the atmosphere is difficult. In addition, for minimization of the hazards associated with energetic materials, these materials are desensitized to reduce the friction, shock, and heat sensitivity by means of stabilizers or binders. The traditional way of desensitizing high energy materials is through a plasticizing technique with the help of a binder and a plasticizer. However, methods such as this and many others, including mixing the energetic material with inert substances such as salts or oil, coating the crystals with waxes, or regulating the oxygen balance of the energetic material with the help of non-explosive nitro compounds entail a considerable amount of spectral contamination that highly compromises their precision and selectivity. The problem is further complicated by low levels of traces of explosives available for detection due to low vapour pressure pose a major difficulty in explosive detection. This calls for a highly sensitive detection technique.

Increased attention in defence and homeland security with improved techniques towards military explosive detection have led to the increased use of non-nitrated explosives [19]. Peroxide-based liquid explosives e.g., triacetone triperoxide (TATP) and hexamethylene triperoxide diamine (HMTD) have been recently involved in terrorism and drug related crimes [19, 20]. Even though these materials are relatively unstable and sublime quickly (resulting in a higher vapour pressure), these materials are significantly different from other classes of explosives, and the best method for their detection is yet to be spotted.

Development of field-deployable stand-off detection techniques has held the highest level of priority among the research communities. Both standoff and point detections of explosives demand different requirements with certain criteria being commonly desirable e.g., lower false alarm rate, better selectivity, higher sensitivity and a wider range of detectable threats. Standoff explosive detection can be defined as a method for sensing the presence of explosives when vital assets and the personnel monitoring, operating and responding to the threat detection are physically separated from the explosive device [21]. The physical separation should ensure to put the personnel and the vital assets outside the zone of severe damage due to an explosion. The zone of severe damage is chosen to be 10 m for a pedestrian suicide bomber to a 100 m for a vehicle based bomb, however it can vary with the scenario and type of bomb (weight, shrapnel *etc.*).

In the current scenario, we are however, faced with another challenge i.e., of Improvised Explosives Devices, commonly known as IEDs. An improvised explosive device (IED) is a destructive device, including explosive but also possibly incendiary or toxic materials, and perhaps with mechanical components to increase fragmentation or armour penetration [22]. Detection of these IEDs poses a serious challenge for the security agencies and therefore, development of any detection technology has to be well equipped to detect IEDs as well.

The invention of lasers spurred activities in various fields of chemistry in general, and spectroscopy in particular. Lasers possess unique properties, such as, long distance propagation of intense energy and high wavelength specificity for molecular and atomic spectroscopy. In contrast to chemical analysis methods, laser-based techniques do not require any sample preparation and therefore can be performed in-situ and often in a standoff configuration. In this chapter we would explore various laser based Raman spectroscopic methods with suitable examples related to detection of explosives. We conclude by comparing different laser based Raman spectroscopic methods which can or have a potential to be used for detection of explosives either for point detection or for stand-off detection. In the following sections, we describe different Raman spectroscopy-based techniques for detection of explosives. Each method described has some advantages over the others for certain classes of explosives; hence the ideal sensing technique should be devised such that it covers up the combination of the merits of the individual techniques.

2. Raman Spectroscopy and Explosives Detection

When light is scattered from a molecule, most photons are elastically scattered. The scattered photons have the same energy (and therefore same frequency and wavelength) as the incident photons. However, a small fraction of the scattered light (approximately 1 in 10^7 photons) is at optical frequencies different from, and usually lower than, the frequency of the incident photons. The process leading to this inelastic scattering is termed the Raman effect.

Raman scattering can be thought of formally as a two-photon absorption-emission process [23]. The difference in energy between the incident photon and the Raman scattered photon is equal to the energy of a vibration of the scattering molecule. An intensity plot of scattered light versus energy difference is a Raman spectrum. The Raman Effect arises when a photon is incident on a molecule and interacts with the electric dipole of the molecule. In classical terms, the interaction can be viewed as a perturbation of the molecule's electric field

[24]. In quantum mechanics the scattering is described as an excitation to a virtual state lower in energy than a real electronic transition with nearly coincident de-excitation and a change in vibrational energy. The virtual state description of scattering is shown in Figure 2.

The energy difference between the incident and scattered photons is represented by the arrows of different lengths in Figure 1. Numerically, the energy difference between the initial and final vibrational levels, $\bar{\nu}$ or Raman shift in wave numbers (cm^{-1}), is calculated through equation

$$\bar{\nu} = 1/\lambda_{incident} - 1/\lambda_{scattered}$$

where $\lambda_{incident}$ and $\lambda_{scattered}$ are the wavelengths (in cm^{-1}) of the incident and Raman scattered photons, respectively. The vibrational energy is ultimately dissipated as heat. Because of the low intensity of Raman scattering, the heat dissipation does not cause a measurable temperature rise in a material. Raman spectroscopy stands out as one of the laser-based non-destructive techniques (depending on laser power and wavelength of the incident laser beam) that can be used for detection of explosives, with a potential drawback of weak signal intensity, as spontaneous Raman scattering has a very low cross section.

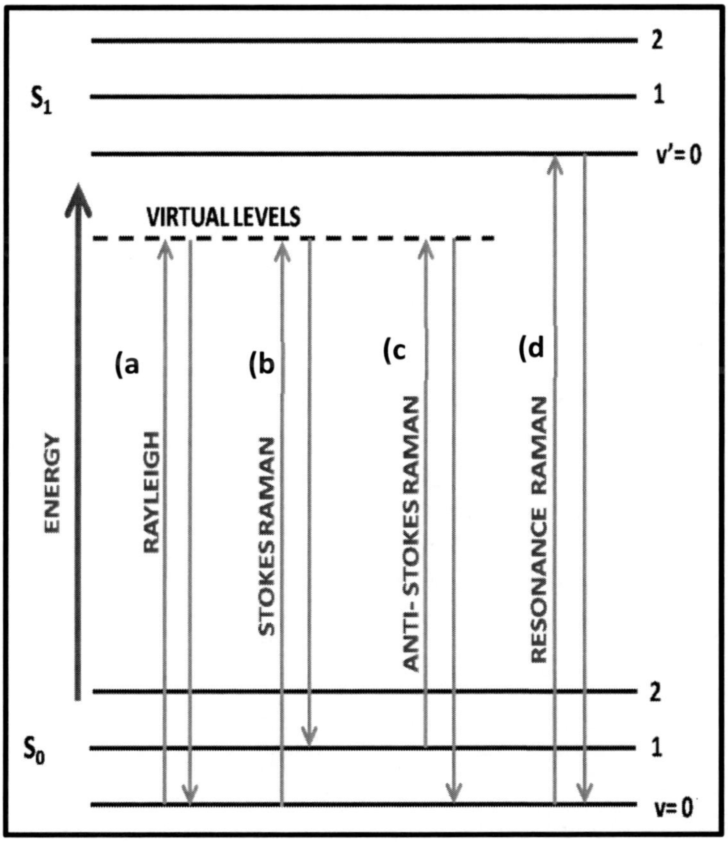

Figure 2. Energy level diagram for Raman scattering; (a) Rayleigh, (b) Stokes Raman scattering (c) anti- Stokes Raman scattering, and (d) Resonance Raman scattering.

Akhavan [25] as early as 1991 reported FT Raman spectra of RDX, PETN, Semtex A, B and C (explosive composition consisting of RDX, PETN and binder). Raman spectra of 32 nitro-containing explosives were studied by Lewis *et al.* [26] and based on the spectral features, they were able to group them into three classes: nitrates esters, nitroaromatics, and nitramines. The same authors reported the results obtained with a no-moving parts dispersive Stokes Raman spectrometer with a Ge array detector and 1064 nm excitation for the study of ten explosives [27]. Gupta *et al.* [28] used Acousto-optic tunable-filter (AOTF) technology based spectrometer from visible to near-IR to measure the Raman spectra of energetic materials like HMX (nitramine cyclotetramethylenetetranitramine), TNT (trinitrotoluene), and NQ (nitroguanidine).

Standoff detection using Raman spectroscopy was theoretically proposed in mid- 1960s, it was hesitantly used owing to the inadequacy of sensitivity due to inherent low signal levels, long distance ranges and low sample concentrations [29]. The performance of standoff detection used to be better during the night than in the daytime. However with tremendous advancements in all features of the spectrometers and the detectors in the last decade, Raman spectroscopy can be used in standoff detection of explosives over considerable distances in daylight and various other environmental conditions [21]. Numerous attempts have been to detect explosives using near- infra red Raman excitation [30-56]. Several groups have demonstrated the ability to detect Raman scattering from explosives at distances as great as 55 m [57, 58].

2.1. Resonance Raman Spectroscopy

Raman scattering has been the subject of intrigue among the vibrational spectroscopic community as it opened the world of molecular structures of the compounds. This led to exploration of various Raman based methods to enhance the efficiency of the process. In this regard, a unique process which became a favorite among Raman community is the resonance Raman spectroscopic technique. When an excitation light source approches an electronic transition in the scattering molecule under investigation, the cross-section, and hence the intensity of the Raman band shows enormous enhancement. This phenomenon is termed as resonance Raman scattering (Figure 2d). The sensitivity of the otherwise weak Raman process can thus be enhanced dramatically using this technique. Resonance Raman spectroscopy has been applied to many organic systems due to its high sensitivity and selectivity [59-63]. In addition, this technique enables fluorescence rejection.

Raman scattering and associated visible luminescence has motivated the exploration of deep ultraviolet (UV) resonance Raman spectroscopy for remote detection [64-69]. Explosives absorb light in the ultra-violet (UV) region and hence the advantage of using deep UV region and increased resonance Raman cross-section is observed. In addition, excitation in this region in condensed-phase samples allows one to avoid fluorescence [70]. Nagli *et al.* [65] have reported observation of 100 to 200 times greater Raman scattering signals with 248 nm excitation than those obtained with excitation in the visible at 532 nm. Furthermore, excitation deeper in the UV allows resonance with the electronic transitions, which should in principle give rise to even larger Raman cross-sections as proposed by Albrecht [71]. Asher's group has been the pioneers in the development of UV resonance Raman spectroscopy for numerous applications, especially biological applications [72-77]. They reported UV Raman studies of explosive molecules and have measured the deep UV Raman cross-sections of several explosives such as trinitrotoluene (TNT), pentaerythritol tetranitrate (PETN), cyclotrimethylene-trinitramine (RDX), another chemically related nitramine (HMX) explosive, and ammonium nitrate. They concluded that deep UV Raman cross-sections for the explosives chosen for the study were roughly three orders of magnitude larger than those for 532 nm excitation. These increased Raman cross-sections enable parts per-billion (ppb) detection level limits for these molecules, indicating promise for the utility of deep UV resonance Raman spectroscopy for stand-off detection of explosives [78].

Asher and co-workers in their recent publication [79] have measured the absolute differential Raman cross-sections of explosives like ammonium nitrate (NH_4NO_3), PETN, TNT, HMX, and RDX in acetonitrile and water solutions between 204 and 257 nm. They observed that the ultraviolet (UV) resonance Raman/differential Raman cross-sections of NH_4NO_3, PETN, TNT, HMX, and RDX increased radically as the excitation wavelength decreased deep into the UV to 204 nm. They noted that NH_4NO_3, PETN, and RDX were best resonance-enhanced by the 204 nm excitation, while the optimum excitation wavelength for TNT and HMX was obtained at 230 nm. In addition, they observed that the excitation profiles for the different Raman bands of each explosive molecule differed, which indicated that multiple-excitation wavelength spectra could provide additional information on the species present. The results obtained by them could form the foundation for the development of standoff deep-UV Raman methods for explosive molecule determinations. The same authors have used DFT methods to obtain UV absorption spectra and UV resonance Raman intensities for explosive molecules like TNT and RDX and for resonant excitations within the electronic absorption bands [80]. Resonance Raman can be ideal for detection of low volatility and low abundance materials, such as explosives due to the availability of new post spectral processing methods and algorithms which enables detection of trace amounts (55 μg/cm) of materials [81].

However, it is important to note that most of the literatures on resonance and UV resonance Raman have been conducted with explosive molecules dissolved in solvents like acetonitrile (ACN), which also serves as an internal standard. In most of the practical applications, explosives are used in their solid state. Moreover, the energy of the photons are very high for resonance Raman experiments, hence being a line-of-sight technique, due consideration should be given to the fact that development of resonance Raman stand-off technique would be challenging for crowded public places like airports, shopping malls etc.

2.2. Surface- Enhanced Raman Spectroscopy

The confluence of Raman Spectroscopy and nanotechnology has given birth to a new hybrid form of Raman Spectroscopy, Surface Enhanced Raman Spectroscopy (SERS) which has the immense potential to overcome the inherent drawback of the weaker signals produced in conventional Raman spectroscopy. With the rapid development in the spectroscopic instrumentation, nanofabrication methods, novel detection schemes and theoretical modeling, renaissance in the field of surface enhanced Raman spectroscopy was brought about. From the first published report on the studies of pyridine on roughened silver electrode, this technique has come a long way and emerged as a state-of-the-art analytical tool for real world sensor applications [82-84]. The enhancement of the signal can be of the order of 10^6 which can go up to 10^{15} for single molecules [85, 86].

The phrase 'surface enhanced Raman spectroscopy' holds the definition of this technique - A surface that enhances Raman signals of an analyte (Figure 3). So, for the instrumentation what one requires is a surface that allows the amplification of the Raman signal of the analyte being probed, an electromagnetic radiation, typically a laser source, an analyzer and a detector. The electromagnetic source, analyzer and the detector forms the spectrometer part, what is essentially introduced in SERS is the 'surface' and the effect that is obtained is, the 'enhancement of Raman signal'. Fleischman *et al.* [87] as early as 1974 observed anomalously high Raman signals for pyridine adsorbed on eletro-chemically roughened silver electrode. This baffling observation led to the origin of a technique which saw an exponential rise of research in SERS in the last four decades resulting in thousands of published papers and patents. This technique has the potential of very high detection sensitivity, which could lead to single molecule level detection [88, 89]. The first question that should be addressed is why some surfaces are capable of giving enhancements while others not, and the second being, what is the genesis of the enhancement.

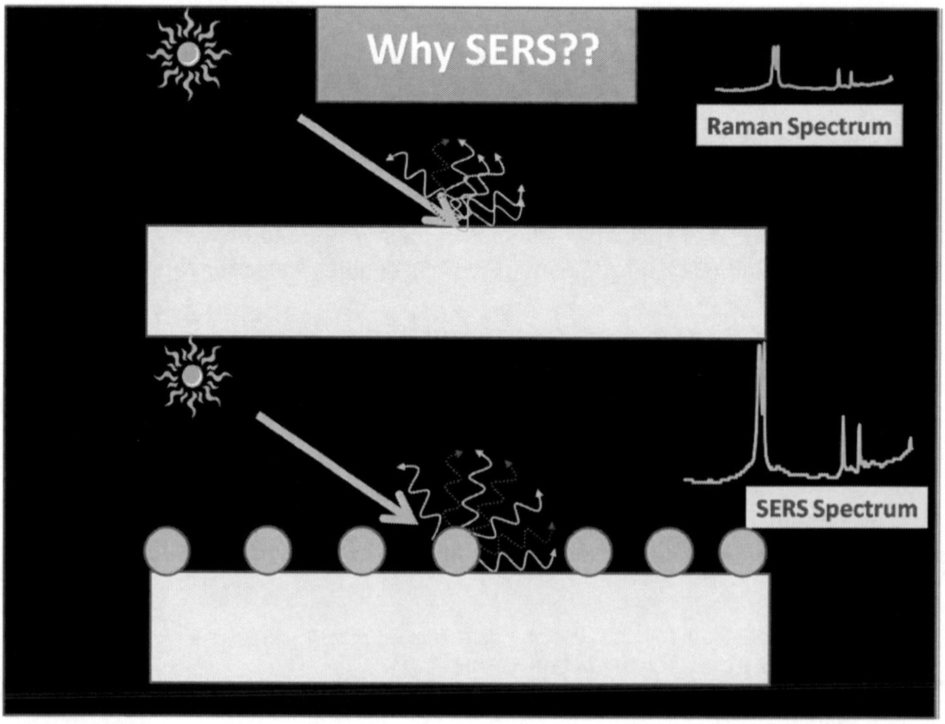

Figure 3. Schematic representation of the surface enhanced Raman spectroscopy.

When electromagnetic radiation is incident on metals, the electrons present in the metal get excited and start oscillating. These oscillations (which can be either longitudinal or transverse) give rise to the plasmons (plasma + mons, Greek for 'plasma' means something molded or shaped like plastic and 'mons' in Latin means mountain), which are collection of interacting particles e.g., electrons on the surface of the metals. Coinage metals such as gold (Au), silver (Ag), copper (Cu), platinum (Pt) support strong localized surface plasmon resonances in the visible region while aluminum (Al) exhibits the plasmonic resonance in the deep ultra violet region and this is the reason why Ag, Au, Cu, Pt and Al nanoparticles (NPs) have been preferred for SERS experiments. When the electric field of the em radiation interacts with the surface electrons of metal NPs, there is a generation of a strong opposing field which is further opposed by the internal depolarization field. The strength of oscillation would therefore, depend upon the shape and orientation of the NP. In addition, NPs with tips and facets would have stronger opposing field, thereby increasing the strength. Hence, the analytes (or molecules) located at the tips/ hot spots have more SERS enhancement compared with the spherical NPs.

The explanation provided by Fleischman *et al.* [87] for the observed Raman signal enhancement was the increased surface area of the Ag electrode due to electro-chemical roughening. This conclusion was however, negated by subsequent reports by Jeanmaire & Van Duyne and Albrecht & Creighton [90, 91] who performed independent experiments and observed that the enhancement was not occurring due to the increase in the surface area of the electrode but some other factor was contributing to the amplification of the electromagnetic radiation. This led to intense research to understand the enhancement mechanisms. Two mechanisms viz. chemical and electromagnetic were proposed for the increase in Raman signals. Moskovits [92] in his report in 1985 gave a detailed explanation of these mechanisms. Chemical enhancement occurs when the compounds adsorbed on the metal surface have lone pair of electrons or π clouds for charge transfer which gives rise to enhancement up to an order of 10^2. On the other hand, electromagnetic enhancement occurs due to the induced dipolar or higher order resonance of the surface plasmon on metal nanoparticles which result in 10^6 order of enhancement. For detailed theory of SERS, the readers are encouraged to refer to the numerous reviews, papers and books that have been published [85, 92, 93-97].

SER spectroscopy has been employed for sensitive and selective molecular identification in biological and chemical sensing. This technique has been applied for the trace detection of pesticides, anthrax, bacteria, glucose, nuclear wastes, and genetic diagnostics to name a few [98]. Detection of explosives in trace quantities, analysis of post blast residues, detection of unexploded ordnances as well as demining, adds to the problem of detection for homeland security. SERS is a potential tool for sensitive and rapid detection of explosive at trace or ultra trace levels. SERS on DNT, TNT, RDX, PETN, NG and TATP has previously been reported [99-112]. Ag or Au NPs were either synthesized by bottom-up approach such as chemical reduction methods or by using top down approach such sputtering, lithographic techniques *etc*.

2.3. SERS Experiments

Our lab has been actively pursuing SERS work for detection of biological analytes as well as for materials. We have successfully synthesized silver nanoparticles (Ag NPs) using natural precursors such as cloves and pepper as reducing agents [113]. We have been successful in obtaining SERS spectra of TNT and CL-20 at sub- ppb levels. In the following paragraphs we describe the synthesis method of silver colloids and would also present the SERS results obtained for explosives.

Silver nanoparticles (Ag NPs) were biosynthesized using previously reported methods using cloves and pepper as natural precursors [114-116]. Both cloves and pepper have eugenol as a component, in addition, pepper also contains piperine. Eugenol can release two electrons at a time which helps in the reduction of $AgNO_3$ (2 Ag^+). Briefly, 1.25g of cloves was soaked in 25ml of MilliQ water for 24h. The extract was collected after filtration and was used for further experiment. Synthesis of Ag NPs was carried out by taking $AgNO_3$ as starting material. 3.33 mM $AgNO_3$ solution was stirred at 300 rpm and the temperature was maintained at 75^0 C, to which measured amount of clove extract was added. $AgNO_3$ to cleave extract ratio was varied from 1:1 to 1:0.01 ratios and the best yield was obtained for 1:0.25 ratio. The reduction was complete in about 2h which was indicated by the formation of greenish yellow colour of Ag NPs. The same protocol was carried out for the biosynthesis of pepper reduced Ag colloids. The synthesized bio Ag NPs were characterized using UV-Vis spectroscopy and scanning electron microscope (SEM) studies (Figures 4 & 5). It was observed that the nano particles were poly-dispersed in nature.

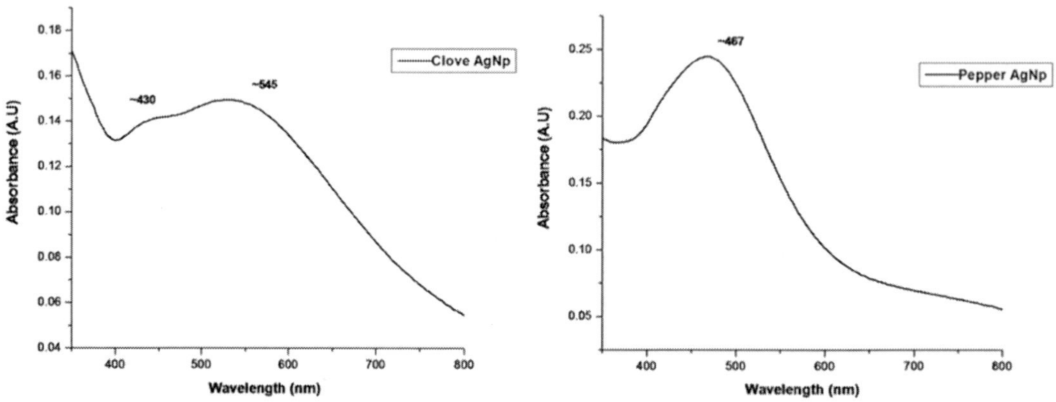

Figure 4. UV-Vis spectra of biosynthesized silver nanoparticles with a) cloves and b) pepper.

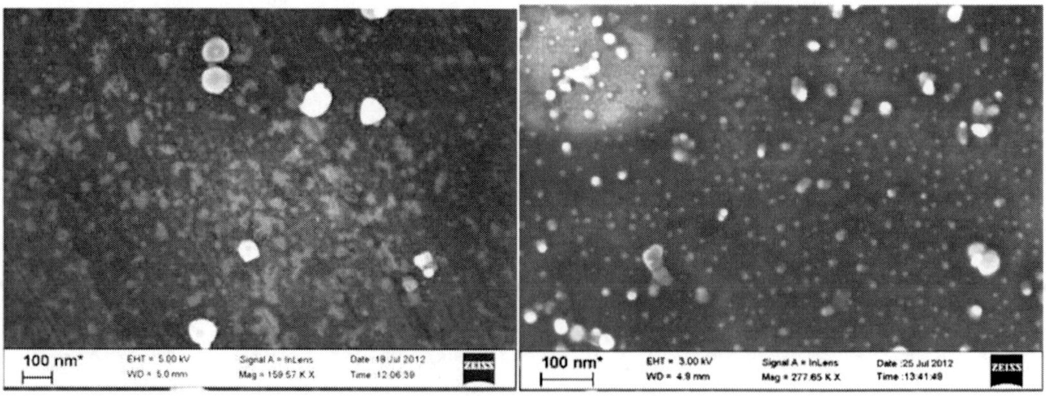

Figure 5. SEM micrograph of clove and pepper AgNPs.

2.4. (a) SERS of Explosives

Raman and SERS spectra were recorded using Renishaw InVia Raman Microscope with a power of 30 mW from an Ar ion laser with the excitation wavelength of 514.5 nm. SERS samples were prepared by mixing equal amounts of biosynthesized Ag colloids described above and the analytes TNT and CL-20 respectively. 50 mM HCl was added for aggregation of colloids. The stock solutions of 1M TNT and 10^{-3} M CL-20 in acetone were prepared and then diluted to obtain desired concentrations. A drop of the colloid mixed with activating agent and analyte was casted on an aluminum slide for recording the SERS spectra.

Raman spectrum of solid TNT was first recorded (Figure 6). The TNT molecule can be identified by the symmetric and the asymmetric NO_2 stretching vibrations, observed at 1366 cm^{-1} and at 1531, 1618 cm^{-1}, respectively [117,118]. The other peaks can be assigned to the NO_2 scissoring modes (790, 824 cm^{-1}), to the CH (ring) bending modes (800–1100 cm^{-1}) and to the ring-breathing at 1209 cm^{-1}. By following the same protocol previously mentioned, SERS of TNT was recorded. In both the cases as shown in Figure 7, the NO_2 stretching vibrations, both symmetric as well as asymmetric were greatly enhanced. In addition, the enhancement of the different vibrational modes was also found to be different.

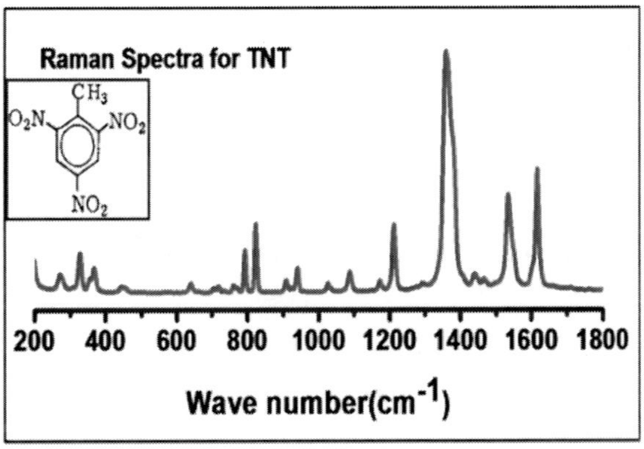

Figure 6. Raman spectrum of pure TNT.

2,4,6,8,10,12-Hexanitro-2,4,6,8,10,12-hexaazaisowurtzitane (HNIW), commonly called CL-20 is one of the most powerful HEM ever synthesized. Infra- red and Raman spectroscopic studies have been reported on the different polymorphs of CL-20 [119,120]. Figure 8 shows the solid Raman spectra of CL-20 molecule. The characteristic Raman bands of CL-20 molecule are 810-870 cm^{-1} which correspond to ring stretch and NO_2 scissoring modes, the band at 1000 cm^{-1} corresponds to N-N stretch, the bands between 1050-1250 cm^{-1} are also attributed to cage stretch from the computational calculations; bands from 1200-1450 cm^{-1} can arise from NO_2 symmetric stretching, while those ranging between 1500-1650 cm^{-1} are mainly attributed to the asymmetric stretching of NO_2. It is evident that only those vibrational modes are enhanced which are either adsorbed on the silver surface or are in close proximity to the surface. Hence, the appearances of peaks as well as their intensities are affected by the extent of interaction of the analyte molecule with the silver nanoparticle. This observation has also been reported in the literature for other molecules [92, 121]. SERS spectra obtained for pM (sub- ppb levels) concentration of CL-20 appeared somewhat similar to the normal Raman spectra of CL-20 (Figure 9). This shows that the molecule although, in the vicinity of the silver nanoparticle did not interact with the silver surface to form chemical bond. The analysis of the spectra revealed that CL-20 molecule possibly interacts to the silver surface through the NO_2 group, keeping the cage almost perpendicular to the metal surface confirming that the NO_2 modes were enhanced for CL-20 SERS spectra in both cloves and pepper colloids. Slight frequency shifts were observed for CL-20 SERS. The observed changes in the spectrum can be attributed to the molecular reorientation upon the surface.

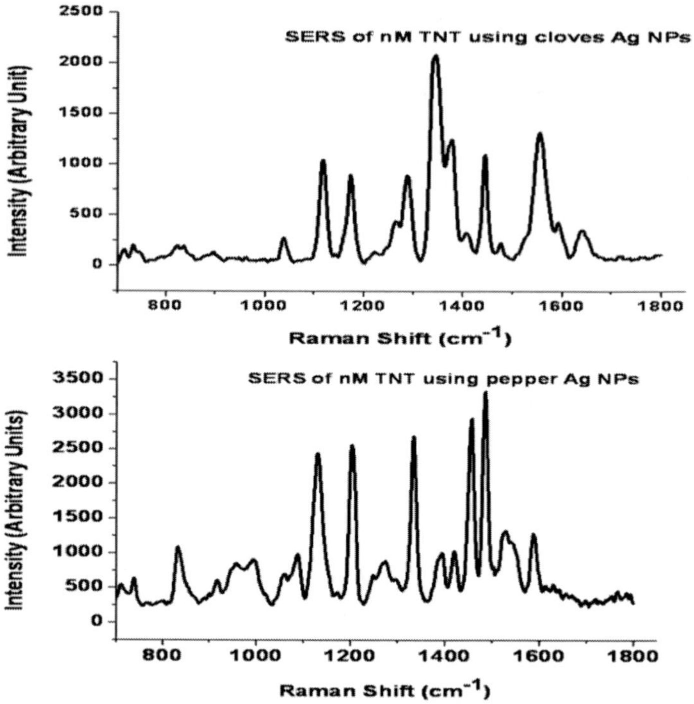

Figure 7. SER spectra of TNT at nM levels using clove and pepper reduced bio Ag NPs respectively.

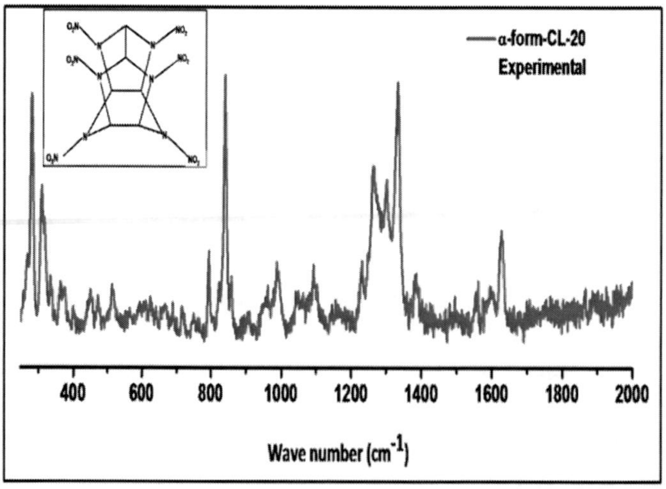

Figure 8. Raman spectrum of pure CL-20.

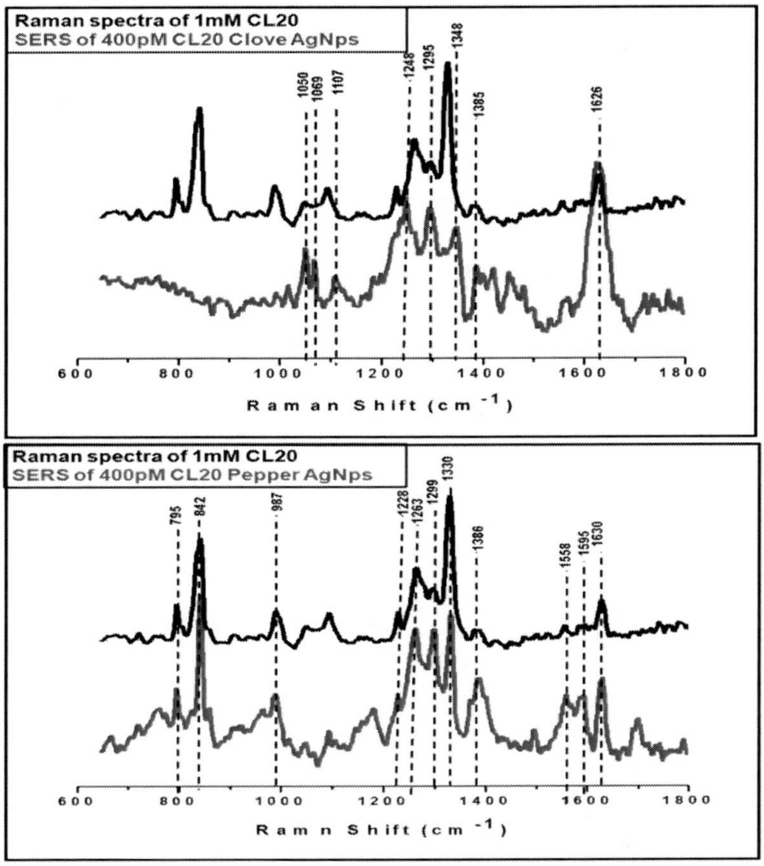

Figure 9. SER spectra of CL-20 at pM level using clove and pepper reduced bio Ag NPs respectively.

We have also initiated work on using carbon based composite substrates for SERS. Raman enhancements using different carbon substrates like chemically reduced graphene, multi-walled carbon nano-tube, graphite, activated carbon and charcoal have been obtained for various analytes [122]. In addition, work on syntheses of bimetallic colloids is also being carried out for the detection of explosives. Efforts are continued to prepare solid SERS substrates for detection of explosives in our laboratory using various physical deposition techniques. The ultimate objective is to prepare robust solid substrates with high sensitivity and reproducibility to detect explosives in solid, liquid or gaseous states. We intend to extend our work in the solid state in the future so that this technique would be useful for field conditions as well, whereby the explosive vapours can be adsorbed on to the solid substrate and then analyzed for obtaining signals at extremely low concentrations.

3. Spatially Offset Raman Spectroscopy

Another off-shoot of conventional Raman spectroscopy is a technique which holds the promise of being used in explosive detection is spatially offset Raman spectroscopy (SORS) [123,124]. Although Raman spectroscopy has proved itself as an invaluable technique to understand the chemical composition of the material, in many of the analytical applications, it is necessary to know the chemical composition of the deep layers in the diffusely scattering samples which are frequently encountered in various spectroscopic applications. Conventional Raman spectroscopy with back scattering collection mode is only capable of probing the shallow depths of semi transparent turbid media. It is therefore imperative to develop a simple and non invasive technique for understanding the chemical composition of different layers of the diffusely scattering media with greater penetration depth. Spatially offset Raman spectroscopy (SORS) is an inelastic scattering based technique in which the Raman photons are collected from the spatially offset regions away from the point of illumination on the sample surface which reduces the Raman signal contribution from the top layer of the sample. The offset spectra contain different relative contributions from the sample layers located at the different depths. SORS allows efficient collection of underlying material spectra by greatly reducing signal from overlying material. The signals from the surface and subsurface have been clearly differentiated within a diffusely scattering sample composed of two layers with the help of SORS [124,125]. Matousek *et al.* [125-129] in their pioneering work have demonstrated SORS for detection of explosives and have also studied human bone *in vivo* using spatially offset Raman spectroscopy.

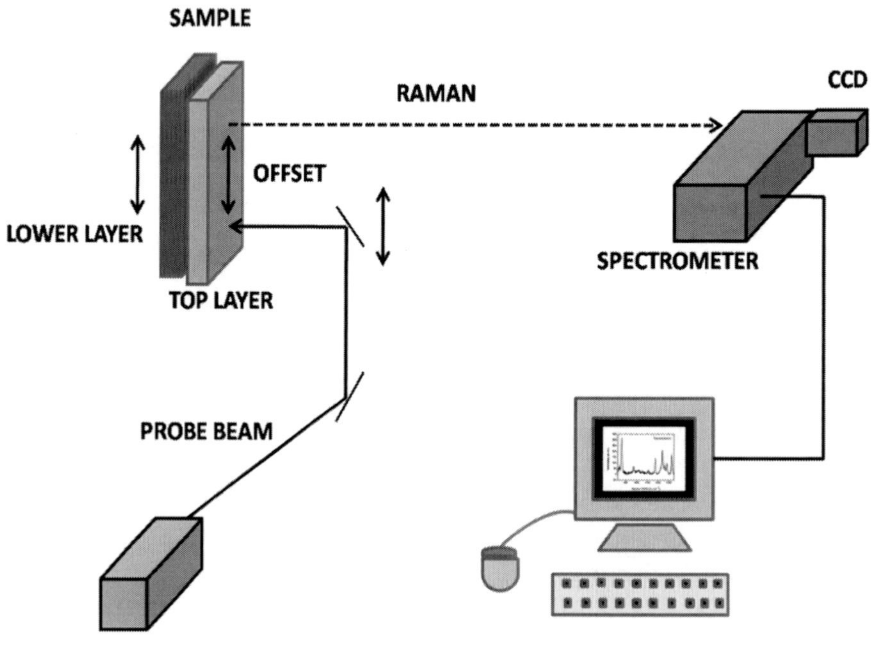

Figure 10. Schematic diagram of the experimental setup.

We have initiated SORS studies in our laboratory and have successfully detected solid powders and explosives concealed in non- metallic containers. The SORS experimental set up is shown schematically in Figure 10. SORS relies upon the focused projection of the laser beam onto the sample where it is reflected, scattered and transmitted through the sample material. The back reflected, diffusely scattered light is then collected by an accessory and then sent to the detector optics as shown in Figure 10. Only part of the beam that is scattered within a sample by generation of secondary photons and returned to the surface is considered to be diffusely reflected. Since a continuous wave laser beam has been utilized in this method, it permits one to use much higher average powers while still remaining within maximum permitted exposure limit of the laser radiation for the non invasive analysis.

This method can provide useful information of the sub surfaces present and hence, the composition of different layers can be obtained. In order to generate the sub surface layer information, it is essential to place the collection optics in suitable geometry to retrieve the scattered photons from the layers and suppress the signal from the overlying or surface materials which may be dominant at a particular geometry (back scattering collection mode). This technique can not only be used for detection of solid explosives but also for liquid chemical precursors and liquid explosives as well.

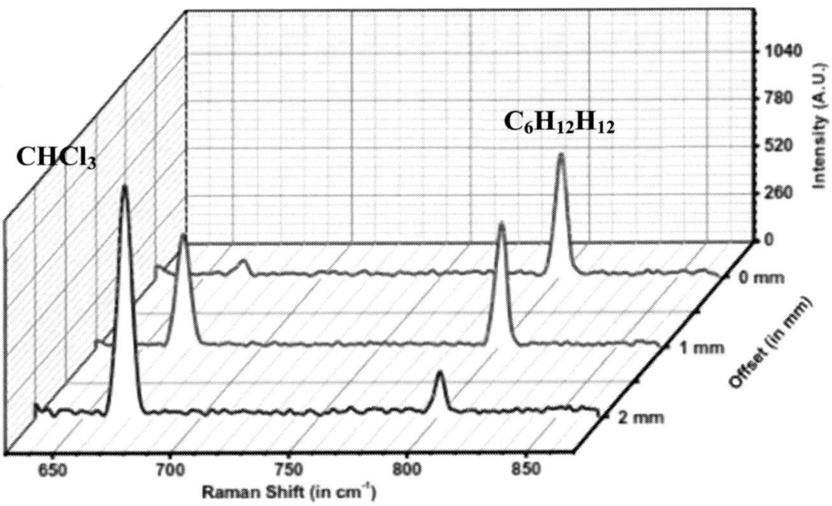

Figure 11. SORS result for transparent liquid-liquid samples taken in a bilayer custom made glass cuvette: 1-mm path cyclohexane and 2-mm path chloroform.

Figure 11 demonstrates SORS with layers of different transparent liquids. For simplicity of comparison, only the marker Raman peaks is shown for different liquid layers. The 667 cm^{-1} peak is for chloroform ($CHCl_3$) and the peak centered at 801 cm^{-1} corresponds to cyclohexane (C_6H_{12}). A bilayer cuvette was used to fill cyclohexane and chloroform in 1-mm and 2-mm path compartments, respectively. Initially, the laser beam was focused on the first layer (cyclohexane) and the Raman spectrum was recorded (spectrum corresponding to 0 mm offset in Figure 11). As the separation between the excitation and collection points was increased, the Raman signal of the second layer, i.e., chloroform could be obtained; At separation >0 mm the second layer contribution began to increase and the magnitudes of Raman signals from both the layers were comparable at 1 mm, beyond which the Raman signal intensity from the first layer diminished quickly. Similar results were obtained with three transparent liquids, benzene/ nitromethane/ acetone, taken in a cuvette of path length 1-mm/1-mm/2-mm, respectively (Figure 12). Here, the laser was focused on 1-mm thick benzene. As the figure demonstrates, the Raman signals from all the three liquid layers show similar trend as we increase the excitation-collection separation distance. These results demonstrate that the SORS technique can easily distinguish and discriminate different transparent, adjacent liquid layers.

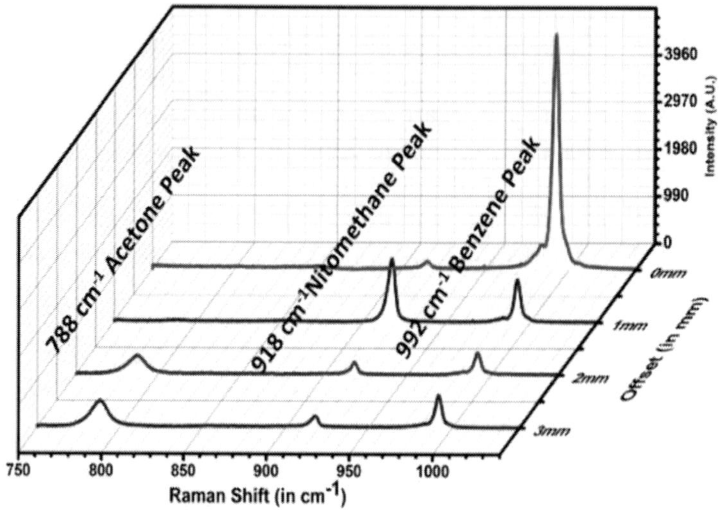

Figure 12. SORS result for trilayer liquid-liquid samples (1-mm benzene/1-mm nitromethane/2-mm acetone). Only the marker Raman peak for each liquid is shown for simplicity of demonstration.

3.1. Detection of Hazardous Chemical Liquids Filled in HDPE Bottles

SORS has been applied for the detection of hazardous chemicals filled in opaque plastic bottles. Plastic bottles are basically high density poly ethylene (HDPE). The most prominent peaks for HDPE are 1061, 1128, 1294, 1415 and 1439 cm^{-1}. 1061 and 1128 cm^{-1} bands were assigned for out-of-phase C–C stretching vibration of all-trans alkyl chains, in-phase (symmetric) C–C stretching vibration of all-trans alkyl chains respectively. 1294 cm^{-1} peak was attributed to the in-phase CH_2 twist; 1415 and 1439 cm^{-1} peaks were assigned for methylene bending vibration of HDPE. The spectra of the test liquids inside the HDPE bottle could be easily obtained by recording the spectra away from the point of illumination. Figure 13 shows the spectra of nitromethane in HDPE bottles. The Figures clearly show the spectra of the hazardous liquid like nitromethane (CH_3NO_2) inside the commercial HDPE bottle acquired at points away from the point of illumination of the sample.

4. Multiply Amplified Raman Spectroscopy (MARS)

Propagation of light in strongly scattering systems like turbid media takes place in the multiple scattering regime and can thus be described in terms of diffusing waves. The study of multiple Rayleigh scattering in diffusely scattering media has led to a powerful technique called the diffusing wave spectroscopy (DWS) [130-132]. In a DWS experiment, the temporal fluctuations of multiply scattered Rayleigh light intensity (speckle pattern) are obtained and these patterns are described by an autocorrelation function that evolves as a result of diffusive

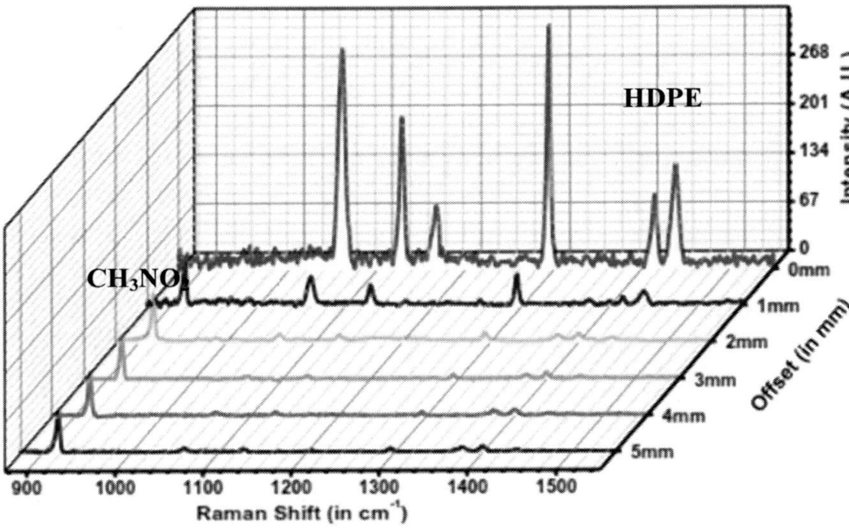

Figure 13. Detection of nitromethane filled in commercial HDPE shampoo bottle.

transport of light in turbid media. The intensity in a given speckle spot fluctuates when the scatterers move with respect to each other. Therefore, the DWS technique can be utilized to study the dynamics of the scattering medium and the inter particle correlations in colloidal suspensions. The idea of DWS has also been utilized in non invasive depth profiling and neuro imaging studies [133,134]. However, all these techniques rely upon the Rayleigh light scattering phenomena and do not yield information about the type of materials present in a sample under study. The multiply scattered light from a turbid media contains both Rayleigh and Raman scattered photons. The sub-surface techniques presently used are spatially offset Raman spectroscopy (SORS) and transmission Raman spectroscopy. The multiple scattering may result in coherent and incoherent scattered fields in different directions and, consequently, coherent backscattering has been observed in various particulate samples; SORS works only with backscattering geometry. We note that similar to the Rayleigh scattered photons, the Raman scattered photons originating at a single subsurface point diffuse and exit the sample not at a single point but distributed over a relatively wide area (Figure 14). We have developed a new Raman based technique in our laboratory that is geometry independent and can probe any type of strongly scattering samples. Thus, we show in our technique that one can record Raman spectra of thick and opaque diffusely scattering samples (both solids and colloidal liquids) in various possible experimental geometries (such as back scattering, sideways, right angle scattering and transmission).

The technique works on the principle of amplification of Raman signal from multiple scattering of light in a randomly distributed scattering medium. The multiply amplified Raman signal enables the technique to probe both strongly and weakly scattering media. The fact that our technique works independent of the experimental geometry, one can easily collect Raman signals from samples lying within other strongly scattering samples using either conventional lenses or optical fibers placed at different positions around the sample. The technique is a potential tool for non- invasive and non- destructive detection or screening of hazardous chemicals packed in different types of containers and also deep imaging of tissues, for example, breast cancer tissue.

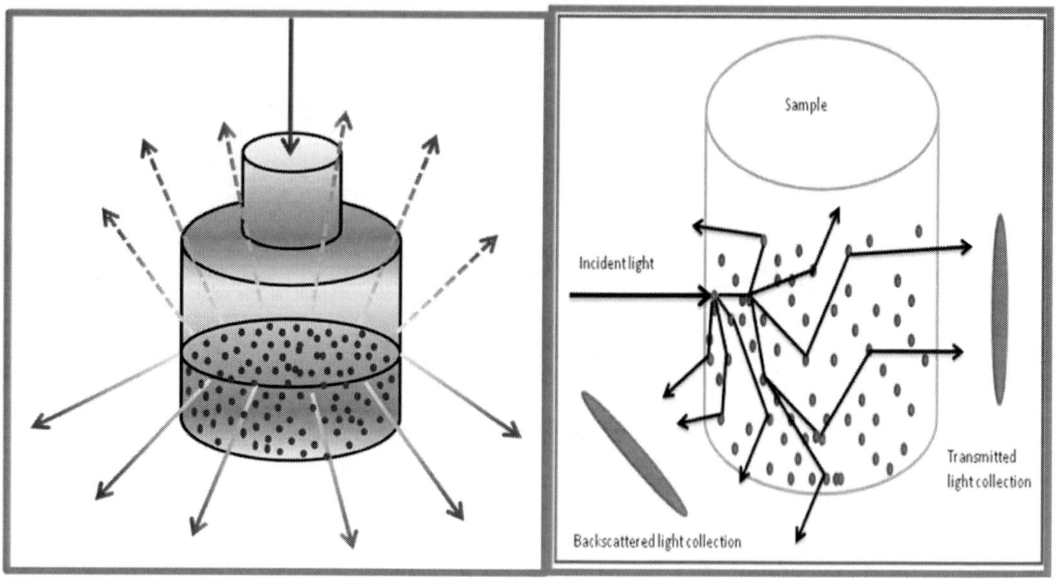

Figure 14. Principle of multiple scattering amplified Raman spectroscopy (MARS).

4.1. Principle of Technique: Multiply-Amplified Raman Signal through Multiple Scattering

In nature, multiple scattering of incident light takes place in turbid media such as cloud, paints, milk and biological tissues. The same phenomenon happens when monochromatic laser light interacts with a system of collection of particles, such as chemical powder. Particles in such a system are electromagnetically coupled, and the scattering results are slightly different from that of a single particle case. Each particle in a collection is excited by the external electromagnetic field and the resultant field scattered by all other particles; but the field scattered by a particle depends on the total field to which it is exposed. When there are many particles and their separation is random, the scattered field will be incoherent, i.e., there is no systematic relation among the phases of the waves scattered by the individual particles. Thus, the total irradiance scattered by the collection in any direction is just the sum of the irradiances scattered by the individual particles in that direction. Even, however, in a collection of randomly separated particles, the scattering can be coherent in forward direction and coherent backscattering is possible for such systems in exact 180° geometry. We note that in a system comprising of less number of particles (low number density) there can be weak localization of light inside the medium which leads to the phenomenon of coherent backscattering, while the photon diffusion process may be absent due to Anderson localization of light inside the medium when the particle number density is significantly increased.

The multiple scattering can be seen as an amplifying process of Raman photons by the medium analogous to the process of multiplicative-noise-induced amplification of electronic signals wherein it is theoretically and experimentally shown that amplification of weak signals is possible in a nonlinear dynamical system with multiplicative noise [135]. An optical signal corresponding to a particular Raman shift from a molecule can be amplified by subsequent addition of similar Raman photons generated by similar molecules by multiple scattering due to the presence of incident and Rayleigh scattered photons. When we see multiple scattering more closely, we find that the scattered photons are not totally lost from the incident beam, i.e., the scattered light can get back into the beam by multiple scattering. The light scattered at other positions in the medium may ultimately contribute to the irradiance at a position of observation. Therefore, the scattered light merely changes direction and is lost from a beam propagating in a particular direction, but contributes to other directions; this fact makes it possible to detect scattered light in all directions with respect to incident light propagation. Clearly, the greater the scattering cross section, the number density of particles and the medium thickness, the greater will be the multiple scattering contributions to the Raman signal at a position of observation. Therefore, this amplified Raman signal can enable us to probe turbid media of thickness exceeding a few tens of mm and probe samples buried deep inside them. Further, scattering by multiple particles in certain directions is incoherent if the separations between the particles are uncorrelated during the measurement time. However, if a converging lens is placed in front of a detector which lies in the focal plane of the lens, the observation distance becomes infinite when compared with the inter-particle separation and in such practical conditions the scattering cross sections of a collection of particles are additive. Therefore, combinations of lenses and optical fibers placed around the dense container and right choice of laser wavelength (for deeper penetration) play vital role in detecting target samples using this method.

4.2. Experimental

The Raman spectroscopic experiment was performed in all geometries (at all angles) (Figure 14).The Raman probe used was the 830 nm output of a cw laser diode. The laser beam was focused onto the sample using a plano-convex lens. The incident beam polarization was perpendicular to the plane of observation of the scattered light. The Raman signals were collected from the samples using a 50-mm diameter biconvex lens (f/4.14) and imaged onto the spectrometer slit. The spectrometer was a high resolution imaging spectrometer (Horiba Jobin Yvon) aided with a LN_2- cooled CCD camera (Symphony, 1024 × 256 pixels, pixel size 26 × 26 μm). The samples taken for study were solid powders filled in envelopes and liquids/ solids filled in coloured glass bottles.

MARS enables detection of diffusely scattering turbid samples, solid powders concealed within layers independent of the collection geometry. It can be noted that using the principle of spatially offset Raman spectroscopy, the samples inside a container can be detected only in backscattering collection geometry. In backscattering collection method, the scatterers close to the surface contribute more to the Raman signal than the scatterers located deeper in the sample. From the statistics of the photon trajectories it is known that the signal is inversely proportional to the square of the depth considered for a depth higher than the transport length l* which implies that the scatterers contained in a layer twice deeper than another give four times less contribution to the signal [136, 137]. This results in dominating contribution from the container wall to the observed signal in backscattering collection geometry. However, in principle, since there is a finite probability of diffusely scattered light from the sample coming out in all directions, one can choose different collection geometries, thereby eliminating the restriction on detection of samples lying inside the opaque container. In the case of solid powders, however, the Raman intensity depends on the particle size and its distribution, and density of packing. In our experiments, the Raman signals were collected from all sides (sideways, 90° geometry or transmission side, etc.) of the container. The same apparatus can also be used for Raman imaging experiments.

4.3. Working Examples

In order to demonstrate the working principle of the technique, the experiment was performed on ammonium nitrate (NH_4NO_3) that was packed in a plastic and concealed inside an envelope (Figure 15) and H_2O_2 mixed with water in varying ratios concealed inside a green coloured bottle (Figure 16). The laser beam was focused onto the top layer i.e., the envelope in the first case and the green bottle in the latter and the signals were collected in different geometries. The objective was to detect the explosives that were concealed and packed within the aforementioned layers. It could be noticed that in both, reflection or back-scattering mode, Raman signal of the explosive concealed in the envelope could be easily obtained (Figure 15). This is also possible in other geometries such as sideways and transmission because of long migration time available to amplification of Raman photons in non-absorbing powder media. However, in the case of hydrogen peroxide mixed with water and concealed in green bottle, Raman spectra could be obtained for 90^{0} as well as for transmission mode (Figure 16). With the present technique, chemical identification of diffusely-scattering samples such as turbid media, colloidal solution, gels, solid powders, biological tissues, blood, layered substances can be performed.

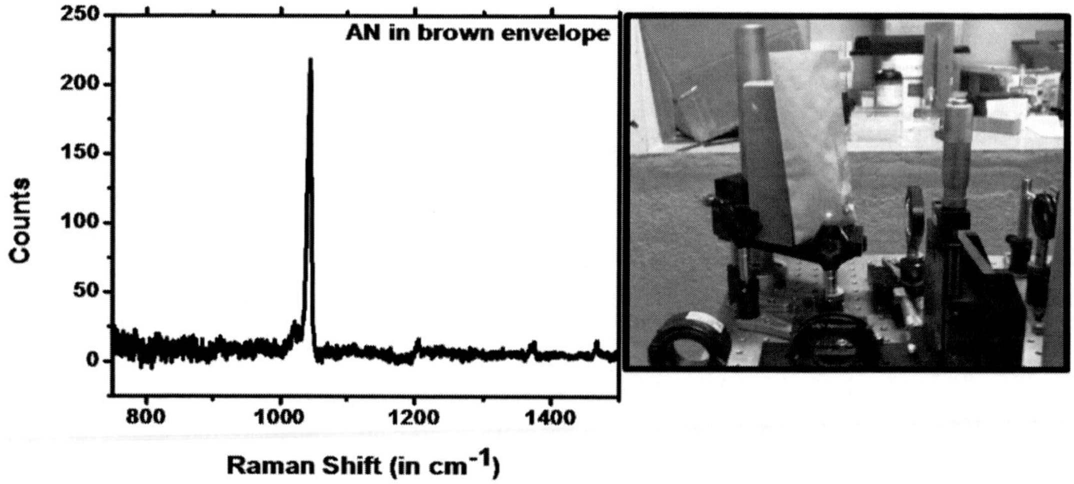

Figure 15. MARS signals of NH_4NO_3 explosive packed inside plastic bag concealed in a brown envelope.

MARS technique can also be used to study liquid chemical precursors as well as liquid explosives. Explosives like Triacetone triperoxide (TATP) and Hexamethylene triperoxidediamine have emerged as more recent explosives that need to be detected. These

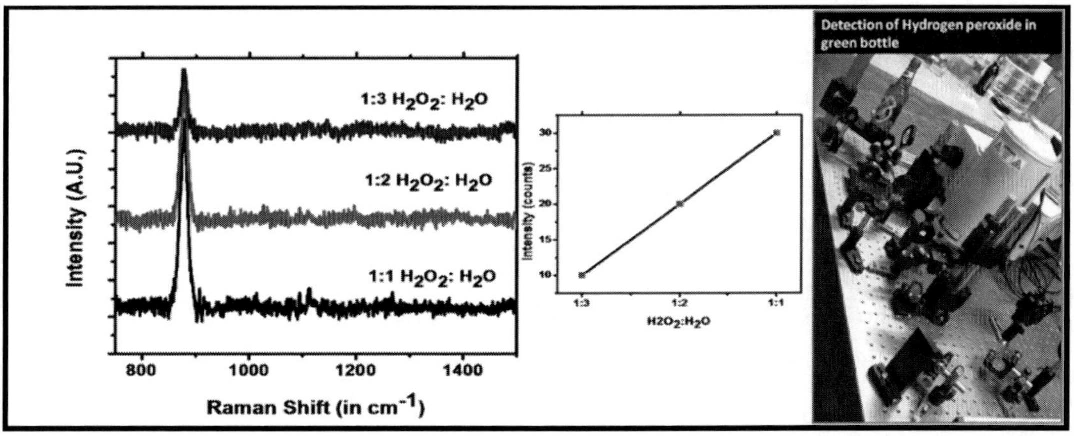

Figure 16. MARS signals of different ratios of H_2O_2 mixed in water concealed in a green glass bottle.

non nitro explosives are synthesized in-situ by mixing hydrogen peroxide and acetone in presence of some acids in the former case and hydrogen peroxide and hexamine in the latter. The precursors are usually common chemicals found in nail polish remover, hair dye, drain cleaner or batteries and are least suspicious items found in general stores. We are extending the work in our laboratory to detect various solid and liquid explosives concealed in different non-metallic containers and packages. We are also working on the instrumentation aspects for miniaturization of the equipment and are working on development of fibre-optic based MARS system. MARS technique holds immense potential to be used for security screening applications at airports or other public places.

From the foregoing discussions it is clear that Raman based spectroscopic techniques hold the promise to be used for detection of explosives (point detection and stand-off detection). There has been immense development in Raman instrumentation in the last few decades which has made it possible for Raman techniques to be used for rapid, non-invasive, sensitive and selective detection of materials including explosives.

However, it is always desirable to incorporate all the aspects in one instrument and develop a fail-proof device. Future research work will be based on keeping in mind the factors that pose challenge for identification of explosives such as low inherent vapour pressure of explosives, packaging or masking, interfering matrices, false positive and negative responses *etc*. Laser based Raman and UV-resonance Raman spectroscopic techniques can provide an answer for explosives detection in field environments at stand-off distances ~ 200 m. SERS based sensors can be useful for detection of explosives at ultra-trace levels in forensics investigation and environmental monitoring. Future sensors would be miniaturized e.g., lab on a chip, which could be produced cost-effectively for rapid explosives detection. In this regard nanotechnology would play a key role. On the other end of the detection spectrum, SORS and MARS can detect explosives in sealed containers such as glass containers, plastic bottles and other non-metallic packages which could be ideal for screening at airports, shopping malls, check-points and other public establishments etc. Development of newer Raman based techniques with novel instrumentation for detection of explosives is still an active area of research owing to the sensitivity of this field. Improving the limit of detection, specificity, selectivity, robustness and reducing the time of detection, size of the instrument and cost continue to be important. One technique alone cannot be an answer to the burning problem of explosives detection. There has to be a multifaceted approach where different techniques can be applied orthogonally. The table provided summarizes the Raman based techniques that have been discussed in this chapter.

5. Conclusion and Outlook

Table. Various Raman based techniques and their features

Technique	Principle	Sample preparation	Sensitivity	Selectivity	Point detection	Standoff detection
Raman Spectroscopy	Inelastic scattering of photons by different molecules when irradiated with a laser source.	No	No	Yes	Yes	Yes
		Remark: Due to low cross section of Raman scattering in conventional Raman scattering, the sensitivity is low.				
Resonance Raman Spectroscopy	Enhanced Raman scattering by matching the incident laser energy to the electronic transition of the system (visible to deep UV range).	No	Yes	Yes	Yes	Yes (UV only)
		Remark: The sensitivity is increased ~ 10^4-10^6 compared to normal Raman spectroscopy.				
Surface Enhanced Raman Spectroscopy	Enhancement of Raman signals by using Ag, Au, Pt etc. nanoparticles or films	No/ Nominal	Yes	Yes	Yes (Analogous to other Analytical Tech-niques)	No
		Remark: Sensitivity as high as 10^4-10^8 compared to normal Raman, can detect upto single molecules.				
Spatially Offset Raman Spectroscopy	Raman signal obtained from spatially offset regions from the point of illumination	No	Yes	Yes	Yes	No
		Remark: Non-invasive detection of explosives concealed in non-metallic containers.				
Multiply Amplified Raman Spectroscopy	Raman signals obtained at all angles and geometry independent.	No	Yes	Yes	Yes	Yes
		Remark: Non-invasive detection of explosives concealed in non-metallic containers.				

Acknowledgement

The authors gratefully acknowledge DST, DBT, DRDO and IISc, Bangalore for their financial support. SU acknowledges the J. C. Bose fellowship from DST.

REFERENCES

[1] J. S. Caygill, F. Davis, S. P. J. Higson, *Talanta*, 88, 14 (2012).

[2] D. S. Moore, *Rev. Sci. Instrum.* 75, 2499 (2004).

[3] G. Majano, S. Mintova, T. Bein, T. M. Klaptke, *J. Phys. Chem. C,* 111, 6694 (2007).

[4] J. F. G. Reyes, et al., *Anal. Chem.* 83, 1084 (2010).

[5] S. D. McGrane, D. S. Moore, *Prop. Explos. Pyrotech.* 36, 327 (2011).

[6] M. Tabrizchi, V. Ilbeigi, *J. Hazard. Mat.* 176, 692 (2010).

[7] E. D. Emmons, J. A. Guicheteau, S. D. Christesen, A. W. *Fountain III,Proc. SPIE 8018, Chemical, Biological, Radiological, Nuclear, and Explosives (CBRNE) Sensing XII, 80181C* (June 06, 2011); doi:10.1117/12.883202.

[8] J. Yinon, S. Zitrin, *Modern methods and applications in analysis of explosives*, 1993 (Wiley, Chichester, UK)

[9] J. C. Oxley, J. L. Smith, K. Shinde, J. Morgan, *Prop. Explos. Pyrotech.* 30, 127 (2005).

[10] U. Willer, W. Schade, *Anal. Bioanal. Chem.* 395, 275 (2009).

[11] M. W. Todd R. A. Provencal, T. G. Owano, B. A. Paldus, A. Kachanov, K.L. Vodopyanov, M. Hunter, S. L. Coy, J. I. Steinfeld, J.T. Arnold, *Appl. Phys. B* 75, 367 (2002).

[12] A.I. Nadezhdinskii, Y. Y. Ponurovskii, D. B. Stavrovskii, *Appl. Phys. B* 90, 361(2008).

[13] D. D. Wu, J. P. Singh, F. Y. Yueh, D. L. Monts, *Appl. Optics*, 35, 3998 (1996).

[14] J. Cabalo, R. Sausa, *Appl. Optics,* 44, 1084 (2005).

[15] D. Heflinger, T. Arusi-Parpar, Y. Ron, R. Lavi, *Opt. Comm.* 204, 327 (2002).

[16] C. Bauer, A. K. Sharma, U. Willer, J. Burgmeier, B. Braunschweig, W. Schade, S. Blaser, L. Hvozdara, A. Müller, G. Holl, *Appl. Phys. B* 92, 327 (2008).

[17] M. P. P. Monterola, B. W. Smith, N. Omenetto, J. D. Winefordner, *Anal. Bioanal. Chem.* 391, 2617(2008).

[18] M. S. Meaney, V.L. McGuffin, *Anal. Chim. Acta.* 610, 57 (2008).

[19] L. R. Schulte, P. Kolla, U. Karst, *Anal. Bioanal. Chem.* 386, 559 (2006).

[20] L. R. Schulte, P. Kolla, U. Karst, *Analyst*, 127, 1152 (2002).

[21] *Committee on the Review of Existing and Potential Standoff Explosives Detection Techniques,* Washington (2004).

[22] *Aspects of Explosive Detection*; Edited by Maurice Marshall & Jimmy Oxley: Elsevier; ISBN: 978-0-12-374533-0.

[23] K. P. J. Williams, G. D. Pitt, B. J. E. Smith, A.Whitley, D. N. Batchelder, I. P. Hayward, *J. Raman Spectro.* 25, 131 (1994).

[24] A. Hédoux, , Y. Guinet, M. Descamps, *Int. J. Pharm.* 417, 17 (2011)

[25] J. Akhavan, Spectrochim. *Acta, Part A*, 47, 1247 (1991).

[26] I. R. Lewis, N. W. Daniel, P. R. Griffiths, *Appl. Spectrosc.* 51, 1854 (1997).

[27] M. L. Lewis, I. R. Lewis, P. R. Griffiths, *Vib. Spectrosc.* 38, 11 (2005).

[28] N. Gupta N.F. Fell Jr, *Talanta*, 45, 279, (1997).

[29] T. Hirschfeld, E. R. Schildkrout, H. Tannenbaum, D. Tannenbaum, *Appl. Phys. Lett.* 22, 38 (1972).

[30] L. C. Pacheco-Londono, W. Ortiz-Rivera, O. M. Primera-Pedrozo, S. P. Hernandez-Rivera, *Anal. Bioanal. Chem.* 395, 323 (2009).

[31] J. C. Carter, S. M. Angel, M. Lawrence-Snyder, J. Scaffidi, R. E. Whipple, J. G. Reynolds, *Appl. Spectrosc*. 59, 769 (2005).
[32] A. Pettersson, I. Johansson, S. Wallin, M. Nordberg, H. Ostmark, *Prop. Explos. Pyrotech.* 34, 297 (2009).
[33] I. R. Lewis, N. W. Daniel, Jr., N. C. Chaffin, P. R. Griffiths, M. W. Tungol, *Spectrochim. Acta, Part A,* 51, 1985 (1995).
[34] D. Sulzle, P. Klaeboe, *Acta Chem. Scand. A,* 42, 165 (1988).
[35] N. F. Fell, J. M. Widder, S. V. Medlin, J. B. Morris, R. A. Pesce Rodriguez, K. L. McNesby, *J. Raman Spectrosc*. 27, 97 (1996).
[36] N. W. Daniel, Jr., I. R. Lewis, P. R. Griffiths, *Appl. Spectrosc*. 51, 1868 (1997).
[37] K. L. McNesby, C. S. Coffey, *J. Phys. Chem*. B 101, 3097 (1997).
[38] M. L. Lewis, I. R. Lewis, P. R. Griffiths, *Appl. Spectrosc*. 58, 420 (2004).
[39] M. L. Lewis, I. R. Lewis, P. R. Griffiths, *Vib. Spectrosc*. 38, 17 (2005).
[40] K. E. Lipinska-Kalita, M. G. Pravica, M. Nicol, *J. Phys. Chem. B*, 109, 19223 (2005).
[41] S. D. McGrane, J. Barber, J. Quenneville, *J. Phys. Chem. A,* 109, 9919 (2005).
[42] J. J. Haycraft, L. L. Stevens, C. J. Eckhardt, *J. Appl. Phys.* 100, 053508 (2006).
[43] J. A. Ciezak, T. A. Jenkins, Z. Liu, R. J. Hemley, *J. Phys. Chem. A*, 111, 59 (2007).
[44] Z. A. Dreger Y. M. Gupta, *J. Phys. Chem. B* 111, 3893 (2007).
[45] V. Vacque, N. Dupuy, B. Sombret, J. P. Huvenne, P. Legrand, *Appl. Spectrosc*. 51, 407 (1997).
[46] F. T. Docherty, P. B. Monaghan, C. J. McHugh, D. Graham, W. E. Smith, J. M. Cooper, *IEEE Sens. J.* 5, 632 (2005).
[47] L. L. Stevens, J. J. Haycraft, C. J. Eckhardt, *Cryst. Growth Design* 5, 2060 (2005).
[48] K. Y. Lee, D. S. Moore, B. W. Asay, and A. Llobet, *J. Energetic Mater*. 25, 161 (2007).
[49] D. S. Moore, K.Y. Lee, *J. Raman Spectrosc*. 38, 1221 (2007).
[50] D. S. Moore, R. J. Scharff, *Anal. Bioanal. Chem*. 393, 1571 (2009).
[51] J. D. Santillan, C. D. Brown, W. Jalenak, *Proc. SPIE-Int. Soc. Opt. Eng*. 6540, 65400P (2007).
[52] R. J. Stokes, W. E. Smith, B. Foulger, C. Lewis, *Proc. SPIE-Int. Soc. Opt. Eng*. 7119, 7119OI (2008).
[53] Y. Fleger, L. Nagli, M. Gaft, and M. Rosenbluh, *J. Lumin*. 129, 979 (2009).
[54] D. S. Moore, J. Fresenius, *Anal. Chem*. 369, 393 (2001).
[55] S. D. McGrane, A. P. Shreve, *J. Chem. Phys*. 119, 5834 (2003).
[56] J. Oxley, J. Smith, J. Brady, F. Dubnikova, R. Kosloff, L. Zeiri, Y. Zeiri, *Appl. Spectrosc*. 62, 906 (2008).
[57] J. C. Carter, S. M. Angel, M. Lawrence-Snyder, J. Scaffidi, R. E. Whipple, J. G. Reynolds, *Appl. Spectrosc*. 59, 769 (2005).
[58] A. Pettersson, I. Johansson, S. Wallin, M. Nordberg, H. Ostmark, Prop. Explos. Pyrotech. 34, 297 (2009).
[59] A. W. Parker, R. E. Hester, S. Umapathy, *J. Chem. Soc. Faraday Trans*. 88, 2649 (1992).
[60] N. Biswas, S. Umapathy, *Chem. Phys. Lett*. 294, 181 (1998).
[61] G. Balakrishnan S. Umapathy, *J. Mol. Struc*. 475, 5 (1999).
[62] M. Puranik, J. Chandrasekhar, J. G. Snijders, S. Umapathy, *J. Phys. Chem. A*, 105, 10562 (2001).

[63] N. Biswas, S. Umapathy, *J. Raman Spectrosc*. 32, 471 (2001).
[64] M. Gaft, L. Nagli, *Opt. Mater*. 30, 1739 (2008).
[65] L. Nagli, M. Gaft, Y. Fleger, M. Rosenbluh, *Opt. Mater*. 30, 1747 (2008).
[66] P. Jander, R. Noll, *Appl. Spectrosc*. 63, 559 (2009).
[67] L. Nagli and M. Gaft, *Proc. SPIE-Int. Soc. Opt. Eng*. 6552, 6552OZ (2007).
[68] R. D. Waterbury, A. R. Ford, J. B. Rose, E. L. Dottery, *Proc. SPIE-Int. Soc. Opt. Eng*. 7304, 73041B (2009).
[69] W. F. Hug, R. D. Reid, R. Bhartia, A. L. Lane, Proc. *SPIE-Int. Soc. Opt. Eng*. 6954, 6954OI (2008).
[70] S.A. Asher, C. R. Johnson, *Science* (Washington D.C.), 225, 311, 1984.
[71] R. J. H. Clark, 'Resonance Raman Spectra of Inorganic Molecules and Ions', *in Advances in Infrared and Raman Spectroscopy* R. J. H. Clark and R. E. Hester, Eds. (Heyden, London, 1975), vol. 1, 151.
[72] A. Asher, *Anal. Chem*. 65, 59A (1993).
[73] A. Asher, *Anal. Chem*. 65, 201A (1993).
[74] S. Bykov, I. Lednev, A. Ianoul, A. Mikhonin, C. Munro, S. A. Asher, *Appl. Spectrosc*. 59, 1541 (2005).
[75] B. Sharma, S. Bykov, S. A. Asher, *J. Phys. Chem. B,* 112, 11762 (2008).
[76] A. Mikhonin, S. Bykov, N. Myshakina, S. A. Asher, *J. Phys. Chem. B,* 110, 1928 (2006).
[77] S. A. Asher, in *Handbook of Vibrational Spectroscopy*, J. M. Chalmers and P. R. Griffiths, Eds. (John Wiley and Sons, New York, 2001), vol. 1, 557–571.
[78] D. D. Tuschel, A. V. Mikhonin, B. E. Lemoff, S. A. Asher, *Appl. Spectrosc*. 64, 455 (2010).
[79] M. Ghosh, L. Wang, S. A. Asher, *Appl. Spectrosc*. 66, 1013 (2012).
[80] W. A. Al-Saidi, S. A. Asher, P. Norman, *J. Phys. Chem*. A 116, 7862 (2012).
[81] P. Jander, R. Noll, *Appl. Spectrosc*. 63, 559 (2009).
[82] N. Weienbacher, B. Lendl, J. Frank, H. D. Wanzenbock, B. Mizaikoff, R. Kellner, *J. Mol. Struc*. 410-411, 539 (1997).
[83] X. Zhang, M. A. Young, O. Lyandres, R. P. Van Duyne, *J. Am. Chem. Soc*. 127, 4484 (2005).
[84] X. Zhang, C. R. Yonzon, R. P. Van Duyne, *Proc. SPIE*, 5221, 82 (2003).
[85] D. Cialla, A. März, R. Böhme, F. Theil, K. Weber, M. Schmitt, J. Pop, *Anal. Bioanal. Chem*. 403, 27 (2012).
[86] C. Le Ru, P. Etchegoin, *Annu. Rev. Phys. Chem*. 63, 65 (2012).
[87] M. Fleischmann, P. J. Hendra, A. J. McQuillan, *Chem. Phys. Lett*., 26, 163 (1974).
[88] S. Nie, S. R. Emory, *Science*, 275, 1102 (1997).
[89] K. Kniepp, Y. Wang, H. Kniepp, L. T. Perelman, I. Itzkan, R. R. Dasari, M. Feld, *Phys. Rev. Lett.,* 78, 1667 (1997).
[90] D. L. Jeanmaire, R. P. Van Duyne, *J. Electroanal. Chem*. 84, 1 (1977).
[91] M. G. Albrecht, J. A. Creighton, *J. Am. Chem. Soc*. 99, 5215 (1977).
[92] M. Moskovits, *Rev. Mod. Phys*. 57, 783 (1985).
[93] K. Kneipp, M. Moskovits, H. Kneipp, *Surface-Enhanced Raman Scattering: Physics and Applications*; Springer. Berlin (2006).
[94] A. Campion, P. Kambhampati, *Chem. Soc. Rev*. 27, 241 (1998).

[95] G. Mcnay, D. Eustace, W. Ewen, S. K. Faulds, D. Graham, *Appl. Spectrosc*. 65, 825 (2011).
[96] B. Sharma, R. R. Frontiera, A. I. Henry, E. Ringe, R. P. Van Duyne, *Materials Today* 15, 16 (2012).
[97] E. C. Le Ru, P. Etchegoin, *Principles of surface enhanced Raman spectroscopy and related plasmonic studies effects. Elsevier, 1st Edition* (2009).
[98] C. L. Haynes, A. D. McFarland, R. P. Van Duyne, ***Anal. Chem.*** 77, 338A (2005).
[99] K. M. Spencer, J. M. Sylvia, J. A. Janni, J. D. Klein, *Proc. SPIE-Int. Soc. Opt. Eng*., 3710 Detection and Remediation Technologies for Mines and Minelike Targets IV, 373 (1999).
[100] A. Baker, D. S. Moore, *Anal. Bioanal. Chem*. 382, 1751 (2005).
[101] S. Hrapovic,E. Majid, Y. Liu, K. Male, J. H.T. Luong, *Anal. Chem*., 78, 5504 (2006).
[102] K. Kniepp, Y. Wang, R. R. Dasari, M. S. Feld, B. D. Gilbert, J. Janni, J. I. Steinfeld, *Spectrochim. Acta,* 51, 2171 (1995).
[103] J. Y. Xu, J. Wang, L. T. Kong, G. C. Zheng, Z. Guo, Z.; J. H. Liu, *J. Raman Spectrosc*., 42, 1728 (2011).
[104] A. Hatab, G. Eres, P. B. Hatzinger, B. Gu, *J.Raman Spectrsc*, 41, 1131 (2010).
[105] S. Botti, L. Cantarini, A. Palucci, *J. Raman Spectrosc*. 41, 866 (2010).
[106] X. Fang, S. R. Ahmad, *Appl. Phys. B*. 97, 723 (2009).
[107] R. Stokes, Raman renaissance in defense and security. *SPIE* 2009, 1-3.
[108] P. R. Sajanlal, T. Pradeep, *Nanoscale*, 4, 3427 (2012).
[109] T. Demeritte, R. Kanchanapally, Z. Fan, A. K. Singh, D. Senapati,M. Dubey, E. Zakarb, P. C. *Ray, Analyst*, 137, 5041 (2012).
[110] B. D. Piorek, S. J. Lee, M. Moskovits, C. D. Meinhart, Anal. Chem. 84, 9700 (2012).
[111] I. Talian, J. Huebner, *J. Raman Spectrosc*. 43, 4237 (2012).
[112] S. Botti, S. Almaviva, L. Cantarini, A. Palucci, A. Puiu, A. Rufoloni, *J. Raman Spectroscop*., 44, 463 (2013).
[113] S. Sil, D. Chaturvedi, B. Keerthi, S. Kumar, S. N. Asthana S. Umapathy, *J. Phys. Chem. A,* Submitted, (2013).
[114] A.K. Singh, M. Talat, D. P. Singh, O. N. Srivastava, *J. Nanopart. Res*. 12, 1667 (2010).
[115] K. Vijayaraghavan, S. P. Kamala Nalini, N. U. Prakash, D. Madhankumar, *Mater. Lett.* 75, 32 (2012).
[116] V. K. Shukla, R. P. Singh, A. C. Pandey, *J. Alloys Comp*. 507, 13 (2010).
[117] I. R. Lewis, N. W. Daniel Jr., P. R. Griffiths, *Appl. Spec*., 51, 1854 (1997).
[118] C. Passingham, P. J. Hendra, H. A. Willis, Spectrochim. *Acta Part A,* 47, 1235 (1991).
[119] P. Goede, N. V. Latypov, H. Őstmark, Propell. *Explos. Pyrot*. 29, 205 (2004).
[120] Y. Kholod, S. Okovytyy, G. Kuramshina, M. Qasim, L. Gorb, J. Leszczynsk, *J. Mol. Struct*. 843, 14 (2007).
[121] M. Moskovits, D. P. DiLella, K. J. Maynard, *Langmuir*, 4, 67 (1988).
[122] S. Sil, N. Kuhar, S. Acharya, S. Umapathy, *Is Chemically Synthesized Graphene 'Really' a Unique Substrate for SERS and Fluorescence Quenching?* Nature Scientific Reports, Submitted (2013).
[123] P. Matousek, N. Everall, M. Towrie, A. W. Parker, *Appl. Spectrosc*. 59, 200 (2005).
[124] P. Matousek, I. P. Clark, E. R. C. Draper, M. D. Morris, A. E. Goodship,N. Everall, M. Towrie, W. F. Finney, A. W. Parker, *Appl. Spectrosc*. 59, 393 (2005).
[125] C. Eliasson, N. A. Macleod, P. Matousek, *Anal. Chem*. 79, 8185 (2007).

[126] P. Matousek, E. R. C. Draper, A. E. Goodship, I. P. Clark, K.L.Ronayne, A. W. Parker, *Appl. Spectrosc*. 60, 758 (2006).
[127] P. Matousek, *Chem. Soc. Rev*. 36, 1292 (2007).
[128] M. D. Hargreaves, N. A. Macleod, V. L. Brewster, T. Munshi, H. G. M. Edwards, P. Matousek, *Journal of Raman Spectroscopy*, 40, 1875 (2009).
[129] P. W. Loeffen, G. Maskall, S. Bonthron, M. Bloomfield, C. Tombling, P. Matousek, "Chemical and explosives point detection through opaque containers using Spatially Offset Raman Spectroscopy (SORS)," in *Chemical, Biological, Radiological, Nuclear, and Explosives (CBRNE) Sensing XII, vol. 8018 of Proceedings of SPIE,* Orlando,Fla,USA,April 2011.
[130] D. J. Pine, D. A Weitz, P. M. Chaikin E. Herbolzheimer. *Phys. Rev*. Lett. 60, 1134 (1998).
[131] J. H. Park, C. Park, H. S. Yu, J. Park, S. Han, J. Shin, S. H. Ko, K. T. Nam, Y. H. Cho, Y. K. Park, *Nature Photonics* 7, 454 (2013).
[132] G. Maret, *Curr. Opin. Coll. Inter. Sc*. 2, 251 (1997).
[133] S. Sasic, Y. Ozaki (Ed.). *"Raman, Infrared, and near-IR chemical Imaging"*. John Willey & Sons Inc. USA. 2010.
[134] G. Strangman, D. A. Boas, J. P. Sutton, *Biol. Psychiatry*. 52, 679 (2002).
[135] O. V. Gerashchenko, S. L. Ginzburg, M. A. Pustovoit, JETP Letters, 67, 997 (1998).
[136] D. A. Weitz, D. J. Pine, in: W. Brown (Ed*.). "Dynamic Light Scattering, the method and some applications"*. Clarendon, Oxford. 1993. p. 179.
[137] A. Brun, H. Dihang, L. Brunel, *Prog. Org*. Coat. 61, 181 (2008).

In: Spectroscopic Techniques for Security, Forensic ... ISBN: 978-1-63117-404-9
Editors: Y. Dwivedi, S. B. Rai and J. P. Singh

Chapter 8

VANADIUM OXIDES FOR ENERGY AND SECURITY APPLICATIONS

Chiranjivi Lamsal and N. M. Ravindra*
New Jersey Institute of Technology, Newark, New Jersey, US

Vanadium oxides have been the active area of research for decades and new possibilities are being revealed day by day due to their well established, yet controversially explained, phenomenon of Insulator Metal phase Transition (IMT). In this study, we have summarized potential solutions to the challenges in the fields of energy, environmental protection, defense and security that are being faced by the world. These challenges include security of energy supply, increasing demand of expensive energy, energy shortage and exploration of new source of energy and minerals, resolving fluctuating energy cost, CO_2 emission associated with energy consumption, climate change, and potential threat regarding the security of people and the country. The role of vanadium oxides in smart materials, their working principle and mechanism, methods for enhancing their performances, parameters controlling their efficiency, spectral range for optimum performance and source of operations have been described. The complicacy of thermal detectors and the use of vanadium oxides as sensing elements in such detectors have been explained and practical examples are presented. The performance parameters of a bolometer detector are illustrated from the data available in the literature. Use of vanadium oxides as a sensing element in stable, high TCR (Temperature Coefficient of Resistance) bolometer operating at room temperature in the flat spectral range without any cooling mechanism has been described and the importance of passivation characteristics, good IR (Infra Red) absorption characteristics and fabrication compatibility of vanadium oxides to modern technology has been highlighted. Recent progress in the dynamic tuning of metamaterials achieved by blending the properties of the VO_2 film during the IMT is described.

* Email: nmravindra@gmail.com.

1.1. Introduction

The ground state electronic configuration of vanadium is $[Ar]3d^3 4s^2$. Being a d-transition metal, vanadium has different oxidation states capable of existing in both single as well as mixed valence state on forming oxides. The vanadium oxides such as VO, V_2O_3, VO_2 and V_2O_5 exist in a single oxidation state whereas many others, for instance: V_3O_5, V_4O_7, V_6O_{11}, V_6O_{13}, V_7O_{13}, V_8O_{15} etc., remain in mixed (two) valence state. However, these oxides can be categorized under the so-called Magnéli (V_nO_{2n-1}) and Wadsley ($V_{2n}O_{5n-2}$) homologous series. Di-, Sesqui- and pentoxides of vanadium (VO_2, V_2O_3 and V_2O_5) are most widely studied oxides of vanadium for technological applications. Vanadium ions in VO_2 and V_2O_3 have $V^{4+}(d^1)$ and $V^{3+}(d^2)$ electronic structures whereas V_2O_5 has V^{5+} ion with no 3d electrons. In these transition metal oxides, d electrons are spatially confined in partially filled orbitals and are considered to be strongly interacting or "correlated" because of Coulombic repulsion between two d electrons of opposite spin on the same ion. In other words, the two conduction electrons with antiparallel spin at the same bonding site repel each other with strong Coulomb force so as to keep them mutually separated and hence spatially localized in individual atomic orbitals rather than behaving as delocalized Bloch functions. Correlated electrons are responsible for the extreme sensitivity of materials for small change in external stimuli such as pressure, temperature or doping [1].

Several vanadium oxides undergo insulator-metal transitions (IMT) at a particular temperature, T_c. The IMT, occurring in these materials, varies over a wide range of temperatures and depends on the O/V ratio, i.e., the transition temperature increases with oxidation states of the vanadium atom [2]. Among them, VO_2 is one of the widely studied materials which undergoes IMT at 340K [3], while V_2O_3 and V_2O_5 exhibit transitions at 160K [4] and 530K [5] respectively. These first order phase transitions are reversible [6] and are accompanied by drastic change in crystallographic, optical and electrical properties. During structural transition, atoms undergo displacement with redistribution of electronic charge in the crystal lattice and hence the nature of interaction changes [7]. Below T_c, the V-O system shows insulating behavior wherein VO_2 and V_2O_3 have monoclinic structure [8, 9] and V_2O_5 has orthorhombic structure [10]. At temperatures greater than T_c, they behave like metal but with crystal structures that are different from their low temperature counterparts [8, 11]. Similarly, the phase transition leads to change in electrical conductivity up to 10 orders of magnitude [12], while optical properties show discontinuity.

The vanadium oxides are chromogenic materials and can change their optical properties due to some external stimuli in the form of photon radiation (photochromic), change in temperature (thermochromic) and voltage pulse (electrochromic); the change becomes discontinuous during IMT. Such properties can be exploited to make coatings for energy-efficient "smart windows" [13], and electrical and optical switching devices [14]. Thin films of VO_2 and V_2O_3 have been found to show good thermochromism in the infrared region [15, 16]. While maintaining the transparency to visible light, a smart window modulates infrared irradiation from a low-temperature transparent state to a high-temperature opaque state [17]. The two oxides, VO_2 and V_2O_5, can change their optical properties in a persistent and reversible way in response to a voltage [18]. V_2O_5 exhibits exceptional electrochromic behavior because it has both anodic and cathodic electrochromism, different from VO_2 which only has anodic electrochromism. These electrochromic materials have four main applications: information displays, variable-reflectance mirrors, smart windows and variable-emittance surfaces.

The V-O systems are widely applicable in technology such as memory devices and temperature sensors [19]. The memory aspect of the material is evidenced from the pronounced hysteresis present in phase transition [20]. Normally the range of operation of a device lies outside the hysteresis region. However, some bolometric devices are operational within the hysteretic transition [21]. Bolometers are thermal infrared (IR) detectors and can be used in infrared imaging applications such as thermal camera, night vision camera, surveillance, mine detection, early fire detection, medical imaging, and detection of gas leakage. A bolometer requires a material with high temperature coefficient of resistance (TCR) and a small 1/f noise constant [22]. Pure, stoichiometric single-crystals of VO_2 and V_2O_5 have high TCR but are difficult to grow. Furthermore, the latent heat involved in IMT is highly unfavorable for the bolometric performance [23]. Since T_c of V_2O_3 is far below room temperature, the resistance and hence the level of noise is low which makes V_2O_3 a good candidate for the fabrication of efficient microbolometers. However, Cole et al. [24] have shown that the thin films of all the three oxides, combined together, can produce a desired material with high TCR and optimum resistance for bolometer fabrication.

Clearly, phase transition in VO_2 is of high technological interest. IMT occurs near to room temperature and T_c can be tuned optically, thermally, electrically [25] and with doping [12]. The phase transition in VO_2 has been used to achieve frequency-tunable metamaterials in the near-infrared range [26, 27]. Recently, Kyoung et al. [28] have extended the study to terahertz range proposing an active terahertz metamaterial, a gold nano-slot antenna on a VO_2 thin film, which transforms itself from transparent to complete extinct at resonance when the VO_2 film undergoes thermo or photoinduced phase transition. Cavalleri et al. [8] showed that the phase transition can be photoinduced within hundreds of femtoseconds which can be an underlying principle for an ultrafast switch.

1.2. Smart Materials

Highly sensitive materials which can change their properties reversibly and persistently as a response to external stimuli such as pressure, temperature, light, electric fields, magnetic fields or chemical stimulus are called smart materials. A number of smart materials have been discovered such as [29] color- and optically changing smart materials (able to change color and or optical properties), adhesion-changing smart materials (able to change the attraction forces of adsorption or absorption of an atom or molecule), light-emitting smart materials (able to emit light by a phenomenon called luminescence), electricity-generating smart materials (able to generate an electric current), energy exchanging smart materials (able to store both sensible and latent energy and exhibit some reversibility), material-exchanging smart materials (able to bind and release matter), and even shape-changing smart materials (able to change shape and/or dimensions). Smart materials are being utilized to meet the increasing demand for expensive energy sources and raw materials, enhanced applications in automation, compact materials and products reacting to sensors and actuators. Since vanadium oxides are chromogenic, our focus will be on the chromogenic smart materials. Chromogenic smart materials are being widely used for architectural glazing, transportation and certain electronic displays. The concept of architectural glazings and the use of smart materials in vehicles (rear view mirrors, sun-roofs, visors, etc), aircrafts, spacecrafts and ships have been very attractive.

Chromogenic technologies are based on various mechanisms such as [30]: electrochromism, electrophoresis, liquid crystal display, thermotropics, photochromism, and thermochromism. However, the common feature of these technologies is the ability of the material to switch optical transmittance and reflectance. For quantifying these optical properties, average transmittance and reflectance quantities are defined based on their transmittance, T(λ), and reflectance spectra, R(λ). Some of such quantities are solar and visible (or luminous) transmittance and reflectance defined as [31]:

$$Tsol{=}0\infty T\lambda\varphi sol\lambda d\lambda 0\infty\varphi so(\lambda)d\lambda\ ,\ Rsol{=}0\infty R\lambda\varphi sol\lambda d\lambda 0\infty\varphi sol(\lambda)d\lambda \qquad (1.1)$$

$$Tvis{=}370770T\lambda\varphi vis\lambda d\lambda 370770\varphi vi(\lambda)d\lambda,$$
$$Rvis{=}370770R\lambda\varphi vis\lambda d\lambda 370770\varphi vi(\lambda)d\lambda \qquad (1.2)$$

where, $\varphi_{vis}(\lambda)$ refers to the visible spectral range of 0.37-0.77 μm and $\varphi sol\lambda$ refers to 0.25 - 3 μm interval of solar radiation. A smart window, for example, capable of transmitting the visible radiation should have high T_{vis} while for the purpose of reflecting heat, it should have low T_{sol} and high R_{sol}.

Smart windows are physical counter parts of biological cell walls, which adjust energy flow in accord with the thermal and optical need of the house such as light, view or privacy and hence can be considered as a salient feature of a home automation system that offers interactive security. The term, smart window, applies to a glass that can manually or automatically change the intensity of light passing through it. The use of switchable glazings in buildings results in reduction in energy used for heating, cooling and lighting. The world is facing two major challenges: security of its energy supply and climate change. Obviously, the most secure energy is saved energy and reducing the energy consumption will help to reduce the CO_2 emission caused by energy generation.

(a) Smart glass "off" **(b) Smart glass "on"**

Figure 1. Smart window in "OFF" and "ON" state [32].

1.2.1. Electrochromism

Electrochromic phenomenon was first invented in tungsten oxide thin films (WO_3) in 1969 [33] and, since then, researchers have focused on a wide range of materials and device structures. A number of alternate optical switching systems have been developed and in the mid-1980s, Granqvist coined a phrase "smart window" to describe electrochromic glazings [31]. Electrochromic material can regulate the transmittance (T), reflectance (R), absorption (A) and emittance (E) upon the application of a voltage between widely separated extrema [34]. In other words, an electrochromic material changes color persistently and reversibly by an electrochemical reaction between a transparent ("bleached") state and a colored state as shown in Figure 1 or between two colored states. This optical change occurs as a result of a small electric current at low dc potentials, well within a few volts [35].

An electrochromic device mainly consists of a configuration of three layers, each with less than one micrometer thick, positioned on a substrate or between two substrates: an ion storage film, an ion conductor (electrolyte) and an electrochromic film sandwiched between electrically conducting films (the electrodes) as shown in Figure 2. On applying an electric field, the ions which are normally small with high mobility such as H^+ or Li^+, move from the ion storage layer into the electrochromic layer via the ion conductor. The external electric circuit provides charge-balancing (an electronic charge equal to ionic charge) counterflow of electrons which creates a variation in the electron density in the electrochromic materials. These electrons also remain in the electrochromic film for the time the ions reside there altering, thereby, the optical properties. If the electrolyte is a pure ion conductor with negligible electronic conductivity, the device acquires the state of open state memory and electric field is not required unless it is desired to change the optical property. The electron injection can alter the transparency depending on the nature of the electrochromic oxide used. Ideally, this process is reversible so that the material returns to a transparent state with the extraction of the ions and electrons. The main requirements of the electrochromic film used in such devices are as follows:

a Coloration efficiency: optical modulation should be commensurate with the finite change in the electron density.

b Ion intercalation and deintercalation: The electrochromic film darkens up as the ions (H^+ or Li^+, etc.) move uniformly into and out of the electrochromic film. Hence it is required that insertion and extraction of ions through the film be easy.

If ion intercalation (insertion) leads to the darkening of the film, the coloration is said to be cathodic, whereas, deintercalation (extraction) led coloration is known as anodic coloration [18]. In other words, charge injection in cathodic electrochromic process leads to a decrease in transmittance while in anodic electrochromism, the injection leads to an increase in transmittance. The ion storage film, in this device, can be chosen with or without electrochromic property. The electrochromic storage film - the counter electrode- in variable light transmission electrochromic devices, such as smart windows, should be chosen in such a way that it works in a mode that is complimentary to the primary electrochromic material. In other words, the counter electrode should darken upon ion intercalation if the primary electrochromic film –the working electrode- darkens upon ion deintercalation and vice versa.

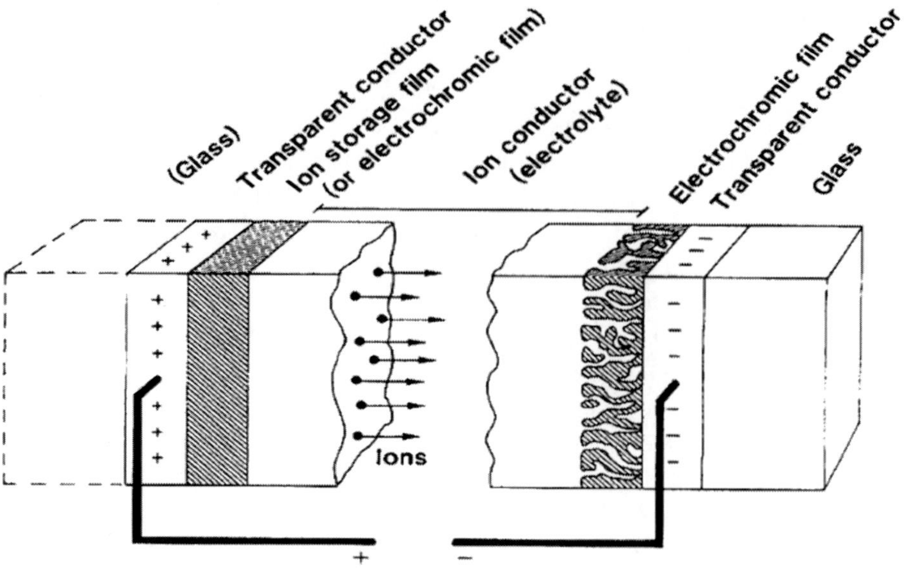

Figure 2. Schematic of an electrochromic device [34] showing the motion of positive ions.

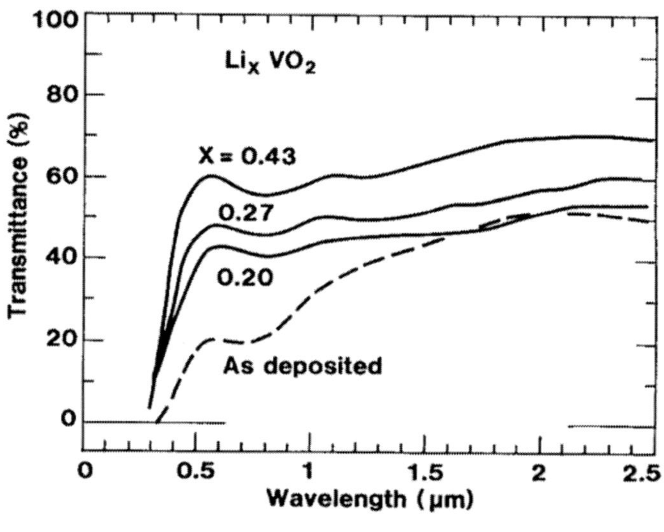

Figure 3. Transmittance spectra of 170 nm thick VO_2 film before and after lithiation [36].

VO_2 and V_2O_5 are electrochromic materials - VO_2 shows anodic coloration whereas V_2O_5 shows exceptional behavior of both types of coloration phenomenon within different wavelength ranges [18]. Other inorganic electrochromic systems are: WO_3, NiO, TiO_2, IrO_2, Nb_2O_5, SnO_2 and Pr_2O_3 among which WO_3 has the most electrochromic efficiency observed so far. In an attempt to see the improved performance of vanadium oxide as an electrochromic film, Khan et al. [36] studied transmittance spectra of Li-intercalated film of VO_2 (Li_xVO_2) at room temperature and found that the intercalation of lithium leads to a large increase in transmittance. Interestingly, such effect is observed over all spectral range with visible region showing the most pronounced change as shown in Figure 3. Achieving high transmittance modulation seems possible by combining anodically coloring Li_xVO_2 film with a cathodically coloring Li_yWO_3 film [36]. In similar studies [37, 38] of Li-intercalated film of V_2O_5 ($Li_xV_2O_5$), thermochromism is found to be weaker and of more complex nature. However, it has been realized that V_2O_5 films have electrochromic properties that are appropriate for counter electrode material for electrochromic devices. In their study, Cogan et al. [37] concluded that V_2O_5 shows a yellow to colorless modulation for a typical choice of film thickness (120nm) with Li^+ insertion level of 10-15mC/cm^2. Hence V_2O_5 film is a good potential counter electrode to WO_3 in electrochromic devices. V_2O_5 is particularly interesting not only because it has an excellent (charge) capacity to incorporate (Li^+) ions [39] but also due to relative ease of production by using simple, inexpensive and non-toxic sol-gel deposition technique. In a WO_3-V_2O_5 electrochromic device, the overall electrochemical reaction occurs as [37]:

$$a{-}LixWO3 + V2\ O5 \Leftrightarrow a{-}WO3 + LiyV2\ O5$$

The absorption in the layers, Li_xWO_3 and $V2\ O5$, occurs in the red and blue spectral region, respectively, while thin film of $a - WO_3$ and $LiyV2\ O5$ are transparent and colorless. Hence the WO_3 and V_2O_5 work synchronously as a complementary pair giving coloring and bleaching action in the electrochromic device.

Electrochromism can be explained theoretically in terms of a band-structure model. For a 0.1 μm thick film of $Li_xV_2O_5$ made with the substrate temperature of 50°C and 300°C, Talledo et al. [40] found the structure of the films to be nanocrystalline and polycrystalline respectively. On changing the doping level from x=0 to 1.5 in the nanocrystalline state, the band gap increased proportionally with x from 2.25 to 3.1 eV. For polycrystalline film, there was a shift in band gap from 2.38 eV at x=0 to 2.75 eV at x=2.2. However, the shift was not proportional to x, unlike in the nanocrystalline structure. The shift was predominant at x=1.0. It was observed that the absorption at $\lambda > 0.5\ \mu m$ increased with increase in the doping level from x=0 to 1.0 and then decreased for x=1.0 to > 2.0. This observation can be understood, at least qualitatively, with band structure effect and polaron absorption.

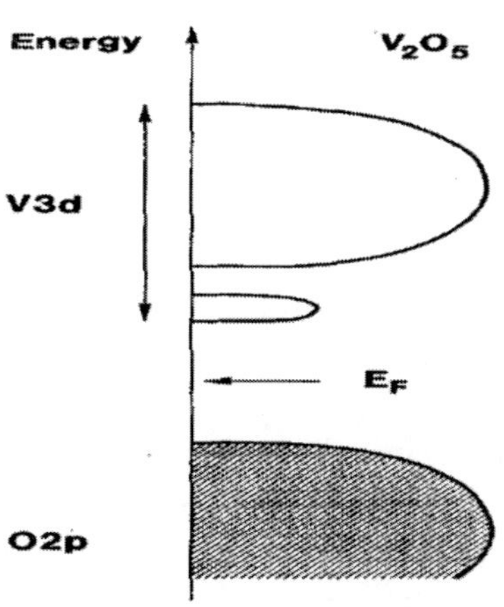

Figure 4. Schematic sketch of density of states for V_2O_5 [40].

The density of states for V_2O_5 can be represented schematically as shown in Figure 4 where O2p states are filled and V3d states are empty. As can be seen in Figure 4, the most bonding part of the d band is split-off from the d band spectrum [40, 41] and the optical band gap is the difference in energy between the top of the O2p band and the bottom of the split-off part of the V3d band, with Fermi level (E_F) lying in the gap. However, the insertion of ions and electrons shifts E_F to the split off band. With the increased ion intercalation, the split-off component gets filled up. Parity selection rule avoids the transition between the two parts of the d-band and hence the optical band gap is redefined as the difference between the top of the O2p band and the bottom of the main part of the V3d band. The observed [18] decrease in absorptance (increase in transmittance) of the blue light can be related to this band gap widening, giving ultimate explanation of the electrochromism based on band structure. Furthermore, V_2O_5 is known to exhibit cathodic coloration in the near-infrared (near IR) region of the spectrum, where it shows increase in absorptance (decrease in transmittance). This increased absorption is thought to arise from oxygen vacancies in the V_2O_5 lattice. Empty 3d orbitals of vanadium atoms, adjacent to such vacancies, localize $3d^1$ electron states within the band gap producing V^{4+} pairs. These V^{4+} states are responsible for creating small polarons resulting in increased near IR absorption.

1.2.2. Thermochromism

Thermochromism in vanadium oxide materials represents another possibility for smart coating in energy efficient buildings. Thin film of VO_2 is one of the most durable thermochromic materials undergoing IMT at 65 °C [42], which is close to its bulk value [3]. However, several techniques such as [43] (a) doping with suitable materials such as tungsten (W), Molybdenum (Mo), Niobium (Nb) and Rhenium (Re) (b) fluorination: replacement of some oxygen atoms by fluorine (c) mechanical stress induced by an over layer, can lower the transition temperature making it quite suitable as a window coating. Figure 5a shows the normal transmittance spectra for VO_2 film at temperatures below and above T_c, measured in the wavelength range of 0.3 to 2.5 μm [44]. For higher wavelengths (the near infrared region, greater than ~0.7 μm), the transmittance is modulated considerably as a function of temperature as compared to the low wavelength region, which is central to an efficient, energy controlling smart window. It means, for wavelengths near infrared, temperature dependent modulation of transmission spectra is observed while maintaining transparency to visible light. At the same time, the near infrared reflectance increases appreciably above T_c in line with the decrease in corresponding transmittance. For use of materials in glazing technology, transmission of light and also the reflectivity (in most applications) are very important. Ideally, a glazing material transmits solar radiation from exterior to interior when the "window" conducts heat out of the building during daytime in the winter and reflects solar radiation when it conducts heat into the building during daytime in the summer.

The transition temperature T_c of W-doped thin film of VO_2 drops almost linearly when the level of doping increases [42]. Sobhan et al. [42] showed that W-doped film, with the composition of $W_{0.032}V_{0.968}O_2$, undergoes IMT at ~32 °C and the normal transmittance for this structure varies as shown in Figure 5b. While doping decreases T_c to a comfort temperature, the thermochromic modulation of infrared transmittance, on the other hand, becomes small for the doped sample making it less useful for the energy control in smart windows. Later, in an attempt to improve the material performance, Granqvist [43] studied the transmittance spectra by replacing some of the oxygen with fluorine at various temperatures as shown in Figure 6. For a 0.13 μm thick VO_xF_y coating, the T_c was found to decrease to 52 °C and the near infrared transmission was found to be strongly temperature dependent. In this case, visible transmittance was found to be 28%, irrespective of the temperature and the solar transmittance was noticed to be ranging from 35% at room temperature to 28% at 70°C. The decrease in T_c and increase in transmittance in the visible spectrum show that VO_2 can be a very good candidate for thermochromic smart window application.

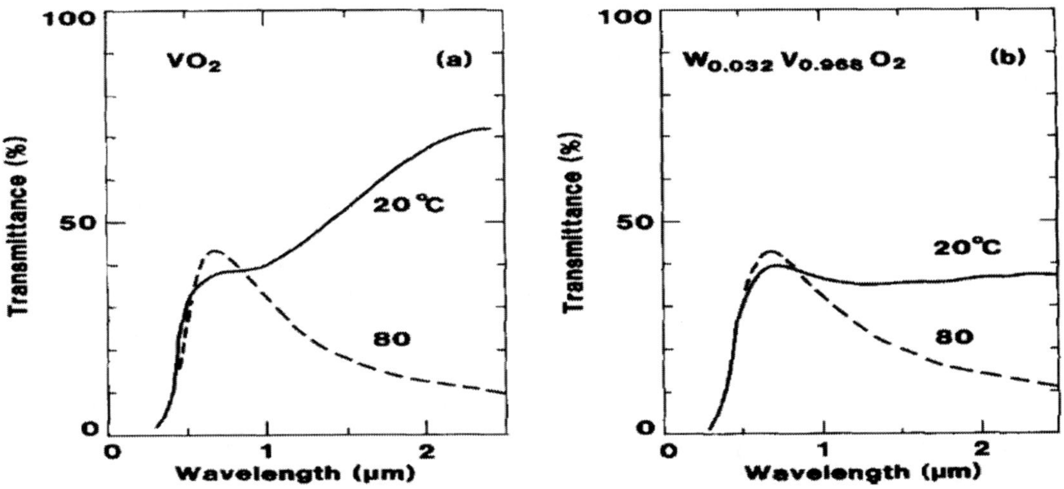

Figure 5. Transmittance spectra for (a) VO_2 and (b) W-doped VO_2 films at temperatures below and above IMT [42].

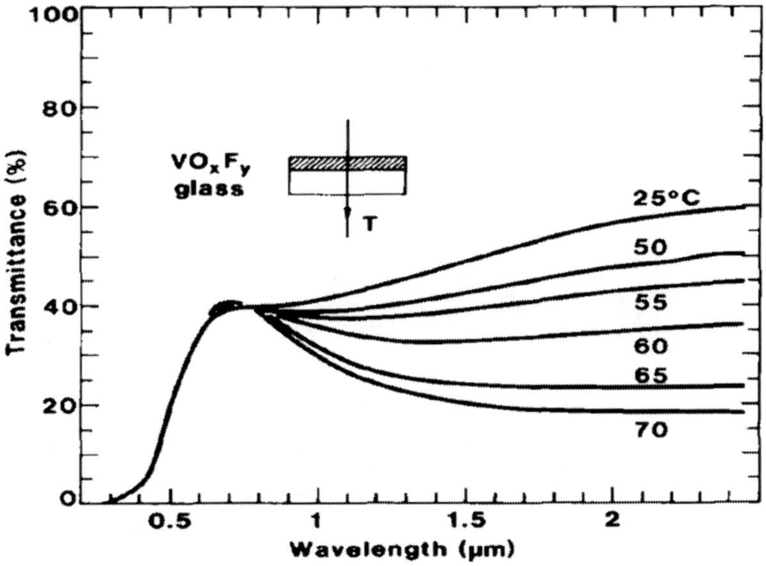

Figure 6. Normal transmittance spectra for vanadium oxyfluoride film at various temperatures [43].

1.3. Thermal Detectors

Any object, from human body to stars, having temperature above absolute zero emits electromagnetic waves depending on its emissivity and its absolute temperature (Stefan-Boltzmann law). In order to detect radiation, we use a photodetector, a device which converts the absorbed photons into a measurable form. There are mainly two types of detectors: photon detectors and thermal detectors. A photon detector is an optoelectronic device which gives rise to an electrical output signal when energy distribution of electrons changes as a result of the interaction of radiation with either free or bound charge carriers in a material. Interaction can be either internal or external. In internal interaction, photons either interact with charge carriers (bound or free) or produce a localized excitation of an electron to higher energy state [45]. However, in external interaction, electrons are emitted as a result of Einstein's photoelectric effect. On the other hand, thermal detectors absorb the photon energy and convert it into heat which, in turn, affects physical or electrical parameters such as electrical conductivity, thermoelectric voltage, and pyroelectric voltage. Hence thermal detectors do not depend on the nature of photon or spectral content of the radiation but depend on radiant power. Since heating and cooling are slower processes compared to interaction between photons and electrons, thermal response is relatively slower than spectral response. Typically, thermal effects occur in millisecond time scale while the effects due to photons are observed on micro or nano second time scale.

Figure 7 shows a schematic representation of a thermal detector. The sensing element, having thermal capacity C, is mounted on a substrate and thermally linked to a heat sink, at a constant temperature T_0, such that the average thermal conductance is G.

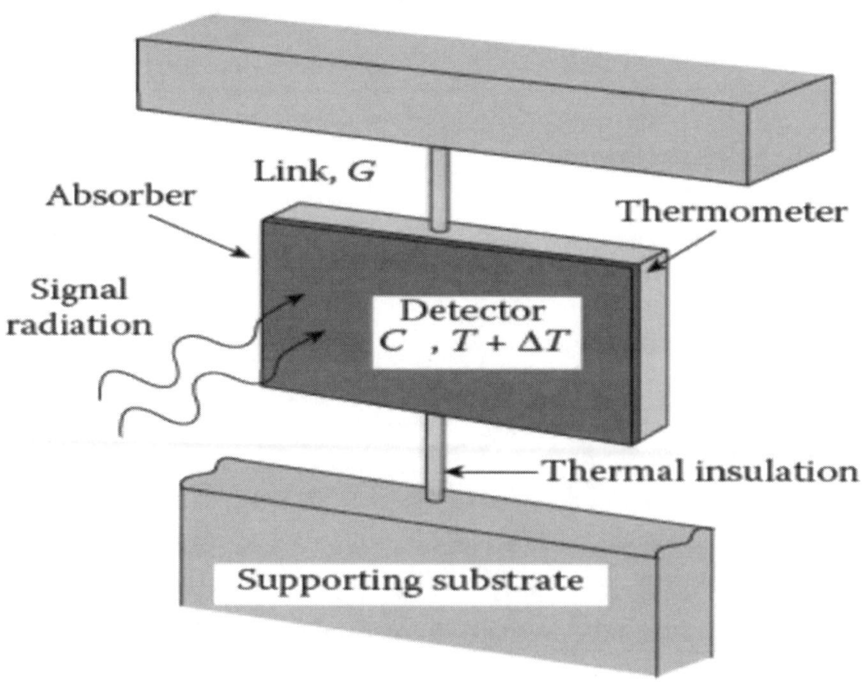

Figure 7. A schematic of a simple thermal detector [46].

The heat capacity C and thermal conductance (G) are given by:

C = mass (m) specific heat (c_p) (1.3)

G= $4\sigma\varepsilon AT3$ (1.4)

where, σ = Stefan-Boltzmann constant

ε = emissivity
A = surface area
T = temperature in Kelvin

The initial temperature of the detector, before the radiation input, is equal to the constant temperature of the sink T_0. However, after the detector absorbs radiation, the temperature of the detector changes by ΔT which is obtained as a solution to the heat balance equation [47, 48],

$(\Delta T)dt{+}G\Delta T{=}\varepsilon\Phi$ (1.5)

Taking radiant power ( ) as a periodic function i.e., $\Phi = \Phi_0 e^{i\omega t}$, where $\Phi 0$ is the amplitude of sinusoidal radiation, the solution to equation (1.5) is given by [49],

$\Delta T{=}\varepsilon\Phi 0G2{+}\omega 2C21/2$ (1.6)

Here, time dependent component of the solution involves the exponential term decaying with time and, hence, is ignored. Root mean square (rms) value of the electrical output voltage of the detector, which is proportional to change in temperature, can be written as:

$\Delta V{=}K\Delta T{=}K\varepsilon\Phi 0G2{+}\omega 2C21/2$ (1.7)

Obviously, a sensing material must be chosen in such a way that we obtain ΔT as large as possible which, in turn, requires, from equation (1.6), that G and C be as small as possible. In other words, while optimizing the interaction between radiation and the detector, the thermal contacts with its surroundings should be minimized. This requires that we choose a small detector mass and fine wire for connecting to the heat sink.

1.3.1. Performance Parameter of a Photodetector

The efficiency of a photodetector is based on several specifications. Normally, the following parameters are used by most manufactures of photodetectors to specify their overall performance.

Responsivity or Sensitivity

Response of a detector is measured in terms of quantities such as current, voltage, wavelength or impulse with respect to incident optical power. For instance, the detector current per unit incident power defines current responsivity (R_I) of the device. In general, responsivity of the detector varies with the wavelength of incident radiation and the responsivity at a particular wavelength is called spectral responsivity. Some other responses such as angular response can also be studied where change in output of the device varies as a function of the angle of incidence.

Quantum Efficiency (η)

Quantum efficiency is the probability that each photon incident on a material produces the charge carrier contributing to the detector current. Since all the incident photons are not absorbed, the quantum efficiency varies within $\mathbf{0 \leq \eta \leq 1}$. In practical situation, many photons impinge on the material and quantum efficiency can be taken as the ratio of the flux of charge carriers contributing to the detector current to the flux of incident photons. In the case of surface defects or reflection of radiation, a photon fails to reach the detector and hence only the absorbed photons should be taken into account. In this sense, we can call this parameter an internal quantum efficiency. Mathematically, quantum efficiency is given by [50],

$$\eta = 1.24\, RI\lambda \times 100\% \tag{1.8}$$

where, λ is the wavelength in μm

α/G *ratio*

The ratio of absorption coefficient to thermal conductance is the fundamental figure of merit for sensing element and determines the detectivity limits of the photodetector.

Dark Current

The detector current observed even when the photodetector is totally shielded from outside radiation is referred as the dark current.

Linear Dynamic Range

In general, a detector responds linearly with the incident optical power. However, an excessively large optical power degrades responsivity of the device and is said to be saturated. The range over which the detector exhibits linearity is called linear dynamic range.

Optical Gain

When electron-hole pairs are generated by incident photons, charge carriers flow in the detector circuit. Normally, each pair (with two charge carriers) is assumed to produce a charge (q=2e) in the external circuit. However, it has been shown [51] that only a charge of e is produced in the circuit. Moreover, some devices produce charge (q) different from e or 2e. The average number of circuit electrons produced per generated pair is called optical gain and can have the value greater or less than unity.

Noise

Normally all detectors suffer from some sort of noise but the source and degree of noise depend on the choice of materials. Major sources of noise are: shot noise generated by random emission of electrons, radiation noise which includes both signal fluctuation and background fluctuation noise, thermal Johnson noise (or thermal noise or Nyquist noise) generated by random motions of charge carriers at finite temperature, amplifier noise, microphonic noise and Flicker or 1/f noise- the noise which increases rapidly with decrease in frequency and is assumed to originate from material and manufacturing defects.

Noise Equivalent Power (NEP)

Noise equivalent power is defined as the incident optical power required for the detector to produce output signal equal to the noise in the frequency bandwidth of 1 Hz. In other words, NEP is the incident power required to produce signal to noise ratio (S/N) equal to one. Mathematically, NEP is given by,

$$NEP = INRI\ , \tag{1.9}$$

where, I_N is the quadrature sum of currents due to all significant noises.

Obviously, it is desired that we obtain a higher value of signal to noise ratio, which requires smaller value of NEP. Hence NEP can be taken as the measure of signal to noise ratio.

Noise Equivalent Temperature Difference (NETD)

Noise equivalent temperature difference (NETD) is the temperature difference between objects in a scene producing a signal-to-noise ratio of 1. The smaller the NETD, higher is the sensitivity and better is the performance of the detector.

Detectivity (D)

This is a figure of merit used for comparing detectors and is defined as inverse of NEP (D=1/NEP).

Specific Detectivity or Area Normalized Detectivity (D*)

This is another figure of merit used to compare detectors of similar spectral response and physical types but with different area.

$$D{*}=(A\,\Delta f)1/2NEP \tag{1.10}$$

where, Δf is the band width of the associated electronics and A is the active area of a detector, the primary light-collecting area of the detector surface, which is different from secondary area in some detectors that absorbs light to generate output signal, called non active area.

Thermal Time Constant or Thermal Response Time

Thermal time constant of a detector shows how quickly a detector responds to the incident IR radiation. It is defined as the ratio of thermal capacity (C) of the sensing element to the thermal conductivity (G). Response time of a detector is typically in the millisecond range for thermal detector which is much longer than for a photon detector. Smaller response time requires small active area of the sensing element but it should be carefully optimized because decrease in area will result in reduction of incident input.

1.3.2. Bolometers

Detection of dangers in advance, superior situation awareness and the use of precise weapon on time (if required) are the main requirements for both military and non military security applications. Infrared detectors, sensitive in both short and long wave infrared region, fulfill most of these requirements [52]. Due to two major performance parameters, excellent signal to noise ratio and fast response time, photon detectors are widely used in IR detector technology. Since the energy of incident photons is comparable to average thermal energies (K_BT) of atoms of the sensing element [53], the noise due to thermal charge carriers is inevitable and hence these photon detectors require cryogenic cooling to 77K or below [54]. Cooling mechanism, included in the photodetectors, make the device not only heavy, bulky and inconvenient but also expensive. Furthermore, photon detectors lack in broad band response i.e., they exhibit selective wavelength dependent response to incident radiation. Lack or difficulty in operating the photon detectors with appropriate spectral response in the IR region [50] is their other drawback. On the other hand, thermal detectors such as thermocouples, bolometers, thermopiles, and pyroelectric detectors are interesting because they are rugged, reliable, light, inexpensive and they can be operated at room temperature. Most of the thermal devices are passive devices for they do not require bias and, most importantly, they provide flatter spectral response. In this section, we will discuss one of the thermal IR detectors - the bolometers (Greek, "bole" meaning- ray). Bolometer was first designed in 1880 by Langley for solar observation; he demonstrated that the device sensitivity increased by 3 orders of magnitude as compared to thermopile and was capable of measuring temperature change as small as 1/100000 of 1°C [55]. Since then, the bolometer has been widely used in civilian and defense utilities. Figure 8 shows a picture taken from a helicopter using such an IR device in Boston, Massachusetts, during a security operation after the recent Boston marathon bombing on April 15, 2013.

The bolometer consists of a sensing element having strong temperature coefficient of resistance [TCR] so that a small temperature change, caused by incident radiation, can be measured. Unlike semiconductor based photon detectors, resistive bolometers are uncooled, based on simple principle and easy to fabricate. It is possible [57] to fabricate thermal bolometric detectors on thermally isolated hanging membranes by utilizing the recent progress in microelectromechanical systems technology (popularly known as MEMS technology). Due to recent developments in MEMS technology, performance levels of cooled infrared photon detectors have now been maintained by these uncooled infrared bolometers [22]. TCR is one of the vital parameters that influence the performance of bolometer and metals, semiconductors, thermistors and superconducting materials, having sufficient TCR, have been studied as possible candidates for sensing element in the bolometric devices.

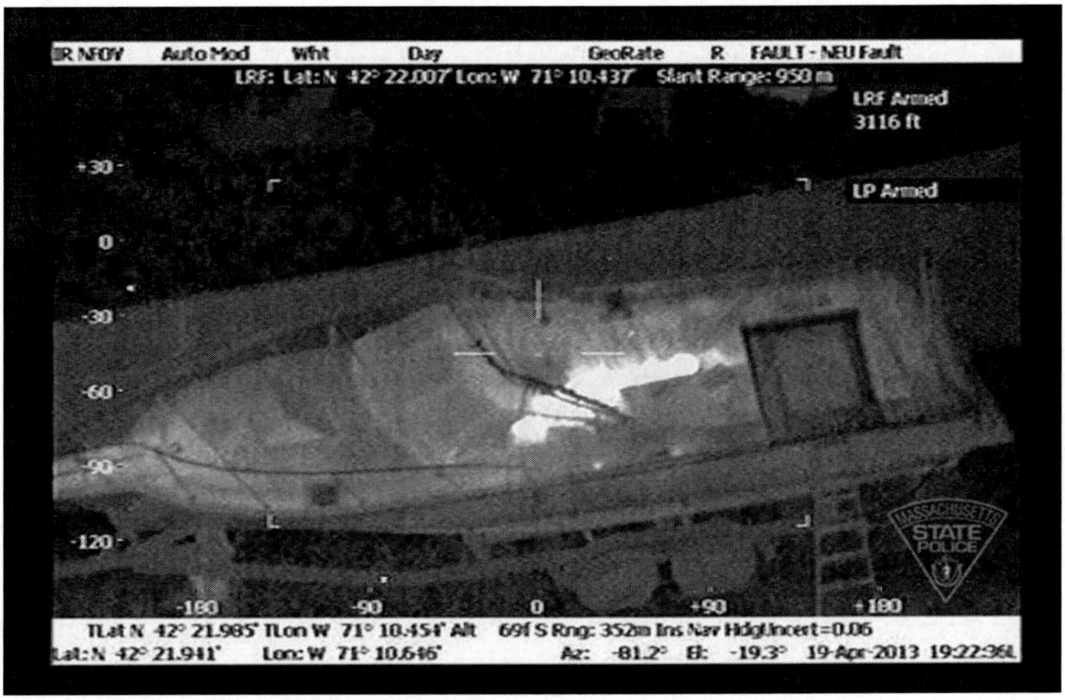

Figure 8. Boston marathon bombing suspect found hiding in a boat [56].

The performance of a thermal detector can be divided into two steps: raising the temperature of a sensing material by input radiation and using the temperature dependent variation of a particular property of the material for signaling in the output circuit. The second step, involving the use of material's property, depends on the type of thermal detector and, for bolometer, we use TCR (α) defined by, $\alpha{=}1RdRdT$. Accordingly, change in voltage of a current biased bolometer becomes,

$$\Delta V = I\Delta R = IR\alpha\Delta T \tag{1.11}$$

Comparing equations (1.7) and (1.11) we get, K=$IR\alpha$ and the equation (1.7) can be rewritten as,

$$\Delta V{=}IR\alpha\varepsilon\Phi 0G2{+}\omega 2C21/2 \tag{1.12}$$

Hence voltage sensitivity of a bolometer is,

$$Rv{=}IR\alpha\varepsilon G2{+}\omega 2C21/2 \tag{1.13}$$

It is seen that the sensitivity is inversely proportional to thermal conductance and thermal capacity. However, the sensitivity is directly influenced by the product of current (I), resistance (R) and TCR (α). Also, high emissivity is desirable in the atmospheric IR window of 8-14 μm bolometric devices.

The simplest representation of a bolometer consists of finding a sensitive material having high α, low thermal mass (C) along with maximum thermal isolation (low G). However, an accurate model of a bolometer requires a complex representation of parameters in the model (some of the important parameters are discussed in section 1.3.1). The performance of bolometers is typically hindered by noises and one of them is 1/f noise, the sources of which are not well understood. The noises vary by several orders of magnitude depending on the sensing material and its composition. 1/f noise has been found to be low for monocrystalline material as compared to amorphous or polycrystalline material. The most common bolometer sensing elements are vanadium oxides (VO_x), silicon diode and amorphous silicon (a-Si). Particularly V_2O_3, which undergoes IMT at low temperature (the critical temperature of transition, T_c=160K), has very low resistance at room temperature and can be used for low noise microbolometers. In addition to V_2O_3, mixed vanadium oxides (VO_x) are becoming popular thermistor materials and are used in the new generation of silicon microbolometers.

Thermistors are known to be stable materials having long life span and resistant to nuclear radiation. Thermistor materials, used in bolometer fabrication, are a mixture of various semiconducting oxides having higher TCR than metals. Their TCR depends on the impurity state, band gap and major conduction mechanism. Since the resistance for a semiconductor varies as $R{=}R0T{-}3/2\exp bT$, the TCR at room temperature can be written as $\alpha = -b/T^2$ [46]. This means TCR varies inversely as square of temperature. The TCR of VO_x exceeds 0.03 per °C (~2% K^{-1} at 25°C) which is sufficiently high for use in infrared imaging applications and is, in fact, five to ten times better than the TCR of most metals. VO_x is also found to have favorable optical properties for enhancing IR absorption. VO_x has a measured responsivity of 250,000 V/W in response to 300K blackbody radiation [46]. Because of a good combination of low 1/f noise, high TCR, high electrical resistivity, fabrication capability, good IR absorption characteristics, good passivation characteristics in conjunction with silicon nitride (Si_3N_4), vanadium oxides are suitable candidates as sensing material in bolometers.

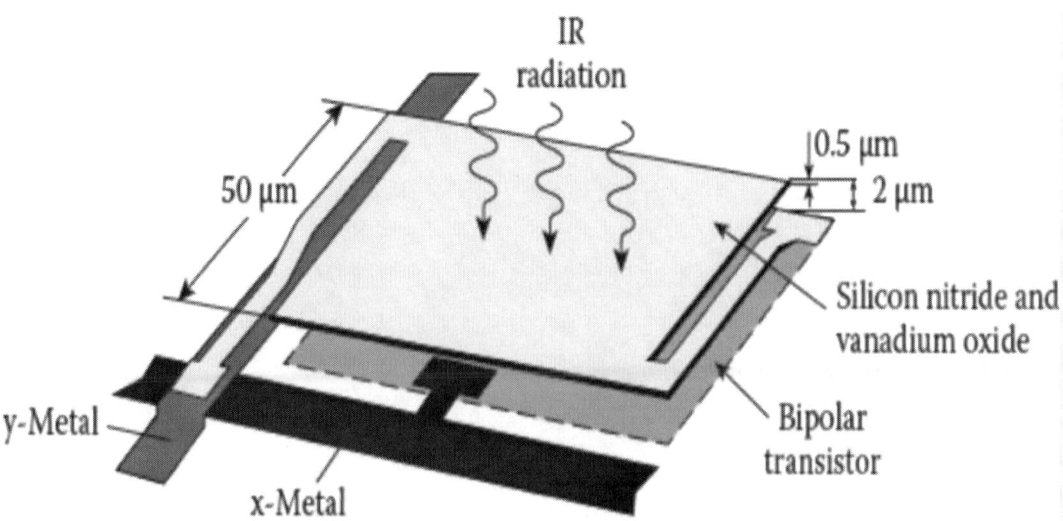

Figure 9. A microbolometer pixel structure [46].

Figure 9 shows the basic design of the Honeywell bolometer, which consists of a two-level structure with a gap of ~ 2 μm between them. The upper level is a square shaped silicon nitride (Si_3N_4) plate, of side 500 μm and thickness 0.5 μm, suspended over an underlying silicon integrated circuit (IC) substrate. A family of fabrication compatible materials with high TCR and well defined resistance can be developed out of mixed oxides of vanadium (VO_2, V_2O_3 and V_2O_5) using low-temperature ion beam sputtering deposition process [24]. Such VO_x films have a resistance of 20K per square film at 25°C which is quite sufficient for microbolometer readout circuits. The resistor material (VO_x) is, hence, formed within each microstructure plate of Si_3N_4. The bridge structure of Si_3N_4 is also supported by two narrow legs of Si_3N_4 that additionally provide the thermal insulation (since microbolometers require high thermal insulation, we do not rely on air insulation but on the microstructure support). The legs also contain a thin metal layer and hence serve as an electrical contact. A bipolar transistor is normally required for amplification purposes. An aluminum layer on the substrate reflects the absorbed IR radiation back to the sensing material and thus maximizes the absorption process. A monolithic control circuit is fabricated in the silicon substrate so that individual leadouts from each microstructure of a large number of individual microstructure arrays can be avoided. The basic purpose of the monolithic circuit is to apply a control voltage to each microstructure in the array and hence measure the microstructure resistance.

Figure 10. A ceramic package of microbolometer arrays operating at room temperature [24].

A prototype of an uncooled IR camera, operating at room temperature, was constructed out of such 240 336 microbolometer arrays as shown in Figure 10. This is a small micromachined structure extremely well thermally isolated from the substrate ($G=10^{-7}$ W/K) and having very small thermal mass ($C=10^{-9}$ J/K and corresponds to a thermal time constant of 10 ms). This thermal isolation is close to the maximum possible physical limit of about 10^{-8} W/K for a 50 μm square detector. With a microbolometer resistance of 50K and TCR of 2%/K, it was shown [24] that a typical IR signal of 1 nW was enough to change the microbolometer temperature by 0.01 K. This resulted in the value of the thermal signal to noise ratio as 275. Furthermore, Honeywell has claimed that the microbridge of Si_3N_4 is such a strong structure that it can tolerate shocks of several thousand g-forces (weight per unit mass). Tables 1 and 2 show the list of performance parameters for the VO_x bolometer.

Table 1. Performance parameters for a typical VO_x bolometer

Parameter	TCR (K^{-1})	Responsivity (V/W)	Resistance (K /square)	G (W/K)	C (J/K)	T (ms)	S/N
Value	~2% @298K	250,000 @300K	20 @ 298K	G=10^{-7}	10^{-9}	10	>275

Table 2. Performance parameter, NETD, for commercial VO_x bolometers (mK @ f-number of infrared optics =1 and frequency= 20-60 Hz) [58]

Company	FLIR USA	L-3 USA	BAE USA	Raytheon USA	DRS USA	NEC Japan	SCD Israel
NETD	35	50	30-50	30-50	35-50	75	50

The monolithic silicon bolometer technology was developed, for the first time in early 1980s, in Honeywell Sensor and System Development Center in Minneapolis, Minnesota. Later, both Honeywell (on VO_x) and Texas Instruments (on a-Si) worked under classified projects sponsored by DARPA and U.S. Army Night Vision and Electronic Sensor Directorate and finally succeeded in producing low-cost night vision systems with NEDT of 0.1°C using f/1 optics. Today, thermistor bolometers are widely used in applications such as fire detection systems, radiometer, space-borne horizon sensors, burglar alarms and industrial temperature measurements. They are also useful in applications requiring flat spectral response. In 2005, John Fluke Mfg. Co. had introduced a number of models of hand-held, portable infrared cameras of commercial standard. A stable, high TCR bolometer material of mixed vanadium oxides was the key to the success of the room temperature, uncooled bolometer at Honeywell.

1.4. METAMATERIALS

The term metamaterials (Greek, "meta" meaning- beyond) was coined by Walser [59] in 1999. Literally speaking, metamaterials are artificially designed materials which have properties that may not exist in nature. According to Munk [60], Walser defined metamaterials as "macroscopic composites having man-made, three-dimensional, periodic cellular architecture designed to produce an optimized combination, not available in nature, of two or more responses to specific excitation". Among the scientific community, metamaterials, today, are popularly recognized as composite materials which simultaneously possess negative permeability (μ) and permittivity (ε) (In order to be consistent with the standard notation, we have chosen ε to symbolize permittivity and readers should not be confused with emissivity of section 1.3). Even when both the constitutive parameters of a material are negative, $\varepsilon\mu$ is positive and hence wave propagation is still possible. However, for the sake of energy conservation, we must choose negative sign in $n{=}\pm\varepsilon\mu$ [61, 62]. Accordingly, such a medium can be characterized by negative refractive index and said to form a "left handed" medium in which Cherenkov radiation, Doppler effect and even Snell's laws are reversed [63]. Depending on the sign of ε and μ, materials can be classified as shown in Figure 11.

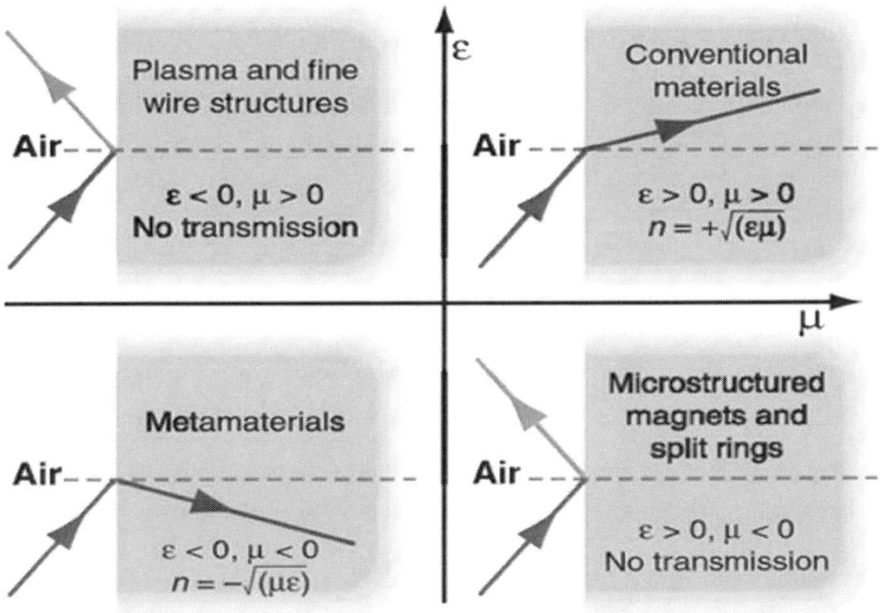

Figure 11. Classification of materials according to μ and ε [62].

Figure 12. A microwave invisible cloak - might be useful for evading radar [64]. It is worthwhile to note here that radar detects the microwave sent from its antenna, after it is reflected back from a remote target.

Normally, the electromagnetic properties of metamaterials are defined by the way they are structured, not by their chemical composition. The graded structure of an inhomogeneous material such as metal-dielectric composites and metamaterials, where materials properties change gradually as a function of position, can provide stronger nonlinear optical response as compared to its homogeneous counterpart [65]. Gradation can either be achieved naturally or be engineered in the manufacturing process. Graded materials can be useful for controlling the physical properties of optical materials. It is the gradations of refractive index of metamaterials that lead to its invisibility. This technology can be used to cause an object, from an individual to spaceship, either partially or totally invisible to parts of the electromagnetic spectrum. A plane coated with a metamaterial can elude detection since metamaterials can result in zero reflectance for all incident angles. Hence, they have enormous applications in the defense and security sectors. Figure 12 shows a cloaking device which is practically invisible when viewing the world in microwaves of a particular wavelength. Similarly, we might have a flat superlens operating in the visible spectrum and then we will have images even smaller than one wavelength of light with ultrahigh resolution [66]. Metamaterials have drawn the attention of the scientific community due to their wide range of potential applications.

Resonance Tuning Using Vanadium Oxides: A Recent Progress

The capacity of a metamaterial to interact with radiation of certain frequencies in a strong and a special manner is known as its so called resonant behavior. Scientists have been able to device metamaterials working at a single frequency [67] or narrow band [68] in the microwave or infrared regions of the spectrum, terahertz domain [69] and even in the lower end of the visible spectrum [70]. Recently, research in metamaterials is progressing by overcoming the limitations of bandwidth and the dynamic control of metamaterial's properties in real time has been achieved either through non linear responses or externally tunable components. Particularly, after Pendry et al. [71] introduced the concept of split ring resonator (SRR), scientists have constructed metamaterials using SRR [63] and further research has been conducted towards the creation of tunable metamaterials using SSR. It is the dynamically controllable metamaterial that allows us to take advantage of the full potential of its peculiar properties. Shadrivov et al. [72] "took the first step towards the creation and study of fully controlled, tunable nonlinear metamaterial systems through the study of the tunability and self-induced nonlinear response of a single SRR." In his study, Shadrivov used a specially doped p-n junction diode: the variation of diode capacitance led to the change in resonance conditions of the SRR in microwave region. Significant progress has been made towards the study of tunable metamaterials [26, 28, 73-98] and has been core research in modern science and technology.

Particularly, Driscoll et al. [26] have achieved dynamic tuning, with a tuning range of at least 20%, in the near infrared region by using a device made of a 100nm thick gold SRR array and 90nm thick VO_2 film grown on sapphire substrate. Both the gold and VO_2 layers are much thinner than the periodicity of SRR array (20μm) and are said to form an effective (single) material layer. This structure is now considered as a hybrid metamaterial since it has the combined property of both SRR and VO_2. It is well established that VO_2 undergoes insulator metal phase transition (IMT), on a sub-picosecond timescale [99], as a result of some external stimuli. At the onset of the phase transition, nanoscale (5 -10nm) metallic grains or "puddles" emerge from the insulating host of VO_2. The percolative nature of the phase transition causes divergent permittivity. This drastic change in optical behavior has a strong effect on the resonance frequency of the SRR metamaterial - the local electromagnetic fields of the SRR gets modified as if a tunable dielectric were inside a capacitor. It is found that the resonance frequency of the hybrid system varies inversely with the change in permittivity of VO_2. Furthermore, the decrease in metallic puddles damps the resonance amplitude.

Wen *et al.*[100] have proposed VO_2 cut-wire fabricated on silica-glass substrate as a terahertz metamaterial. The advantage of cut-wire resonator over SRR is that it exhibits simple yet broad band response [101-105]. Furthermore, cut-wire resonators are easy to design and fabricate. Motivated by the 4 orders of magnitude change observed in the dielectric properties of VO_2 during IMT, researchers are rigorously busy to find out whether the same order of magnitude change in extinction can be obtained in metamaterials [106]. Furthermore, due to the pronounced hysteresis present in phase transition [20], scientist are interested in designing memory metamaterials where the metamaterial tuning persists even if triggering stimulus for IMT disappears [25]. A wide variety of structures based on VO_2 phase transition have been studied in recent years due to the versatile properties of VO_2 [28, 106-111]. More interestingly, some researchers have identified VO_2 as a natural disordered metamaterial [112].

REFERENCES

[1] G. Kotliar, D. Vollhardt, Physics Today 57 (2004) 53-59.
[2] C. Lamsal, N. M. Ravindra, J. Mat. Sci. 48 (2013) 6341-6351.
[3] A. Zylbersztejn, N. F. Mott, Phys. Rev. B 11 (1975) 4383-4395.
[4] F. Morin, Phys. Rev. Lett. 3 (1959) 34-36.
[5] G. S. Nadkarni, V. S. Shirodkar, Thin Solid Films 105 (1983) 115-129.
[6] D. M. Lamb. Semiconductor to Metalic Phase Transitions from Vanadium and Titanium Oxides Induced by Visible Light. Missouri State University (2009).
[7] C. Batista, R. M. Ribeiro, V. Teixeira, Nanoscale Res. Lett. 6 (2011) 301 (7 pages).
[8] A. Cavalleri, C. Tóth, C. W. Siders, J. A. Squier, F. Ráksi, P. Forget, J. C. Kieffer, Phys. Rev. Lett. 87 (2001) 237401 (4 pages).
[9] C. Wessel, C. Reimann, A. Müller, D. Weber, M. Lerch, T. Ressler, T. Bredow, R. Dronskowski, J. Comput. Chem. 33 (2012) 2102-2107.
[10] E. E. Chain, Appl. Opt. 30 (1991) 2782-2787.
[11] P. D. Dernier, M. Marezio, Phys. Rev. B 2 (1970) 3771-3776.
[12] F. Chudnovskiy, S. Luryi, B. Spivak, Switching device based on first-order metalinsulator transition induced by external electric field, in: S. Luryi, J. M. Xu, A. Zaslavsky, (Eds.) Future Trends in Microelectronics: the Nano Millennium, Wiley Interscience (2002) pp. 148-155.
[13] C. G. Granqvist, Phys. Scripta 32 (1985) 401-407.
[14] V. S. -Milosevic, N. Nilius, H. -P. Rust, H. -J. Freund, Phys. Rev. B 77 (2008) 125112.
[15] J. B. K. Kana, J. M. Ndjaka, P. O. Ateba, B.D. Ngom, N. Manyala, O. Nemraoui, A. C. Beye, M. Maaza, Appl. Surf. Sci. 254 (2008) 3959-3963.
[16] M.S. Thomas, J.F. DeNatale, P. J. Hood, Materials Research Society Symposium Proceedings 479 (1997) 161-166.

[17] Z. Zhang, Y. Gao, H. Luo, L. Kang, Z. Chen, J. Du, M. Kanehira, Y. Zhang, Z. L. Wang, Energy & Environ. Sci. 4 (2011) 4290-4297.

[18] C. G. Granqvist, Handbook of Inorganic Electrochromic Materials. Elsevier Science, Amsterdam, Holland (1995).

[19] A. Gupta, R. Aggarwal, P. Gupta, T. Dutta, R. J. Narayan, J. Narayan, Appl. Phys. Lett. 95 (2009) 111915 (3 pages).

[20] T. Driscoll, H. -T. Kim, B. -G. Chae, M. D. Ventra, D. N. Basov, Appl. Phys. Lett. 95 (2009) 043503 (3 pages).

[21] L.A.L. D. Almeida, G.S. Deep, A.M.N. Lima H. Neff, Appl. Phys. Lett. 77 (2000) 4365 (3 pages).

[22] F. Niklaus, A. Decharat, C. Jansson, G. Stemme, Infrared Physics & Technology, 51 (2008) 168-177.

[23] R.T.R. Kumar, B. Karunagaran, D. Mangalaraj, S. K. Narayandass, P. Manoravi, M. Joseph, V. Gopal, Sens Actuat. A: Physical, 107 (2003) 62-67.

[24] B.E. Cole, R. E. Higashi, R. A. Wood, Proceedings of the IEEE, 86 (1998) 1679-1686.

[25] T. Driscoll, H.-T. Kim, B. -G. Chae, B. -J. Kim, Y. -W. Lee, N. M. Jokerst, S. Palit, D. R. Smith, M. D. Ventra, D. N. Basov, Science 325 (2009) 1518-1521.

[26] T. Driscoll, S. Palit, M.M. Qazilbash, M. Brehm, F. Keilmann, B. -G. Chae, S. -J. Yun, H.-T. Kim, S. Y. Cho, N. M. Jokerst, D. R. Smith, D. N. Basov, Appl. Phys. Lett. 93 (2008) 024101 (3 pages).

[27] M.J. Dicken, K. Aydin, I. M. Pryce, L. A. Sweatlock, E. M. Boyd, S. Walavalkar, J. Ma, H. A. Atwater, Opt. Exp. 17 (2009) 18330-18339.

[28] J.S. Kyoung, M. A. Seo, S. M. Koo, H. R. Park, H. S. Kim, B. J. Kim, H. T. Kim, N. K. Park, D. S. Kim, K. J. Ahn, Phys. Status Solidi C 8 (2011) 1227-1230.

[29] A. Ritter, Smart Materials in Architecture, Interior Architecture and Design. Basel, Switzerland: Birkhäuser (2007).

[30] C.M. Lampert, Materials Today, 7 (2004) 28-35.

[31] J. Bell, M. Schwartz, Smart Materials. Boca Raton, Florida: CRC Press (2009).

[32] http://en.wikipedia.org/wiki/Smart_glass (Date: 06/27/2013: 1:10 pm).

[33] S.K. Deb, Appl. Opt. Supp. 1 (1969) 192-195.

[34] C.G. Granqvist, E. Avendaño, A. Azens, Thin Solid Films, 442 (2003) 201-211.

[35] P. R. Somani, S. Radhakrishnan, Mat. Chem. Phys. 77 (2002) 117-133.

[36] M.S.R. Khan, K.A. Khan, W. Estrada, C. G. Granqvist, J. Appl. Phys. 69 (1991) 3231-3234.

[37] S. F. Cogan, N. M. Nguyen, S. J. Perrotti, R. D. Rauh, J. Appl. Phys. 66 (1989) 1333-1337.

[38] A. Talledo, A. M. Andersson, C. G. Granqvist, J. Mat. Res. 5 (1990) 1253-1256.

[39] A. M. Andersson, C. G. Granqvist, J. R. Stevens, Appl. Opt. 28 (1989) 3295-3302.

[40] A. Talledo, C.G. Granqvist, J. Appl. Phys. 77 (1995) 4655-4666.

[41] D.W. ullett, J. Phys. C: Solid State Physics, 13 (1980) L595-L599.

[42] M.A. Sobhan, R. T. Kivaisi, B. Stjerna, C. G. Granqvist, Solar Energy Mat. Solar Cells, 44 (1996) 451-455.

[43] C.G. Granqvist, Thin Solid Films, 193-194 (1990) 730-741.

[44] C. G. Granqvist, Solar Energy Mat. Adv. Mat. 15 (2003) 1789-1803.

[45] J. Art, Photon Detectors for Confocal Microscopy, in: Pawley, J.B. (Ed.) Handbook of Biological Confocal Microscopy, Springer Science + Business Media, LLC, New York, 2006.
[46] A. Rogalski, Infrared detectors. Boca Raton: CRC Press (2011).
[47] A. Rogalski, Infrared detectors. Amsterdam: Gordon and Breach (2000).
[48] E.L. Dereniak, G.D. Boreman, Infrared Detectors and Systems. New York: Wiley (1996).
[49] A. Rogalski, A., Quantum Well Infrared Photoconductors in Infrared Detectors Technology, in: Ryzhii, V. (Ed.) Intersubband Infrared Photodetectors, World Scientific New Jersey, pp. 1-66 (2003).
[50] S.C. Stotlar, Infrared detector, in: Waynant, R. & Ediger, M. (Eds.) Electro-Optics Handbook McGraw-Hill, New York, 200, pp. 17.11-17.24.
[51] B.E.A. Saleh, M. C. Teich, Fundamentals of Photonics. New Jersey: Wiley-Interscience (2007).
[52] R. Breiter, M. Münzber, IR modules and devices for security applications, in: Beyerer, J. (Ed.) Future Security: 2nd Security Research Conference, Germany, pp. 96-99 (2007).
[53] G. Hyseni, N. Caka, K. Hyseni, Infrared Thermal Detectors Parameters: Semiconductor Bolometers Versus Pyroelectrics. Wseas Transactions on Circuits and Systems, 9, 238-247 (2010).
[54] R.B. Darling, S. Iwanaga, Sadhana, 34 (2009) 531-542.
[55] S. P. Langley, The Bolometer. Nature 25 (1881), 14-16.
[56] http://photoblog.nbcnews.com/_news/2013/04/20/17841024-infrared-police-chopper-images-show-boston-marathon-suspect-hiding-in-boat?lite (Date: 07/28/2013: 3:05 pm).
[57] R. K. Bhan, R. S. Saxena, C. R. Jalwania, S. K. Lomash, Defence Science Journal, 59 (2009) 580-589.
[58] F. Niklaus, C. Vieider, H. Jakobsen, MEMS-based uncooled infrared bolometer arrays: a review. Proc. SPIE, 6836 (2008) 68360D (15 pages).
[59] R. Walser, Metamaterials: An Introduction, in: Weiglhofer, W.S. & Lakhtakia, A. (Eds.) Introduction to Complex Mediums for Optics and Electromagnetics, SPIE-The International Society for Optical Engineering, Bellingham, Washington, pp. 295-316 (2003).
[60] B.A. Munk, Metamaterials : critique and alternatives. Hoboken, New Jersey: John Wiley & Sons, Inc. (2009).
[61] J.B. Pendry, hys. Rev. Lett. 85 (2000) 3966-3969.
[62] M.C.K. Wiltshire, Science 292 (2001) 60-61.
[63] D.R. Smith, W. J. Padilla, D. C. Vier, S. C. N. -Nasser, S. Schultz, Phys. Rev. Lett. 84 (2000) 4184-4187.
[64] A. Cho, Science 314 (2006) 403.
[65] J.-P. Huang, K. W. Yu, New Nonlinear Optical Materials: Theoretical Research. New York: Nova Science Publishers, Inc. (2013)
[66] C. M. Soukoulis, Optics & Photonics News, 17, 16-21 (2006).
[67] D. Schurig, J. J. Mock, B.J., Justice, S. A. Cummer, J. B. Pendry, A. F. Starr, D. R. Smith, Science 314 (2006) 977-980.

[68] T. Driscoll, D. N. Basov, A. F. Starr, P. M. Rye, S. N. -Nasser, D. Schurig, D. R. Smith, Appl. Phys. Lett. 88 (2006) 081101 (3 pages).
[69] S. Linden, C. Enkrich, M. Wegener, J. Zhou, T. Koschny, C. M. Soukoulis, Science 306 (2004) 1351-1353.
[70] C.M. Soukoulis, S. Linden, M. Wegener, Science 315 (2007) 47-49.
[71] J. B. Pendry, A. J. Holden, D. J. Robbins, W. J. Stewart, IEEE Transactions on Microwave Theory and Techniques 47 (1999) 2075-2084.
[72] I.V. Shadrivov, S.K. Morrison, Y.S. Kivshar, Opt. Exp. 14 (2006) 9344-9349.
[73] P.V. Kapitanova, S.I. Maslovski, I.V. Shadrivov, P.M. Voroshilov, D.S. Filonov, P.A. Belov, Y.S. Kivshar, Appl. Phys. Lett. 99 (2011) 251914 (3 pages).
[74] I.V. Shadrivov, A.B. Kozyrev, D.W.V.D. Weide, Y. S. Kivshar, Opt. Exp. 16 (2008) 20266-20271.
[75] I.V. Shadrivov, A.B. Kozyrev, D.W.V.D. Weide, Y. S. Kivshar, Appl. Phys. Lett. 93 (2008) 161903 (3 pages).
[76] Y. Wang, J. Yin, G. Yuan, X. Dong, C. Du, Appl. Phys. A, 104 (2011) 1243-1247.
[77] D. Shrekenhamer, S. Rout, A. C. Strikwerda, C. Bingham, R. D. Averitt, S. Sonkusale, W. J. Padilla, Opt. Exp. 19 (2011) 9968-9975.
[78] Z. Wang, Y. Luo, L. Peng, J. Huangfu, T. Jiang, D. Wang, H. Chen, L. Ran, Appl. Phys. Lett. 94 (2009) 134102 (3 pages).
[79] H.-T. Chen, J. F. O'Hara, A. K. Azad, A. J. Taylor, R. D. Averitt, D. B. Shrekenhamer, W. J. Padilla, Nature Photonics 2 (2008) 295 - 298.
[80] D. Mittleman, Nature Photonics 2 (2008) 267 - 268.
[81] Q. Zhao, L. Kang, B. Du, B. Li, J. Zhou, H. Tang, X. Liang, B. Zhang, Appl. Phys. Lett. 90 (2007) 011112 (3 pages).
[82] D.H. Werner, D.-H. Kwon, I.-C. Khoo, A.V. Kildishev V. M. Shalaev, Opt. Exp. 15 (2007) 3342-3347.
[83] R. Pratibha, K. Park, I.I. Smalyukh, W. Park, Opt. Exp. 17 (2009) 19459-19469
[84] Y. Poo, R.-X. Wu, G.-H. He, P. Chen, J. Xu, R. -F. Chen, Appl. Phys. Lett. 96 (2010) 161902 (3 pages).
[85] G. He, R. -X. Wu, Y. Poo, P. Chen, J. Appl. Phys. 107 (2010) 093522 (5 pages).
[86] J. Han, A. Lakhtakia, C. -W. Qiu, Opt. Exp. 16 (2008) 14390-14396.
[87] M. Lapine, D. Powell, M. Gorkunov, I. Shadrivov, R. Marqués, Y. Kivshar, Appl. Phys. Lett. 95 (2009) 084105 (3 pages).
[88] M. Lapine, I. V. Shadrivov, D. A. Powell, Y. S. Kivshar, Nature Materials 11 (2012) 30-33.
[89] W.M. Zhu, A. Q. Liu, X. M. Zhang, D. P. Tsai, T. Bourouina, J. H. Teng, X. H. Zhang, H. C. Guo, H. Tanoto, T. Mei, G. Q. Lo, D. L. Kwong, Adv. Mat. 23 (2011) 1792-1796.
[90] Y.H. Fu, A. Q. Liu, W. M. Zhu, X. M. Zhang, D.P. Tsai, J. B. Zhang, T. Mei, J. F. Tao, H. C. Guo, X. H. Zhang, J. H. Teng, N. I. Zheludev, G. Q. Lo, D. L. Kwong, Adv. Func. Mat. 21 (2011) 3589-3594.
[91] D.A. Powell, K. Hannam, I.V. Shadrivov, Y.S. Kivshar, *Phys. Rev. B* 83 (2011) 235420 (6 pages).
[92] H. Liu, Y. M. Liu, T. Li, S. M. Wang, S.N. Zhu, X. Zhang, *Phys. Status Solidi* (b) 246 (2009) 1397-1406.

[93] F. Hesmer, E. Tatartschuk, O. Zhuromskyy, A. A. Radkovskaya, M. Shamonin, T. Hao, C. J. Stevens, G. Faulkner, D. J. Edwards, E. Shamonina, Phys. Status Solidi (b) 244 (2007) 1170-1175.
[94] J.Y. Ou, E. Plum, L. Jiang, N. I. Zheludev, Nano Letters 11 (2011) 2142-2144.
[95] H. Tao, A. C. Strikwerda, K. Fan, W. J. Padilla, X. Zhang, R. D. Averitt, Phys. Rev. Lett. 103 (2009) 147401 (4 pages).
[96] X.G. Peralta, M.C. Wanke, C.L. Arrington, J.D. Williams, I. Brener, A. Strikwerda, R. D. Averitt, W. J. Padilla, E. Smirnova, A. J. Taylor, J. F. O'Hara, Appl. Phys. Lett. 94 (2009) 161113 (3 pages).
[97] H. Tao, A. C. Strikwerda, K. Fan, C. M. Bingham, W. J. Padilla, X. Zhang, R. D. Averitt, J. Phys. D: Appl. Phys. 41 (2008) 232004 (5 pages).
[98] I.E. Khodasevych, C.M. Shah, S. Sriram, M. Bhaskaran, W. Withayachumnankul, B.S.Y. Ung, H. Lin, W.S.T. Rowe, D. Abbott, A. Mitchell, Appl. Phys. Lett. 100 (2012) (3 pages).
[99] A. Cavalleri, T. Dekorsy, H.H.W. Chong, J. C. Kieffer, R. W. Schoenlein, Phys. Rev. B 70 (2004) 161102(R) (4 pages).
[100] Q.-Y. Wen, H.-W. Zhang, Q.-H. Yang, Y.-S. Xie, K. Chen, Y.-L. Liu, Appl. Phys. Lett. 97 (2010) 021111 (3 pages).
[101] L.V. Panina, A. N. Grigorenko, D. P. Makhnovskiy, Phys. Rev. B 66 (2002) 155411 (17 pages).
[102] A. N. Lagarkov, A. K. Sarychev, Phys. Rev. B 53 (1996) 6318-6336.
[103] V.A. Podolskiy, A.K. Sarychev, E.E. Narimanov, V.M. Shalaev, J. Opt. A: Pure and Appl. Opt. 7 (2005) S32-S37.
[104] L. Fu, H. Schweizer, H. Guo, N. Liu, H. Giessen, Phys. Rev. B 78 (2008) 115110 (9 pages).
[105] G. Dolling, C. Enkrich, M. Wegener, J. F. Zhou, C. M. Soukoulis, S. Linden, Opt. Lett. 30 (2005) 3198-3200.
[106] M. Seo, J. Kyoung, H. Park, S. Koo, H.-S. Kim, H. Bernien, B. J. Kim, J. H. Choe, Y. H. Ahn, H. -T. Kim, N. Park, Q. -H. Park, K. Ahn, D.- S. Kim, Nano Lett. 10 (2010) 2064-2068.
[107] M. Liu, H. Y. Hwang, H. Tao, A. C. Strikwerda, K. Fan, G. R. Keiser, A. J. Sternbach, K. G. West, S. Kittiwatanakul, J. Lu, S. A. Wolf, F. G. Omenetto, X. Zhang, K. A. Nelson, R. D. Averitt, Nature 487 (2012) 345.
[108] T. Paik, S. -H. Hong, T. Gordon, A. Gaulding, C. Kagan, C. Murray, (2013) http://meetings.aps.org/link/BAPS.2013.MAR.C19.13.
[109] Q.-Y. Wen, H.-W. Zhang, Q.-H. Yang, Z. Chen, Y. Long, Y.-L. Jing, Y. Lin, P. -X. Zhang, J. Phys. D: Appl. Phys. 45 (2012) 235106-235110.
[110] W.-X. Huang, X.-G. Yin, C.-P. Huang, Q.-J. Wang, T.-F. Miao, Y.-Y. Zhu, Appl. Phys. Lett. 96 (2010) 261908 (3 pages).
[111] K. Appavoo, F. Richard, J. Haglund, Nano Letters 11 (2011) 1025-1031.
[112] M.A. Kats, R. Blanchard, S. Zhang, P. Genevet, C. Ko, S. Ramanathan, F. Capasso, Vanadium dioxide as a natural disordered metamaterial: perfect thermal emission and large broadband negative differential thermal emittance. arXiv:1305.0033 [physics.optics] (2013).

ISBN: 978-1-63117-404-9

Chapter 9

SPECTROSCOPIC TECHNIQUES FOR ATMOSPHERIC SENSING

***Bomidi Lakshmi Madhavan**[*1]* **and Yonghua Wu**[2]*
[1]Leibniz Institute for Tropospheric Research (TROPOS), Leipzig, Germany
[2]NOAA CREST Center at the City College of New York (CUNY), New York, US

A wide variety of spectroscopic techniques in remote sensing and in-situ measurements are used in understanding the atmospheric composition along with its susceptibility to changes as a result of natural and anthropogenic forcings. Improvements in laboratory spectroscopic measurements are essential to meet the needs of these endeavours, particularly in light of the increasing demands of current sensing techniques with respect to increased accuracy and precision required to address various atmospheric science issues. This chapter provides a review of various laboratory molecular spectroscopic techniques in application to probe the atmospheric processes. Though there are many spectroscopic techniques related to atmospheric sensing that are common to all wavelengths, we focus on specific wavelength regions due to instrumentation limitation, concentrating on spectroscopic principles. In particular, we present the most notable techniques like DOAS, FTIR, and lidar spectroscopy along with other existing and promising ones.

1. INTRODUCTION

The electromagnetic (EM) radiation in the Earth-atmosphere system is crucial to understand the atmospheric energetics, including Earth's radiation budget. Historically, the EM radiation includes gamma rays (< 0.00002 μm), X-rays (10^{-5}-10^{-2} μm), ultra-violet (UV; 10^{-2}-0.4 μm), visible (VIS; 0.39-0.75 μm), infrared (IR; 0.75-10^{3} μm), microwaves (10^{3}-10^{6} μm), and radio-waves (10^{3}-10^{14} μm), extending between different frequency or wavelength intervals. It is well known that this radiation propagates in the form of harmonic waves, represented by either wavelength or frequency to distinguish between the various forms of radiation. From spectroscopic point of view, these regions may be associated to different

[*] Email: blmadhavan@gmail.com, madhavan.bomidi@tropos.de.

properties and to different motions of molecules, but the different wavelengths and frequencies comprising various forms of EM radiation are fundamentally similar in that they all travel at the same speed equal to the speed of light (~ 3×10^8 m s^{-1}). The energy of an electromagnetic wave is directly proportional to its frequency and inversely proportional to its wavelength. A *spectrum* is thus defined as the intensity of radiation received by a detector at each wavelength or frequency, which can be shown on a photographic plate or a graph of intensity versus frequency. In general, any spectrum depends on the characteristics of a particular light source and on the nature of the matter between the source and the detector. Instruments like spectrometers use prisms or other devices to split radiation into various wavelengths. A spectrum of the atmosphere can be recorded by an instrument on the ground or an instrument carried by a satellite, balloon, or aircraft. The types and amount of molecules present in the atmosphere can be determined at the time and place where the spectrum was recorded. This is certainly not an easy task and one has to examine the laboratory measurements (act as a *reference)* of the spectrum for each molecule to compare with the observed spectrum of the atmosphere, which is a mix of various molecules. In order to understand the composition of our atmosphere, the most widely used tool is *spectroscopy.*

Spectroscopic studies of the Earth's atmosphere started sometime in the 19th century have contributed largely to the understanding of atmospheric processes; especially they have played an important role in recognizing the importance of atmospheric ozone. The detection of the *ozone hole* in 1985 not only created public awareness about the earth's atmosphere, but also proved that our environment is continuously being affected by various anthropogenic activities on large-scales. The propagation of radiation in the atmosphere is a very complex process, governed by several interaction processes of radiation with matter as listed below:

- **Absorption** (e.g., by trace gases, remove photons from the atmosphere and converts their energy into some other form of energy i.e., heat).
- **Elastic scattering** (i.e., Rayleigh and Mie scattering, possible in any given direction changing the direction of propagation of the photon).
- **Inelastic scattering** (i.e., Raman scattering, changes the direction of photon and its energy).
- **Thermal emission** - from molecules and aerosols.

The electromagnetic radiation is represented as a continuum of wavelengths (Figure 1), from long radio waves to short x-rays and gamma rays. Although a range of electromagnetic wavelengths to which Earth's atmosphere is largely or partially transparent, most spectral regions are affected to some extent by absorption in the atmosphere. Note that the visible portion (detectable with our eyes) constitutes a very small portion of the electromagnetic spectrum. While the optical and radio windows are nearly transparent ranges, there also exist a narrow partial infrared window. Radio light was the first window after the optical one to be used for astronomy because certain portions of the radio spectral region can pass through the atmosphere and be detected from cosmic sources.

Molecules in the gaseous phase absorb or emit radiation mainly at frequencies from the microwave to the visible. Each molecule has a unique spectrum ascertained by the type and arrangement of atoms in the molecule. Since molecules are merely building blocks for substances, the spectrum of a molecule can identify the type of substance to which it belongs.

The total energy of a molecule can be seen as the sum of the kinetic, electronic (visible/ultraviolet), vibrational (thermal/near-infrared), and rotational (microwave/far-infrared) energies. Vibrational transitions (higher energy) are usually followed by rotational transitions (lower energy) and thus we see groups of lines that comprise a vibration-rotation band.

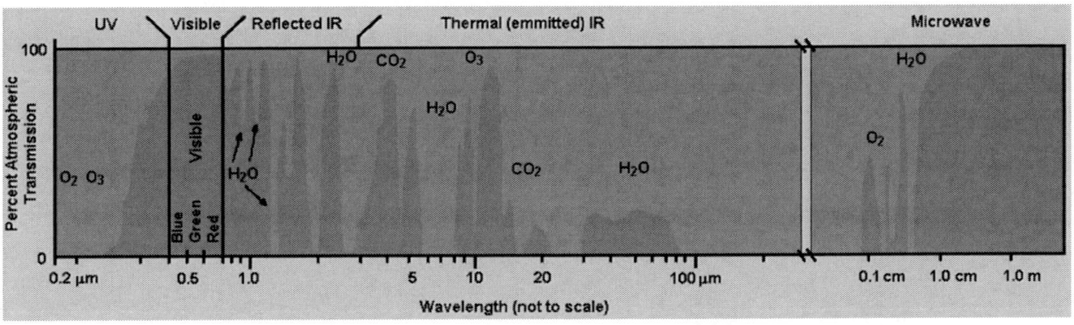

Figure 1. The electromagnetic spectrum showing the ultraviolet-to-microwave wavelength range and the atmospheric transmission of various wavelengths (note the "windows" or areas of high transmission, in the visible and microwave regions). Chemical notation (CO_2, O_3 etc.) indicate the gas responsible for blocking sunlight at the representative wavelength.

All objects whose temperatures are greater than absolute zero tend to emit electromagnetic radiation. This electromagnetic radiation is commonly called as *thermal emission*. The temperature of the radiating body determines the intensity and characteristic of the radiation it emits. The relationship between a radiating body's temperature and the radiation it emits is described by: (a) Stefan-Boltzmann's law (i.e., hotter objects emit more total energy per unit area than colder objects), and (b) Wein's displacement law (i.e., the hotter the radiating body, the shorter the wavelength of maximum radiation). To visualize this, we see our Sun as yellow hot (~ 6000 K) object emitting most of its energy in the visible portion of the electromagnetic spectrum (around 0.5 µm). Basing on the Earth's temperature (~ 300 K), most of Earth's emitted radiation (from land, ocean and clouds) is in the longer thermal IR portion of the electromagnetic spectrum (around 10 µm). Though our eyes are not able to detect this radiation, satellite sensors can measure this and from the amount of emitted radiation will be able to infer the temperature of the surface (e.g., land, ocean) and atmosphere. Both the Sun and Earth emit energy in the microwave portion of the electromagnetic spectrum. Since our Earth is colder than Sun, more radiation will arrive at the satellite directly from Earth than from solar emissions reflected by Earth's surface. Different natural bodies and surfaces emit radiation differently based on their chemical composition and structure, even if they are at the same temperature. The emissivity (i.e., the rate at which a natural body emits radiation relative to the rate of a perfect emitter or black body) of a body or surface is independent of temperature and is a physical property of it. As the emissivity is different for different bodies and surfaces, they can be distinguishable from space. By applying the Rayleigh-Jeans approximation, the microwave radiation emitted by a surface can be expressed as a brightness temperature ($T_B = eT_s$, where T_B is the brightness temperature of a black body at the same temperature as the surface temperature, T_s, and e is the emissivity of the surface). The ability to use brightness temperatures at microwave wavelengths for distinguishing ice from water provides greater scope to probe the single/mixed phase clouds in the atmosphere.

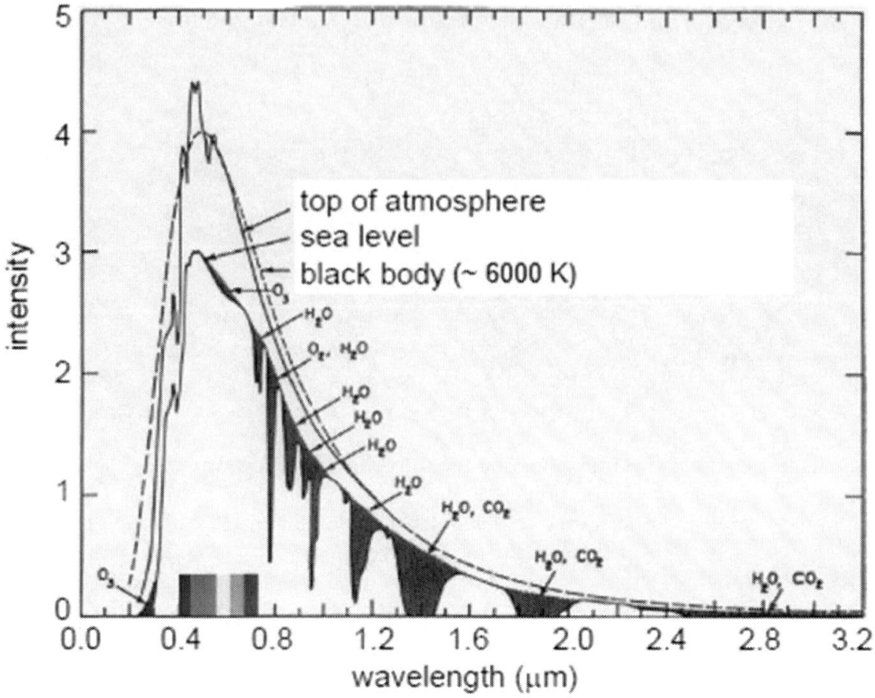

Figure 2. Solar spectrum at the top of the atmosphere and at the Earth's surface [1].

Absorption of visible and near-infrared in the gaseous atmosphere is primarily due to H_2O, O_3 and CO_2 (Figure 2). Water vapor absorbs strongly from the microwave through the visible and is a significant contributor to the radiation budget of the atmosphere, convective transport and atmospheric photochemistry. As a result, accurate spectroscopic parameters for water are important. However, the problems like the low pressures at the lower atmospheric temperatures, the high centrifugal distortion values which confuse the calculations for line strengths, the difficulty in determining the quantity of the water in the gas cell, and the non-Lorenz line shapes that give a continuum opacity over a broad spectrum, have prevented the spectroscopy of water vapor to be represented with sufficient accuracy to address the scientific questions being asked about it.

The infrared techniques for measuring trace gases in the air have been evolving for many years for almost more than three decades. Air pollution research mostly relies on these measurements for mixing ratios as low as 10^{-10}. The infrared method has its success in the study of the upper atmosphere using balloon-borne spectrometers aimed at the sun. In the upper atmosphere, measurements have less chance for interference from water vapor. However, at ground level, the water vapor absorbs nearly everywhere in the spectrum and thus it is necessary to detect the trace gases by the small perturbations they make on the water vapor spectrum. The Fourier transform (FT) spectrometer has been responsible for recent progress in terms of obtaining superb spectra with no influence signal fluctuations (caused by air turbulence) introducing noise into the spectrum.

To simulate the changes in the solar radiation due to atmospheric scattering and absorption processes, radiative transfer models, also referred to as *forward models*, are widely used in atmospheric research community. These models play an important part in any retrieval process as they provide the synthetic radiances that can be measured by the sensor for a specific state of the atmosphere. The main characteristic of the radiation field in the atmosphere is the radiance (I), also referred to as '*intensity*', which is defined as the flux of energy in a given direction per unit time per unit wavelength range per unit solid angle per unit area perpendicular to a given direction [2]. All interactions between the radiation and the atmosphere are classified by the sign of the change in the radiation intensity as a result of the interaction. Processes which reduce the intensity in the direction under consideration by absorption and scattering processes from the original direction into other directions are commonly termed as *extinction*. Processes which increase the intensity in the direction under consideration, for example by scattering into the beam from other directions, by thermal or other emission processes within the volume, are referred to as *source function*.

The general form of the radiative transfer equation describing all processes affecting the radiation field as a result of its interaction with a medium, taking energy conservation into account is given by: $\frac{dI}{ds} = -\alpha(I - J)$, where I is radiance (or intensity) in a given direction, s is the path length, α is the extinction coefficient describing the fraction of the energy which is removed from the original beam by absorption and scattering, and J is the source function which describes the increase of radiance (I) in the original direction due to scattering and/or emission.

If the amount of light traveling in a certain direction through the atmosphere can only be increased due to the scattering processes, the source function depends on the intensity falling on the elementary volume from all directions: $J = \frac{\omega}{4\pi} \int p(\varphi) I d\Omega$, with φ being the scattering angle, i.e., the angle between the directions of the incident and scattered radiation, and ω is the single scattering albedo representing the probability that a photon, which interacts with a volume element, will be scattered rather than being absorbed. The term $\frac{p(\varphi)d\Omega}{4\pi}$ denotes the probability that the radiation is scattered into a solid angle dΩ about a direction forming an angle φ with the direction of the incident radiation. The quantity p(φ) is called the *phase function*.

The total radiation field can be split into two components, $I = I_{dir} + I_{dif}$ with I_{dir} being the direct radiation, which is never scattered in the atmosphere or reflected from the Earth's surface, and the diffuse radiation I_{dif}, which is scattered or reflected at least once. Since there is no relevant process in the atmosphere which increases the intensity of the direct solar radiation, the radiative transfer equation for the direct radiation leads to the homogeneous differential equation $\frac{dI_{dir}}{ds} = -\alpha I_{dir}$ with a solution described by Lambert-Beer's law $I_{dir} = I_0 \exp\left(-\int \alpha(s) ds\right)$, where I_0 being the incident radiance at the top of the atmosphere, which is described by the solar irradiance multiplied by the Dirac delta-function [3]. The integral $\int \alpha(s)ds$ along the photon path defines the optical depth τ(s). Integration is performed along the direct solar beam from the surface to the top-of-atmosphere. If the diffuse radiation is not negligible, it has to be considered in addition to the direct one. The corresponding radiative transfer equation for diffuse component using the scattering source function can be written as:

$$\frac{dI_{dif}}{ds} = -\alpha\left[I_{dif} - \frac{\omega}{4\pi}\int p(\varphi)I_{dif}d\Omega - \frac{\omega}{4\pi}p(\varphi_0)I_{irr}\exp\left(-\int \alpha(s)ds\right)\right]$$

where φ_0 denotes the scattering angle between the direct solar beam and the direction of observation, and I_{irr} is the extraterrestrial solar irradiance. Various standard methods to solve the above equation are presented in [4] or [2].

Light is scattered by atoms and molecules, as well as by various types of aerosols and clouds. Molecular scattering cross-sections are characterized by the Rayleigh law (λ^{-4}), whereas aerosol scattering typically shows a much less pronounced dependence on wavelength with about λ^{-1} (Mie scattering). For cloud-free scenes, molecular scattering usually dominates in the UV spectral range, while aerosols become more important in the VIS and NIR (Near-IR) range. The molecular scattering consists of two parts: the elastic Rayleigh component which accounts for about 96% of scattering events, and 4% inelastic Rayleigh component which is responsible for the *Ring* effect, the 'filling in' of solar Fraunhofer lines in the registered earth-shine spectra [5]. Depending on the scientific application, several radiative transfer models are available to simulate and interpret the atmospheric spectra (for various measurement geometries) obtained from microwave through ultraviolet.

Remote sensing and in-situ measurements play an important role in developing an understanding of the physics and chemistry of the Earth's atmosphere along with its susceptibility to changes as a result of natural and anthropogenic forcings. Improvements in laboratory spectroscopic measurements are essential to meet the needs of these endeavors, particularly in light of the increasing demands of current sensing techniques with respect to increased accuracy and precision required to address various atmospheric science issues. This chapter provides a detailed view on various laboratory molecular spectroscopic techniques to probe atmospheric processes and define the needs of the atmospheric sensing community. Though there are many spectroscopic issues related to atmospheric sensing that are common to all wavelengths, we focus on specific wavelength regions, due to instrumentation limitation, concentrating on spectroscopic aspects.

2. Spectroscopic Techniques in Satellite Remote Sensing

Most remote sensing instruments on aircraft or space-based platforms operate in one or more of atmospheric windows by making their measurements with detectors tuned to specific frequencies (or wavelengths) that pass through the atmosphere. When a remote sensing instrument has a line-of-sight with an object that is reflecting sunlight or emitting heat, the instrument collects and records the radiant energy.

Ultraviolet and visible satellite remote sensing spectroscopic analysis of electromagnetic radiation allows retrieving information on the processes that control its release and transfer. Spectroscopy has been a key method of remote sensing. About 200 years ago Joseph von Fraunhofer (1787 – 1826) was the first to be able to produce spectroscopic gratings of so far unrivalled quality. With these gratings, he found numerous strong and weak vertical lines, which are darker than the remaining part of the spectrum while some of them are almost weak. He was able to determine the exact wavelength of many '**Fraunhofer lines**' as we call them today. The discovery of the physical principles of absorption and emission in 1859 with the collaboration of the physicist Gustav R. Kirchhoff (1824-1887) and the chemist Robert W. Bunsen (1811-1899) resulted in assigning specific emission lines to different elements. These discoveries constituted the foundation for two basic kinds of spectroscopy: spectroscopy by absorption and emission.

Spectroscopic techniques were applied to the earth's atmosphere, but not focused on atmospheric absorption. It was possible to identify excited atoms and molecules as sources of the observed radiation. The spectroscopy of atmospheric absorption using Sun light is complicated by the fact that most atmospheric absorption structures are usually by far weaker than the solar Fraunhofer lines. Thus, initial absorption spectroscopy of the earth's atmosphere was focused only on strong absorptions like those of ozone (O_3), which can be identifiable clearly in absorption spectra of solar light. In 1925, Dobson developed a new stable photo-spectrometer (double monochromator using quartz prisms) for the quantification of the atmospheric vertical column density of O_3. The thickness of atmospheric O_3 layer is expressed in Dobson units (DU), which can be defined as the thickness of the atmospheric O_3 vertical column density under normal conditions. Dobson spectrometers use a very simple and stable spectroscopic method: the direct or scattered solar intensity is measured in different narrow (~ 1 nm) spectral intervals which are located either in or outside of O_3 (Huggins) absorption bands. From the ratio of the radiation intensity of such wavelength pairs (and a geometric correction factor for the solar zenith angle) the vertical column density of O_3 is determined. A possible influence on the intensity ratio due to the absorption of atmospheric aerosols can be accounted for by the combination of several wavelength pairs, for which the influence of O_3 and aerosol is different. Following this method, Alan Brewer in 1973 measured the atmospheric NO_2 column density from ground [6]. He measured light in different narrow wavelength intervals around 450 nm, for which the NO_2 absorption showed strong differential structures. Although the atmospheric NO_2 absorptions are by far smaller than those of O_3, it was possible to clearly identify these absorptions with this new spectrometer and confirm the permanent existence of stratospheric NO_2.

A major step towards measuring weak atmospheric absorbers was introduced by J.F. Noxon [7] in 1975 when he first measured the solar spectrum over a continuous interval in the spectral region where the atmospheric absorption appears (440 nm for NO_2).Then the strong structures of the solar Fraunhofer lines were removed by dividing a measured spectrum (during sunset or sunrise) by another spectrum measured during noon. Due to longer atmospheric light path at low sun this ratio contains strong absorption structures (the difference between twilight and noon) of the atmospheric absorber. First, the Fraunhofer lines are removed by dividing the measurements by a spectrum of direct sun light. The logarithm of this ratio is simultaneously fitted by the absorption cross-sections of the trace gases which show differential absorption structures in the selected spectral range. Also included in the fitting procedure is the so called '*Ring spectrum*', which corrects the filling-in of solar-

Fraunhofer lines by Raman scattering [5]. Usually, the logarithm of the sun light spectrum is included in the fitting procedure.

In 1976, U. Platt and D. Penner introduced the DOAS (Differential Optical Absorption Spectroscopy) method [8-10]. The key concept of DOAS is the simultaneous fit of several trace gas absorption spectra to a measured atmospheric spectrum [11], which is the best possible way to avoid spectral interference between different absorption structures. Several improvements to this concept were added (see also 12).

Ultraviolet and visible measurements of the atmosphere are often made in the following three basic measurement geometries:

a). Nadir – measure the backscattered light in geometry when the line of sight intersects the Earth's surface. Here, the light source is solar radiation that is backscattered from a combination of surface reflectance and aerosol, cloud and Rayleigh scattering.
b). Occultation – direct illumination by the Sun acts as the light source (or, less frequently, the Moon or stars).
c). Limb scattering – Rayleigh scattering of Sun light in the limb geometry.

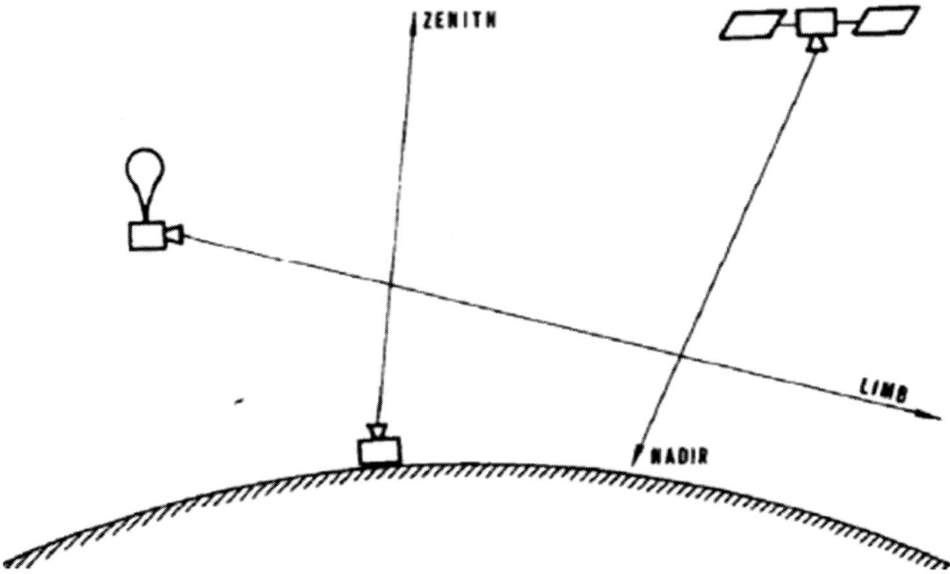

Figure 3. Geometry of observations.

A large number of observations (both in space and spectral domain) improve the conditioning of the inversion problem. In nadir sounding, the good conditioning can be exploited for attaining 1 km vertical resolution. In limb sounding the good conditioning can be exploited for attaining 50 km horizontal resolution. The vertical and horizontal resolution depends on the geometry of observation. Typical resolution of nadir measurements is 10 km horizontal and 10 km vertical. Typical resolution of limb measurement is 2 km vertical and 700 km horizontal. Polar platforms are usually the preferred choice because of global coverage requirements. Meteorological and now-casting applications in general, require frequent observations that can be obtained with geostationary platforms. Figure 4 shows the geometries of the satellite orbits.

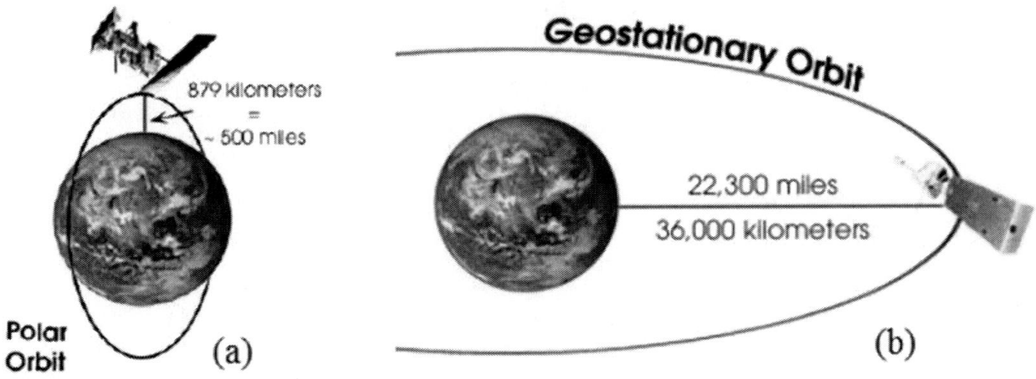

Figure 4. Geometries of satellite orbits: (a) Polar and (b) Geostationary.

The first spectroscopic UV/VIS observations started in 1970 on board of the US research satellite Nimbus 4. These measurements (Backscatter Ultraviolet, BUV, also called Solar BUV, SBUV) operated in nadir geometry. Like the Dobson instruments, the BUV/SBUV instruments measure the intensity in different narrow spectral intervals and thus determine information on the atmospheric O_3 profile since the penetration depth into the atmosphere strongly depends on wavelength. Due to instrumental instabilities, the big breakthrough in UV/VIS satellite remote sensing of atmosphere took place in 1979 with the launch of TOMS (Total Ozone Mapping Spectrometer) on Nimbus-7. TOMS is similar to BUV/SBUV instrument but is more stable and measures light at longer wavelengths. In addition, TOMS is sensitive to the total O_3 column instead of O_3 profile. Further, several TOMS instruments have been launched on other satellites. The first DOAS instrument is operational from space since April 1995. The Global Ozone Monitoring Experiment (GOME) was launched on the European research satellite ERS-2 [13]. Like SBUV and TOMS, GOMS is a nadir-viewing instrument while it covers a large spectral range (240 – 790 nm) at a total of 4096 wavelengths arranged in four channels with a spectral resolution between 0.2 and 0.4 nm. Its normal ground pixel size is 320 × 40 km^2; the global coverage is achieved after three days. For O_3 measurements the intensities at short wavelengths are observed (BUV/SBUV instruments); for the determination of the total atmospheric O_3 column the intensities at larger wavelengths are used (TOMS instruments). In contrast to the limitations of BUV/SBUV and TOMS instruments, GOME spectra yield a surplus of spectral information. By applying the DOAS technique to these measurements it is thus possible to retrieve a large variety of atmospheric trace gases, the majority of which are very week absorbers (O_3, NO_2, BrO, OClO, HCHO, H_2O, O_2, O_4, and SO2). In addition, other quantities like aerosol absorptions, the ground albedo or indices characterizing the solar cycle can be analyzed. Because of the high sensitivity of GOME, it is in particular possible to measure various tropospheric trace gases (NO_2, BrO, HCHO, H_2O, and SO2). Further, the GOME spectra can be analyzed with respect to a spectrum of direct sun light, which contains no atmospheric absorption. Table 1 provides the details of absorption wavelengths in UV/VIS solar spectrum for different atmospheric trace gases.

Table 1. Wavelengths of absorption in the solar spectrum (UV + visible) by several atmospheric gases

Gas	Absorption wavelength (µm)
N_2	< 0.1
O_2	< 0.245
O_3	0.17 – 0.35 0.45 – 0.75
H_2O	< 0.21 0.6 – 0.72
H_2O_2 (Hydrogen peroxide)	< 0.35
NO_2 (Nitrogen oxide)	< 0.6*
N_2O (Nitrous oxide)	< 0.24
NO_3 (Nitrate radical)	0.41 – 0.67
HONO (Nitrous acid)	< 0.4
HNO_3 (Nitric acid)	< 0.33
CH_3Br (Methyl bromide)	< 0.26
$CFCl_3$ (CFC11)	< 0.23
HCHO (Formaldehyde)	0.25 – 0.36

* **NO_2** absorbs at $\lambda < 0.6$ µm, but photo-dissociate at $\lambda < 0.4$ µm.

A second DOAS satellite instrument, SCIAMACHY (SCanning Imaging Absorption SpectroMeter for Atmospheric ChartographY) [14 -16] was launched in March 2002 on board European research satellite ENVISAT. SCIAMACHY measures over a wider wavelength range (240 – 2380 nm) including the absorption of several greenhouse gases (CO_2, CH_4, N_2O) and CO in the infrared. It also operates in additional viewing modes (nadir, limb, occultation), which allows to derive stratospheric trace gas profiles. Additional advantage is that the ground pixel size for the nadir viewing mode was significantly reduced to 30 × 60 km² (in a special mode even to 15 × 30 km²). Especially for the observation of tropospheric trace gases this is very important because of the strong spatial gradients occurring for such species. The first tropospheric results of SCIAMACHY show that it is now possible to identify pollution plumes of individual cities or other big sources. Several additional space borne DOAS instruments are planned for future missions. A large challenge addresses the interpretation of the slant atmospheric column densities derived from the DOAS fit of satellite spectra, which includes the numerical modeling of the atmospheric radiative transport.

2.1. Nadir Trace Gas Retrieval Scheme with Differential Optical Absorption Spectroscopy

Due to the specific properties of a given molecule its absorption spectrum (or emission spectrum) in the UV/VIS spectral range is a characteristic combination of electronic, vibrational and rotational transitions. While in the microwave and infrared region individual lines are resolved by typical instrumentation, typical DOAS instruments are usually not designed to resolve individual lines in the UV/VIS spectral range. For example the full-width-half-maximum (FWHM) of the GOME and SCIAMACHY instruments in the UV range is about 0.2 nm, and thus much larger than the atmospheric widths of the absorbing molecules. Many trace gases are identified by means of their individual absorption structure. The Differential Optical Absorption Spectroscopy (DOAS) technique is successfully adapted to nadir and partly limb measurements from GOME, SCIAMACHY, OMI and GOME-2 [17]. The DOAS approach is based on the following aspects:

- The isolation of high frequency structures of molecular absorbers from broadband scattering features (Rayleigh, Mie) by a high pass filter.
- The separation of spectroscopic retrievals and radiative transger calculations using the Air Mass Factor (AMF) concept.

Given a measured spectrum $I_{abs}(\lambda)$ containing the absorption of the trace gas of interest, a reference spectrum without the absorption of the trace gas of interest $I_{ref}(\lambda)$and the absorption cross-sections $\sigma_i(\lambda)$ of all relevant species, the optical depth can be written according to Lambert-Beer's law.

The basis for early spectroscopic measurements and present quantitative trace gas analytical methods in the atmosphere and the laboratory is *Lambert-Beer's law* (also referred as *Bouguer-Lambert law*) written as:

$$I(\lambda) = I_0(\lambda)\exp(-\sigma(\lambda)cL)$$

Here, $I_0(\lambda)$ denotes the initial intensity of a light beam emitted by a suitable source of radiation, while $I(\lambda)$ is the radiation intensity of the beam after passing through a layer of thickness L, where the absorber is present at a uniform concentration of *c*. The quantity $\sigma(\lambda)$ represents the absorption cross-section at wavelength, and is a characteristic property of any species. The determination of the light path length (L) is essential for active DOAS applications. Once these quantities are known, the average trace gas concentration (*c*) can be calculated from the measured ratio $\frac{I_0(\lambda)}{I(\lambda)}$: $c = \frac{\ln\left(\frac{I_0(\lambda)}{I(\lambda)}\right)}{\sigma(\lambda)L} = \frac{D}{\sigma(\lambda)L}$

The expression $D = \ln\left(\frac{I_0(\lambda)}{I(\lambda)}\right)$ is called the optical density of a layer of a given absorber. Above equation is the basis for most absorption spectroscopic applications in the laboratory where the intensities $I(\lambda)$ and $I_0(\lambda)$ are determined by measurements with and without the absorber in the light beam. However, the application of *Lambert-Beer's law* to open atmosphere is more challenging as the true intensity, $I_0(\lambda)$, as it would be received from the Sun in the absence of any atmospheric absorber is difficult to determine. While this seems to present a dilemma rendering atmospheric absorption spectroscopy useless, the solution lies in measuring the so-called '*differential*' absorption, i.e., the difference between the absorptions at two different wavelengths. This principle was used by Dobson in 1930s to determine the total column of atmospheric ozone. The Dobson spectrometer compares the intensity of direct solar light of two wavelengths (λ_1, λ_2) with different ozone absorption cross-sections given by, $\sigma_1 = \sigma(\lambda_1)$, $\sigma_2 = \sigma(\lambda_2)$ [18].

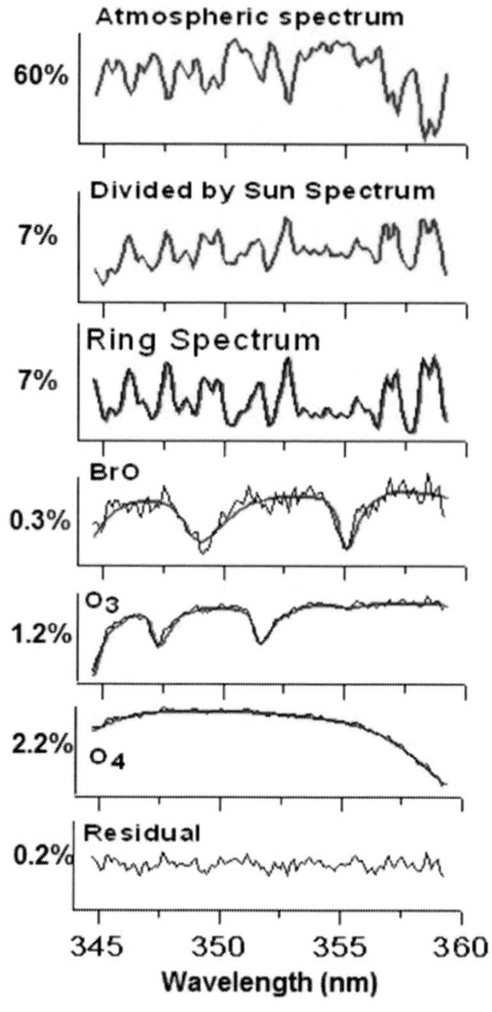

Figure 5. Example of DOAS analysis of spectrum [Reference: http://joseba.mpch-mainz.mpg.de/doas_history.htm]

This spectroscopic technique is used for the detection of various trace gases as listed in Table 1, such as nitrogen dioxide (NO_2), bromine oxide (BrO), ozone (O_3), formaldehyde (HCHO) and nitrous acid (HONO).

3. Spectroscopic Techniques in Lidar Remote Sensing

Lidar ('LIght Detection and Ranging') is an active optical remote sensing technique that measures properties of scattered light to find the range and/or information of a distant target. It uses the same principle as radar except that it uses a laser instead of radio waves. In contrast to other passive remote sensing, lidar has a high vertical and temporal resolution and works in the day- and night-time. The information on the atmospheric composition or thermodynamic state and their physical properties can be derived from the scattering and propagation properties of light by molecules (gases), particles (liquids and solids) and hydrometeors. Lidar has largely contributed to the observations and applications of the Earth's atmosphere during the past decades. It is particularly useful for the investigation of highly variable range-resolved atmospheric parameters. Lidar can measure air density, wind speed and direction, air temperature, the concentrations of ozone, carbon dioxide, and water vapor, and the density, basic shape and size of aerosols, ice crystals, and water droplets, etc. Lidar helps monitor emission rates and concentration levels of trace gases. Meteorological phenomena such as frontal passages, hurricanes, and mountain lee-waves and PBL-dynamics were studied. Lidar contributes to our knowledge of the climatic effects of aerosols. The stratospheric perturbation after major volcanic eruptions has been studied and the intercontinental transport of air pollution, desert dust, and forest-fire smoke has been detected. In the mesosphere, lidar has proven the existence of layers of metallic atoms and ions and of gravity waves therein. Comprehensive reviews of the various existing lidar techniques for atmospheric remote sensing can be found, for example, in ref. [19].

Lidar instruments can operate from the ground or from aircraft, one system has been flown on the Space Shuttle, and in the near future satellite-based lidar instruments will carry out global observations of atmospheric constituents and wind from space. These and many more aspects of lidar are presented in this chapter giving an overview on the state of the art of the basic lidar techniques used in the investigation of the Earth's atmosphere.

3.1. Lidar Measuring Principle and System Prototype

According to the physical process or interactions of laser-beam with atmosphere, the following types of lidar are widely applied to atmosphere such as elastic-scattering lidar, Raman-scattering Lidar, Polarization Lidar, Differential-Absorption Lidar and High Spectral Resolution Lidar [20]. Briefly, the **elastic-scattering lidar** makes use of the Mie-scattering by aerosol-particle and Rayleigh-scattering by molecules in which the scattering wavelength is same as the incident laser wavelength. The elastic-scattering lidar usually has strong scattering intensity thus has strong signal-to-noise ratio. This type of lidar usually is capable to measure aerosol and optically thin clouds (cirrus) from the surface to stratosphere, thus including tropospheric aerosol pollution and volcanic aerosol in the stratosphere. In case of an elastic-scattering (Rayleigh-Mie) lidar system with the assumption of no multiple scattering, the portion of the emitted power that is backscattered and detected by the receiver unit, $P(\lambda_L,r)$, as a function of range, r, is given by the lidar equation:

$$P(\lambda_L,r) = P_0(\lambda_L)O(r)\frac{A}{r^2}C(\lambda_L,r)(\beta_{mol}(\lambda_L,r)+\beta_{aer}(\lambda_L,r))\exp\left[-2\int_0^r \alpha(\lambda_L,r')dr'\right]+P_{bg}$$

$P_0(\lambda_L)$ is the emitted power at laser wavelength λ_L, $O(r)$ is the overlap function, A is the area of the telescope, $C(\lambda_L,r)$ is a system constant accounting for the transmission of the receiver unit, β_{mol} and β_{aer} are the backscatter coefficients of air molecules and aerosols, respectively, $\alpha(\lambda_L,r)$ is the extinction coefficient, and P_{bg} is the background power (straylight or detector noise). To retrieve aerosol backscatter or extinction coefficient, a constant lidar-ratio (e.g., extinction-to-backscatter ratio) and a far-end boundary are usually assumed in the Fernald-Klett inversions [20, 21]. The simultaneous measurements of aerosol optical depth by sunphotometer can be used to constrain the lidar-ratio and thereby reduce the uncertainty of retrievals [22].

Raman lidar uses the Raman-scattering properties of molecules such as water vapor, nitrogen, oxygen and trace gases. Raman-scattering is an inelastic scattering process since the scattering molecule changes it's vibrational and/or rotational energy state and hence changes the wavelength of the scattered photon. The change in wavelength depends on the two involved energy levels and is specific for the scattering molecule. Figure 6 gives the spectrum of Raman backscatter coefficient for a stimulation wavelength of 355 nm. The Raman-scattering cross section or intensity is usually a few orders weaker than the elastic-scattering intensity, thus Raman-lidar often has the small signal-to-noise ratio. It has been used to measure the profiles of water vapor, aerosol, other species and temperature in the field campaigns such as the Atmospheric Radiation Measurement (ARM) [23] for water vapor and aerosols, the Raman Lidar for Meteorological Observations RALMO [24], operated by the Swiss Weather Service for water vapor, aerosol and temperature profiling.

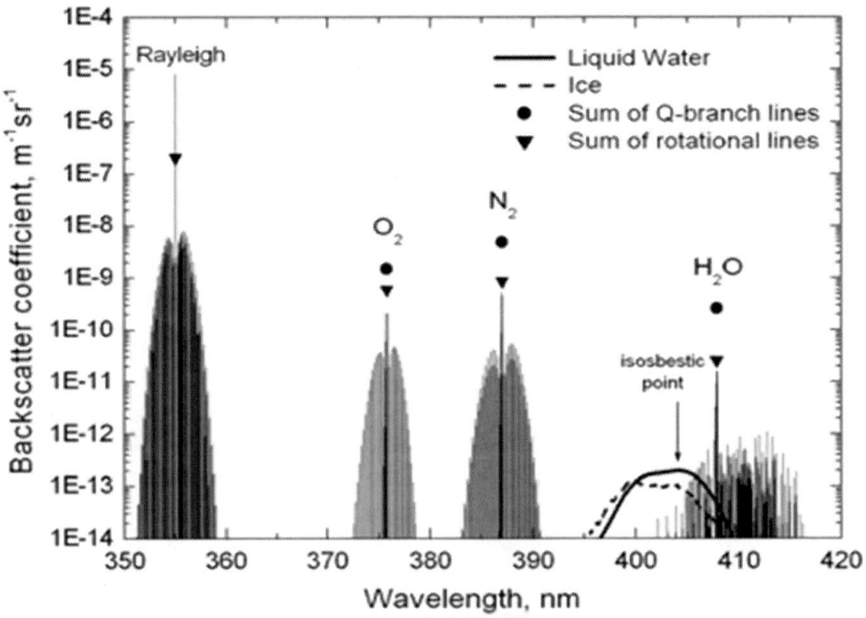

Figure 6. Spectrum of Raman backscatter coefficient for a stimulation wavelength of 355 nm (*source*: http://lidar.tropos.de/en/research/raman.html).

In the case of a Raman lidar system (inelastic scattering) the backscattered power at wavelength λ_X that has been shifted from the laser wavelength due to inelastic Raman scattering by molecular species X is given by the following expression:

$$P(\lambda_X,r)=P_0(\lambda_L)O(r)\frac{A}{r^2}C(\lambda_X,r)\beta(\lambda_X,r)\exp\left[-\int_0^r(\alpha(\lambda_L,r')+\alpha(\lambda_X,r'))dr'\right]+P_{bg}$$

The molecular backscatter coefficient is given by the number density of the scattering medium, N, and by the backscattering cross section σ. Thus, the extinction coefficient of particulates (aerosol or cloud) can be directly derived from the N_2-Raman return:

$$\alpha_p(\lambda_0,z)=\frac{1}{1+\left(\frac{\lambda_N}{\lambda_0}\right)^{-v}}\left[\frac{d}{dz}[\ln\frac{N(z)}{P(\lambda_N,z)z^2}]-\alpha_m(\lambda_0,z)[1+(\frac{\lambda_N}{\lambda_0})^{-4}]\right]$$

Combing the Raman-scattering signals of water vapor and nitrogen, the water vapor mixing ratio q (mass of water vapor to mass of dry air) can be expressed as follows:

$$q=k_q\frac{O_N(r)\sigma(\lambda_N)}{O_H(r)\sigma(\lambda_H)}\Delta\tau(\lambda_N,\lambda_H,r)\frac{P(\lambda_H,r)-P_{bg,H}}{P(\lambda_N,r)-P_{bg,N}}$$

where $\Delta\tau(\lambda_N,\lambda_H,r)=\exp\left[-\int_0^r(\alpha(\lambda_H,r')-\alpha(\lambda_N,r))dr'\right]$.

P(z) is Raman-scattering signal and k_q is the calibration constant related to the lidar system configurations, and can be determined based on a comparison with an external measurement like a radiosonde.

Differential-Absorption Lidar (DIAL) often makes use of the absorption spectrum of trace gas at two adjacent wavelengths, one has the strong absorption (λ_{on}) and other one has the weak absorption (λ_{off}). The ratio of lidar-returns at the two wavelengths can cancel out the aerosol influences, thus obtaining trace gas concentration, such as H_2O, Ozone and industrial emissions for trace gas like SO_2 and NO_2. Figure 7 show the schematic diagram of water vapor spectrum and returns at wavelength λ_{on} and λ_{off}.

Polarization lidar receives the depolarized information from the non-spherical particles such as dust and ice-clouds, so it helps to discriminate the dust aerosol type and cloud phase (ice clouds or water phase clouds). According to the exact Lorenz–Mie theory, spherical particles that are homogeneous in content (with respect to the refractive index) always backscatter linearly polarized electromagnetic radiation in the same (incident) plane of polarization. A variety of approximate scattering theories predict that non-spherical or inhomogeneous particles will introduce a depolarized component into the backscattering. Thus, polarization lidar is unique among remote sensors in that it has the potential to unambiguously identify the thermodynamic phase of clouds. The strength of the depolarization process in non-spherical particles depends on the amount and complexity of the particles' deviation from spherically symmetrical shape, but also on the particle size relative to the wavelength. In addition, the multiple-scattering between droplets in dense water clouds can result in significant depolarization effects, particularly for space-borne lidar with the large laser-beam footprint in clouds. The experiments revealed that the depolarization was related to laser beam size, cloud droplet concentration and size distribution, cloud-height, and field of view of lidar [25].

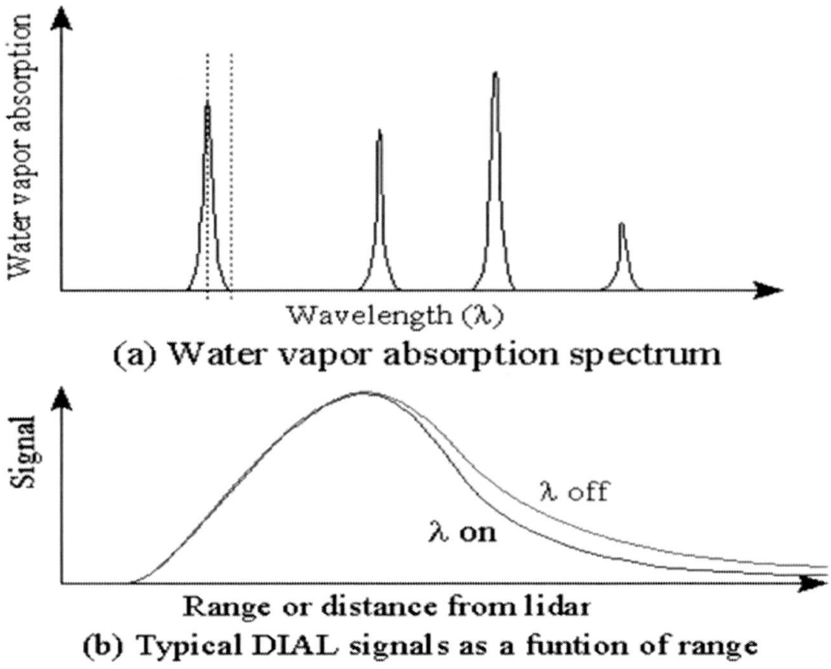

Figure 7. Water vapor absorption spectrum and lidar returns.

Doppler Wind Lidar (DWL) systems fall in two categories: coherent (or heterodyne) detection and direct detection (or incoherent) systems. A number of lidar detection techniques for measuring the atmospheric wind-fields have been developed or proposed over the past several years. These techniques include aerosol and molecular measurements. The aerosol-based wind measurements offer the possibility of high sensitivity measurements in those areas where the aerosol backscattering is high. On the other hand, the molecular-based measurements could provide wind measurements even in those areas where the aerosol concentration is near zero. A DWL is a rugged instrument that meets the requirement of unmanned and unattended operation, with a very high spatio-temporal resolution.

In a lidar system, the emitted radiation of laser beam encounters diffusion by atmospheric particles and molecules along the line of sight. A part of the radiation is scattered backward and collected by an opto-electronic device into the LIDAR reception system. The optical signal is then translated into a voltage over time and distance by multiplication of the speed of light. Therefore, a lidar basically consists of a laser source, generating light pulses at a specific wavelength, of a transceiver to modify the beam diameter and divergence, and of a receiver unit (telescope, spectrometer, acquisition system), to detect the backscattered light as a function of wavelength. Figure 8 gives a schematic diagram of lidar system.

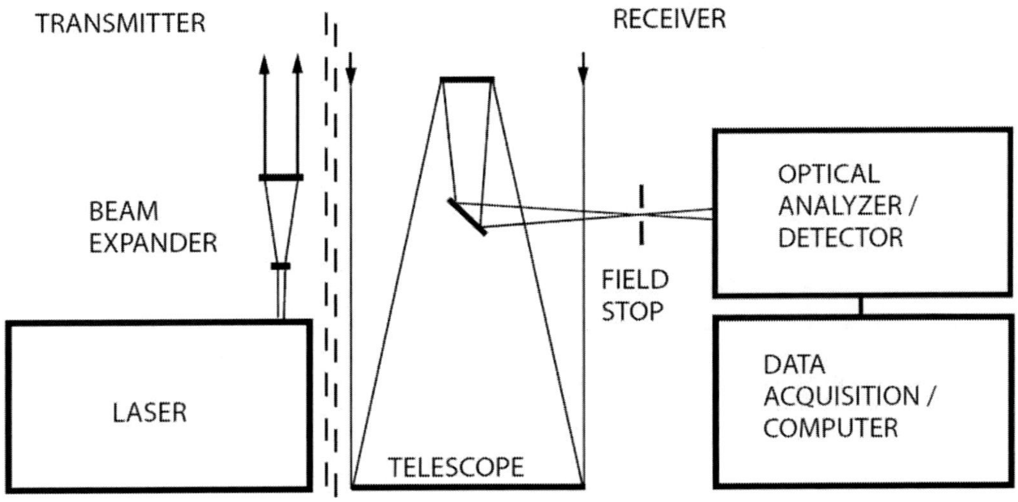

Figure 8. Block diagram of a typical lidar system in parallel [adapted from ref. 19].

3.2. Lidar Observations of Aerosol Plumes, Water Vapor, Cloud and PBL-Height

A ground-based multi-wavelength elastic-Raman scattering lidar system operational at CCNY [26, 27] emits three laser beams at 1064 nm, 532 nm and 355 nm with a pulsed Nd:YAG laser (a repetition rate of 30 Hz), and receives the atmospheric returns at three elastic-scattering channels and two Raman-scattering by nitrogen (387 nm) and water vapor (407 nm) molecules excited by the 355-nm laser pulse. The signals are acquired by a transient recorder (LICEL TR40-160) with 1 min interval and a 3.75 m range resolution. The multi-wavelength measurement provides some level of particle size information by quantifying the wavelength dependence of the scattering parameters (i.e., Angstrom exponent). Below the typical observations of aerosol, water vapor, cloud and PBL-height are illustrated.

3.2.1. Smoke and Dust Plumes Optical Properties and Type Classification

A heavy plume intrusion event was observed by the CCNY-lidar on August 14-15, 2007. Figure 9 shows the time-height cross section of the range-corrected lidar backscatter returns at 1064 nm on August 15, 2007. The colorbar indicates the return intensities with warm color (yellow and red) representing aerosol layer, blue for the clean air, and dark red for clouds. Two separate aerosol plume layers were observed at 2-4 km and 4.5-6.5 km altitude, respectively. The lower layer subsequently mixed downward into the PBL at ~14:00 (local time). To quantitatively analyze the event, we plot the vertical distribution of aerosol extinction coefficients and Angstrom exponents for the 1064-532 nm pair. The upper layer of smoke at 4-6 km has the extinction peak value of 0.3 km^{-1} at 532 nm, the relatively large Angstrom exponents of 1.7-2.0 in the aerosol plume layer at 2.0-6.0 km indicate the plumes are fine mode dominated. An aerosol multi-layer structure can be identified but the stable Angstrom exponent implies all the different plume layers are reasonably of similar type. The column averaged lidar-ratios were then obtained with values of 42±2 sr and 69±9 sr at 1064 and 532-nm, respectively, which are consistent with the existing calculations for smoke types of aerosol. With the combined HYSPLIT-model and satellite observations, these dense plumes are demonstrated to be transported from the episode of Montana/Idaho forest fire smoke in August, 2007.

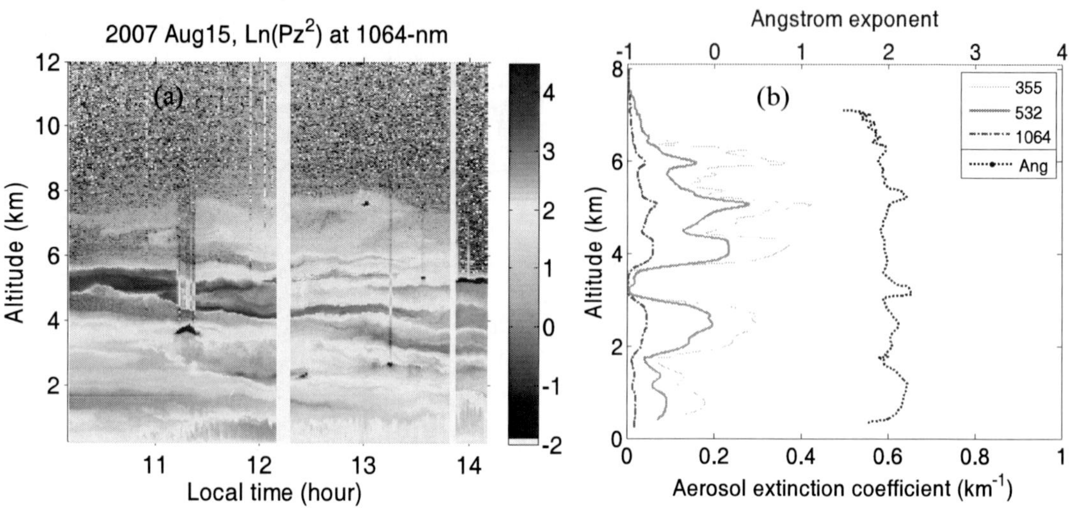

Figure 9. (a) Time-height cross section of range-corrected lidar returns at 1064-nm and (b) Aerosol extinction and Angstrom exponent profiles at 13:35-13:50 on August 15, 2007.

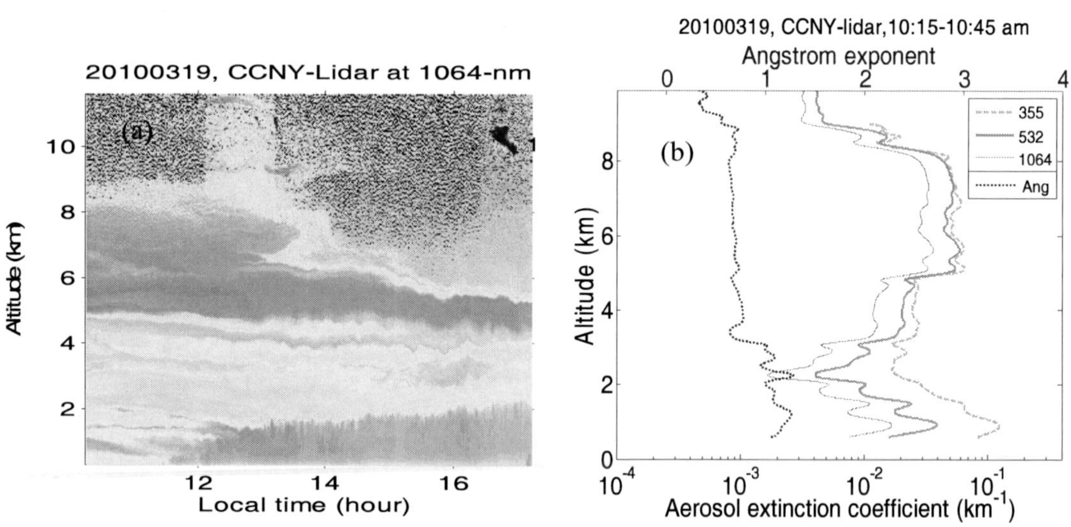

Figure 10. (a) Range-corrected lidar returns, (b) aerosol extinction and Angstrom exponent on March 19, 2010.

Figure 10 gives a dust event observed by CCNY-lidar on March 19, 2010. The range-corrected lidar returns show the thick plume layers at 3.5-8.5 km altitude in Figure 10(a). The extinction-related Angstrom exponents are about 0.5-1.0 in these aerosol layers indicating the coarse-mode dominated. The aloft aerosol plumes contribute 80% to the total AOD according to the lidar-derived vertical extinction profile. Meanwhile, the simultaneous AERONET sunphtometer measurements shows that the total AODs vary from 0.4 to 0.15 with the Angstrom exponents of 0.7-1.0, which are consistent with the lidar retrievals. In particular, the aerosol size distributions from AERONET-inversions clearly indicate the coarse-mode dominance. Additionally, the linear depolarization ratios of lidar are estimated to be 0.135 from the above inversions. The properties with coarse-mode and non-sphere in shape imply that these aloft plumes layers are likely dust particles. With the NOAA-HYSPLIT model and MODIS level-1B image analysis, these aloft aerosol plumes are demonstrated to originate from the **Taklamakan** and Gobi deserts in East Asia. It takes about 5-6 days to arrive east US from the dust source regions.

3.2.2. Raman-Elastic Lidar Observing Water Vapor and Clouds

As an example, Figure 11 gives the elastic-scattering returns at 1064-nm that clearly shows the PBL below 1.1 km and clouds at 5.5-6.5 km altitudes. Water vapor mixing ratios are given in Figure 11(b). First, the rich water vapor can be seen in the PBL below 1.1 km altitude. Second, the cloud layers also show the high mixing ratio of water vapor and some multiple-structures of moisture layers appear below clouds in the free troposphere. To verify the Raman-lidar results, Figure 12 gives the comparisons of water vapor mixing ratio profiles between the Raman-lidar and radiosonde measurements. They generally show good consistency, in particular for those multiple-layers structures of water vapor distributions. The advantage from the Raman lidar is its continuous even long-term measurement with the high spatial-temporal resolution (Figure 11(b)) against only twice regular measurements per day by the radiosonde site. Meanwhile, the simultaneous measurements of H_2O, aerosol and clouds by Raman lidar have the enhancements on investigating the aerosol hygroscopic properties, aerosol-cloud interactions, PBL-dynamics and atmospheric transport mechanism of aerosol/H_2O, etc.

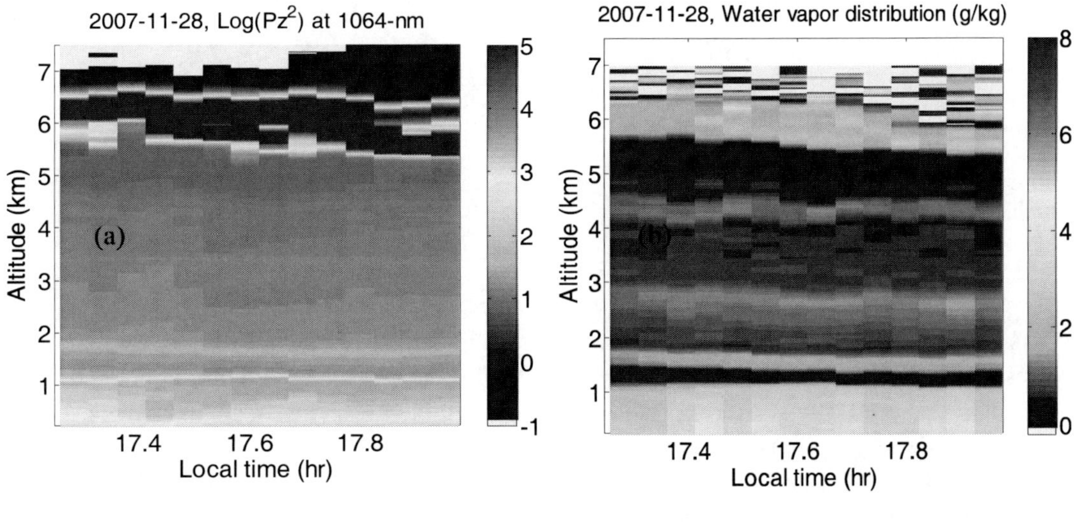

Figure 11. (a) Elastic-scattering returns at 1064-nm channel, (b) Water vapor mixing ratios on November 28, 2007.

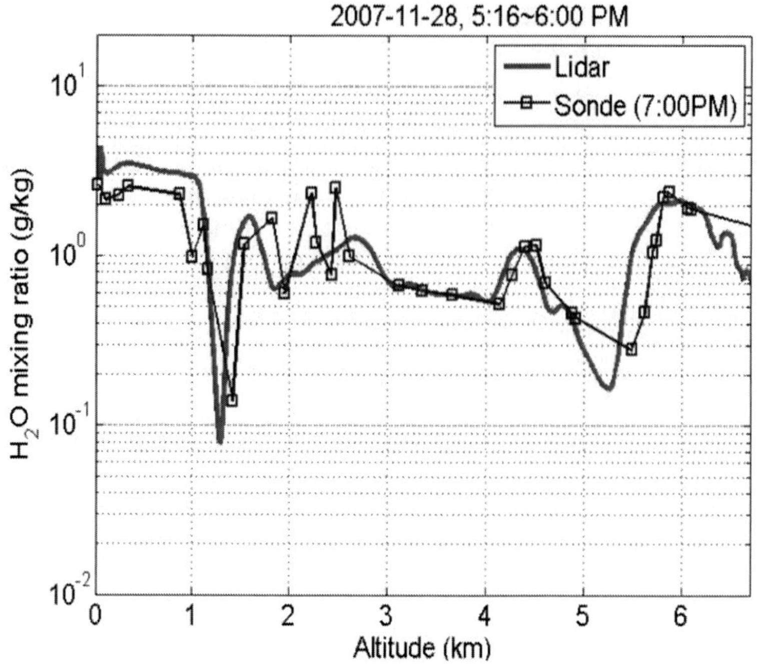

Figure 12. H_2O mixing ratio profiles from Raman-lidar and radiosonde.

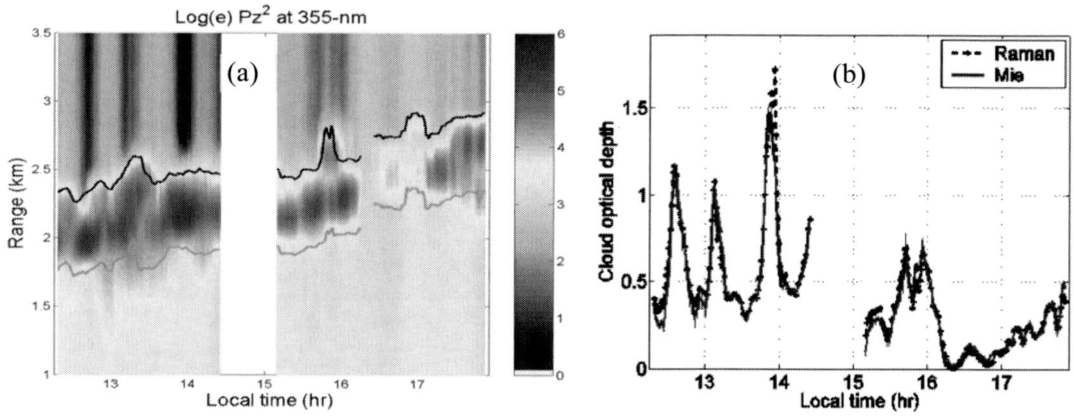

Figure 13. (a) Elastic-scattering returns at 1064-nm channel and (b) cloud optical depth at 355-nm on March 15, 2006.

A representative example for optically thin cloud measurements is shown in Figure 13. First, range-square corrected elastic returns in Figure 13(a) indicate the cloud heights of 1.8-3.0 km marked by the two lines. Complementary radiosonde data is used to identify the cloud is most likely water phase dominated. Second, cloud optical depths (COD) are obtained from the Raman lidar and Mie-scattering returns. The COD undergoes significant change as shown in Figure 13(b). After aerosol contamination is eliminated (Mie-method), the two retrievals are nearly coincident with each other and cloud optical depths vary from 0.1 to 1.7 at 355 nm wavelength. They show a good correlation between the retrievals with R^2=0.959. However, we do note that discrepancies become larger at higher CODs as expected. The mean lidar-ratios in cloud layers are observed that they mostly fluctuating around 20 sr line with standard deviation of 6.3 sr indicating the dominance of water phase in the cloud.

3.2.3. Lidar Measuring Urban PBL-Height

Lidar can detects the planetary-boundary-layer (PBL) height according to the sharp gradient of elastic scattering signal caused by the aerosol or the structure of water vapor profile. Herein, the PBL height or top is calculated from the elastic returns at 1064-nm with the wavelet transform technique (or wavelet covariance technology, WCT). PBL-height plays an important role in weather forecasting and air quality modeling that reflects turbulence and influences the dispersion for air pollutants. Figure 14 gives an example of lidar observing PBL-height. Figure 14 (a) shows the profile of lidar returns which indicate the sharp gradient between 1.5-2.0 km altitudes. Applying the WCT method, a maximum value of WCT is clearly located at ~1.8 km altitude which corresponds to the aerosol gradient well and represents the PBL-height. In Figure 14 (b), the evolution of PBL-height is shown and added into the image of range-corrected lidar returns. With the enhancements of solar radiances, the PBL grows up gradually and reach the maximum around local noon time (14:00 – 15:00), then decreases after 16:00. Most aerosols are trapped into the PBL-layer. Therefore, lidar-derived PBL-height and its temporal variation are important to be assimilated into or validate the forecast of weather research and air quality models.

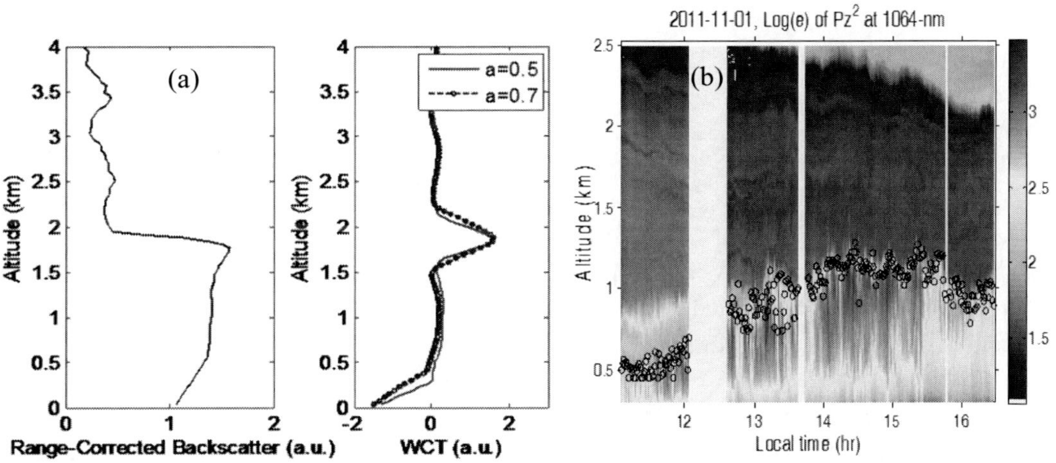

Figure 14. (a) Single Lidar signal (log scale) profile and resulting wavelet covariance transform, (b) PBL-height evolution from lidar measurement on November 1, 2011.

3.3. Laser Induced Fluorescence Spectroscopy

Laser-based spectroscopic techniques have been applied to the analysis of trace gases in the atmosphere. After recognizing the importance of OH^* in the atmosphere, laser induced fluorescence (LIF) measurements on OH^* were performed in 1972. LIF measurements are typically performed in reduced pressure chambers and the method id also known as flourescence assay by gas expansion (FAGE). Specific experimental setups for LIF OH^* measurements are numerous, but all modern instruments possess a gas expansion region, a high repetition rate laser system, an electronically-gated detector (photomultiplier or micro-channel plate) that is operated in photon counting mode, and a calibration system. Since the lifetime of OH^* is roughly around 1 second, LIF offers the advantages of in-situ measurement, fast analysis time and excellent sensitivity. In addition, selectivity is very good. Fluorescence from other species at 308 nm, such as SO_2 and formaldehyde are discriminated against by delayed gating. The only major interference is O_3. However, LIF instruments are costly and have sufficient size and weight to make field measurements cumbersome (i.e., the instrument is usually housed within a small flat-bed trailer with a second trailer for the roots pump). The biggest disadvantage of this technique is that it requires calibration and the measurement accuracy is typically limited by the certainty of the OH^* calibration source.

4. Air Monitoring by IR/UV Spectroscopy

Active optical remote sensing was explored to understand the threats to public health and the environment. Open path Fourier transform infrared (FTIR) spectrometers and ultraviolet (OPUV) spectrometers are currently being used as alarms for accidental releases of toxic gases at industrial plants. A non-dispersive infrared remote sensor was used to determine that about 10% of the motor vehicles are causing most of urban smog in our cities. An ultraviolet differential absorption lidar (DIAL) was used to map the vertical structure of ozone, a major component of urban smog. Ozone accumulation was shown to be a regional phenomenon that occurs on geographical scales extending far beyond the well-known urban situation. Thus, high ozone concentrations are found in urban areas, where it is a public health threat, as well as in rural areas, where it causes plant stress.

The FT-IR and OPUV open-path spectrometers measure path averaged concentrations of multiple gases simultaneously over path lengths of between 50 to 1000 m. During field measurements, the FT-IR and OPUV monitors direct beams of IR and UV energy towards retromirrors that return each beam to its receiver. Gases that pass through these beams and absorb some of the transmitted energy will appear in the measured absorbance data. The separation between the monitors and the retromirrors is chosen according to the pollutants being observed, expected concentrations, and the physical layout of the monitoring site. Once setup, these systems can provide continuous observations of pollutant fluctuations 24-hours a day. The FT-IR and OPUV spectrometers offer several advantages: (1) they are capable of monitoring multiple gases simultaneously and in real time, (2) no canister sampling is required; and (3) data from open-path monitors provide a spatial resolution that is more appropriate for model comparisons than that of point monitors.

4.1. Fourier Transform Infrared Spectroscopy

Infrared radiation spans a section of the electromagnetic spectrum having wavenumbers from roughly 13000 to 10 cm^{-1}, or wavelengths from 0.78 to 1000 μm. It is bounded by red end of the visible region at high frequencies and the microwave region at low frequencies.

The IR region is commonly divided into three small areas: near IR: (0.78–2.5 μm), mid-IR: (2.5-50 μm), far-IR: (50-1000 μm). IR spectroscopy determines the chemical functional groups in the sample. Different functional groups absorb characteristic frequencies of IR radiation. IR spectroscopy is an important and popular tool for structural elucidation and compound identification.

IR absorption positions are generally presented as either wavenumbers or wavelengths. Wavenumber defines the number of waves per unit length. Thus, wavenumbers are directly proportional to frequency, as well as the energy of the IR absorption. The wavenumber unit is more commonly used in modern IR instruments that are linear in the cm^{-1} scale. IR absorption information is generally presented in the form of a spectrum with wavelength or wavenumber as the x-axis and absorption intensity or percent transmittance as the y-axis. Transmittance, T, is the ratio of radiant power transmitted by the sample (I) to the radiant power incident on the sample (I_0). Absorbance (A) is the logarithm to the base 10 of the reciprocal of the transmittance (T).

$$A = \log_{10}\left(\frac{1}{T}\right) = -\log_{10}(T) = -\log_{10}\left(\frac{I}{I_0}\right)$$

The transmittance spectra provide better contrast between intensities of strong and weak bands because transmittance ranges from 0 to 100% T where absorbance ranges from infinity to zero. If the same sample is analyzed for the IR spectrum, it will give quite different profiles which are linear in wavenumber. It will appear as if IR bands have been contracted or expanded.

In recent years, Fourier transform infrared (FTIR) spectroscopy has been the dominating technique used for measuring the infrared absorption and emission spectra of most materials, with substantial advantages in signal-to-noise ratio, resolution, speed and detection limits. The major advantages of FTIR technique over other spectroscopic techniques is that practically all compounds show characteristic absorption/emission in the IR spectral region and based on this property they can be analyzed both quantitatively and qualitatively have recently replaced dispersive instruments for most applications due to their superior speed and sensitivity. For studies involving lower atmosphere, one can obtain superb spectra by combining a Fourier transform instrument of about 0.1 cm^{-1} resolution with a white cell made from mirrors of modest size. Nitrogen cooled detectors should be used.

Many situations do not allow the measurement of a background spectrum prior to the measurement of pollutants in order to perform background removal. After a radiometric calibration of the FTIR spectrometer with IR reference sources the spectral radiance of the environment can be measured. With the inverse function of Planck's radiation law, the spectrally resolved brightness temperature is computed. The temperature spectrum has a constant baseline for many natural materials that serve as the background in field measurements because their emittance is high and almost constant in the spectral range 800-1200 cm^{-1}. With an FT instrument, signal fluctuations caused by air turbulence do not appear to introduce noise into the spectrum. Pollutants that have been measured by infrared in ambient air include: HNO_3, HNO_2, NH_3, O_3, CO, CO_2, SO_2, NO, NO_2, H_2CO, HCOOH, hydrocarbons peroxy nitrates, HCl and HF.

In an active remote sensing measurement the infrared light is emitted by a hot infrared source, and is received by the spectrometer optics. Thus, the emitted infrared light propagates in direction of an open optical path, which is defined by the distance and position of the IR-source and the spectrometer. The received infrared light is modulated by an interferometer, detected by an infrared detector, converted and subsequently stored as an interferogram by help of the spectrometer electronic and the connected data system. The infrared spectrum is then computed from the interferogram by a mathematical operation (Fourier-Transformation).

If there are IR-active molecules (i.e., molecules with a significant IR absorption cross section in the observed frequency range) in the optical path characteristically absorption features (lines or bands) will appear in the IR-spectrum. From the transmittance spectrum the composition and quantity of the species in the optical path can be evaluated *(Lambert-Beer law*) with high accuracy and reliability. The technique is commonly applied, if low detection limits are required.

Generally for a given concentration (partial pressure) of the target species the sensitivity of this detection technique increases with the:

– Optical throughput of the sensor,
– Transmittance and modulation efficiency of the FTIR spectrometer,
– Sensitivity of the IR-detector element,
– Optical path length,
– Absorption cross section of the target molecule i.e., absorption coefficient, and
– Number of coadded spectra i.e., measurement time.

In passive remote sensing by FTIR spectroscopy, the only difference is that the ambient infrared radiation is detected instead of the light of the artificial IR source. This results in the advantage of mobile and fast operation (man-held system), remote sensing distance upto several kilometers and easy handling. Passive FTIR spectroscopy is less sensitive than the active configuration, because the sensitivity decreases with the temperature difference between the background and the target compounds. In case of a small temperature contrast ΔT the detection sensitivity is poor. For an effective temperature difference of 1 K we can estimate that the sensitivity is about 1000 times smaller compared to the sensitivity of the active setup.

Typical applications of passive remote sensing include:

– Reliable identification of industrial released hazardous gaseous compounds,
– Observation of stack gas plumes,
– Civil defense and fire brigade operation purpose (chemical accidents, wildfires)
– Remote sensing of aircraft emissions, and
– Detection of warfare agents under battlefield conditions.

Passive remote sensing is the only detection method which allows mobile, fast, man-held and standoff (wide range) detection of hazardous chemical agents.

FTIR spectrometers have been used to measure gas concentrations in both the stratosphere and the troposphere [28, 29]. In stratosphere, infrared spectrometers are designed with a fine resolution (0.01 cm^{-1}) because atmospheric pressure is low. However, a lower resolution of between 0.5 cm^{-1} and 2 cm^{-1} is commonly used in the troposphere due to pressure broadening effects that result in broadened absorption lines. Infrared spectroscopic techniques in the troposphere are complicated by water vapor concentrations that are much higher than those in the stratosphere. The strong interference of water vapor in the troposphere is overcome by detecting chemical species in narrow bands of the infrared spectrum where water absorption is very weak. The FTIR can detect over a hundred volatile organic compounds (VOCs) emitted from industry and biogenic VOC emissions such as isoprene and a-pinene.

4.2. OPUV Spectroscopy

UV spectroscopic techniques complement IR measurements very well especially in cases when the IR absorbance features of a particular species is lacking or masked by another compound. For examples, oxides of nitrogen are quickly masked by strong water vapor lines in the IR as the monitoring path length increases beyond about 75 m. Strong absorbance features appear between 340 and 380 nm for NO_2, between 220 and 230 nm for NO, between 280 and 330 nm for SO_2, and between 240 and 300 nm for O_3. In addition, the OPUV can detect several aromatic hydrocarbons, such as benzene and toluene, oxides of nitrogen and sulfur, and formaldehyde.

A scanning-slit technique has been applied to measuring molecular species in the UV relating to atmospheric photochemistry and smog formation [30]. An alternative to the scanning slit approach is to use a grating or prism spectrometer with a photodiode array as a detector [31]. An important advantage of simultaneous measurement of the desired spectral interval (as opposed to scanning) is the elimination of time dependent changes due to atmospheric scintillation effects that occur during the scan. A xenon arc lamp is commonly used as the source and a 1024 element photodiode array as the detector. An instrument resolution of around 0.3 nm is quite adequate for resolving pollutant absorbance peaks. The concentration of each species is determined using a least-squares fit similar to that described for the FT-IR. Note that the fit is carried out over a large portion of the spectral region, instead of at a single pixel, in order to reduce the effect of interfering absorbance features due to other gases.

4.3. Emission Spectroscopy

By emission spectroscopy techniques, the molecules are excited to a higher electronic state and the intensity of the emitted radiation as it returns to the ground state is monitored. Electronically excited species can be produced chemically (chemiluminesence), by absorption of light (fluorescence) or by introducing the substance of interest into a flame (flame photometry).

Chemiluminesence is the production of light from the energy released in a chemical reaction and is used to monitor a variety of pollutants, including O_3 and NO. A number of chemical reactions generate products not in their lower energy state, but in their upper levels. The energy released from the reaction is channeled internally into electronic, vibrational or rotational energy of one or more of the products, rather than being released as heat. The excited product molecules may emit this energy as light. By chemiluminesence technique, air containing the pollutant to be measured is drawn into the monitoring instrument where it is mixed with a species with which it reacts to produce light. The emission is proportional to the pollutant concentration. In case of O_3, two reactions take place:

(a) Reaction with ethylene (CH_2=CH_2), where electronically excited formaldehyde (HCHO*) is produced which is the light emitting species. At low pressures all simple olefins emit light upon reaction with O_3. The emitting species have been identified as electronically excited formaldehyde and in some cases, dicarbonyl compounds and vibrationally excited OH in its ground state. At atmospheric pressure in commercial instruments based on ozone (O_3) – ethylene (CH_2=CH_2) reactions all emissions, except that of due to HCHO, are effectively quenched.

$$O_3 + CH_2{=}CH_2 \text{ -> } HCHO^* + \text{other products}$$

$$HCHO^* \text{ - } HCHO + h\nu \ (300 < \lambda < 550 \text{ nm})$$

(b) Reaction with nitric oxide (NO) where electronically excited nitrogen dioxide (NO_2^*) is produced, which is the emitting species. The nitric oxide is monitored using the below reactions where excess O_3 is used.

$$O_3 + NO \text{ -> } NO_2 + O_2$$

$$NO_2^* \text{ -> } NO_2 + h\nu \ (590 < \lambda < 3000 \text{ nm})$$

Also, converting them first to NO, several other important gaseous nitrogen containing pollutants can be monitored. It is clear that any compounds in the air other than NO_X that will be converted to NO in the converter will interfere with the measurement of NO_X. Gaseous nitrogenous species, such as PAN (peroxyacetyl nitrate), organic nitrates, NH_3, HNO_3, N_2O_5, $ClNO_X$, and HONO are measured since these can be reduced to NO.

Nitric oxide (HNO_3) can be measured by the chemiluminiscence method using a nylon filter, which removes quantitatively the HNO_3 from the air sample. The HNO_3 concentration is determined by the difference between the NO_X/NO concentration with and without the nylon filter in the incoming air stream. The HONO may act as interference in the HNO_3 measurements by this technique since it is also removed quantitatively by the nylon filter. Since HONO rapidly photolyzes during the day, it is more likely to present a problem at night when significant HONO concentrations may accumulate.

Chemiluminiscence techniques are more specific than wet chemical techniques (e.g., KI method) because of the relatively few reactions that give off light. In addition, filters or monochromators can be used to isolate the wavelengths of interest in order to discriminate against interfering emissions. However, in the NO_X measurement some selectivity is lost in the NO_X reduction because of the number of compounds that can be reduced to NO and hence give positive responses.

4.4. Aerosol Chemical Composition with Mass Spectrometry

Mass spectrometry is a powerful analytical technique used to quantify known materials, to identify unknown compounds within a sample, and to elucidate the structure and chemical properties of different molecules. The complete process involves the conversion of the sample into gaseous ions, with or without fragmentation, which are then characterized by their mass to charge ratios (m/z) and relative abundances. This technique basically studies the effect of ionizing energy on molecules. It depends upon chemical reactions in the gas phase in which sample molecules are consumed during the formation of ionic and neutral species.

A mass spectrometer generates multiple ions from the sample under investigation; it then separates them according to their specific mass-to-charge ratio (m/z), and then records the relative abundance of each ion type. The first step in the mass spectrometric analysis of compounds is the production of gas phase ions of the compound, basically by electron ionization. This molecular ion undergoes fragmentation. Each primary product ion derived from the molecular ion, in turn, undergoes fragmentation, and so on. The ions are separated in the mass spectrometer according to their mass-to-charge ratio, and are detected in proportion to their abundance. A mass spectrum of the molecule is thus produced. It displays the result in the form of a plot of ion abundance versus mass-to-charge ratio.

Ions provide information concerning the nature and the structure of their precursor molecule. In the spectrum of a pure compound, the molecular ion, if present, appears at the highest value of m/z (followed by ions containing heavier isotopes) and gives the molecular mass of the compound. Laboratory studies show that distinctive mass spectra are obtained when aerosol particles consisting of single chemical components are measured with the Aerosol Mass Spectrometer (AMS) [32 – 34]

5. MICROWAVE SPECTROSCOPY

Currently ground-based, air-, balloon-, and space-borne sub-millimeter radiometers (e.g., MLS, ASUR, SLS, ODIN, SMILES) measure a long list of tropospheric and stratospheric molecules (H_2O, O_3, HCl, CO, N_2O, HNO_3, HCN, H_2CO, CH_3CN, H_2O_2, HNO_4, ClO, OH, etc.). The principal spectroscopic deficiencies in microwave spectral parameters are line-widths and non-resonant absorption continuum (i.e., H_2O). For both, the laboratory data are sparse and theoretical models imprecise. Remote sensing of air temperature and relative humidity in the lower atmosphere is of great interest in understanding the evolution of the boundary layer and convective systems. The ground based radiometric profiling technique provides continuous measurements under a wide range of meteorological conditions. In contrast to satellite based microwave remote sensing, having coarse spatial and temporal resolution, ground based radiometric observations with very high temporal resolution are useful for studying localized convective events more effectively.

Ground based microwave profiling methods make use of atmospheric radiation measurements in the 20 to 200 GHz region. The atmospheric absorption spectrum at several altitudes for a typical mid latitude atmosphere is shown in Figure 15. The features near 22 and 183 GHz are water vapor resonances that are pressure broadened according to the pressure altitude of the water vapor distribution. The feature near 60 GHz is an assemblage of atmospheric oxygen resonances. Clouds are still major uncertainties in atmospheric models, whereby their role in linking radiation and dynamics is ill represented [e.g., 34]. The cloud liquid water emission spectrum has no resonances in this frequency range, and increases approximately with the second power of frequency in this region.

Temperature profiles can be obtained by measuring the radiation intensity, or bightness temperature at points along the side of the oxygen feature at 60 GHz. By sampling the brightness temperatures outward from line center, where opacity is so great that all signal originates from the atmosphere close to the antenna, on to the "wing of the line", where the radiometer sees further into the atmosphere, the instrument can obtain altitudes information. Emission at any altitude is proportional to local temperature and density of oxygen; thus the temperature profile can be retrieved.

Water vapor profiles can be obtained by observing the intensity and shape of emission from pressure broadened water vapor lines. The line near 22 GHz is suitable for ground based profiling in relatively moist areas. The more sensitive 183 GHz line is well suited for ground based water vapor profiling in arid environments. The emission from water vapor is in a narrow line at high altitudes and is pressure broadened at low altitudes. The intensity of emission is proportional to vapor density and temperature. Scanning the spectral profile and mathematically inverting the observed data can therefore provide water vapor profiles.

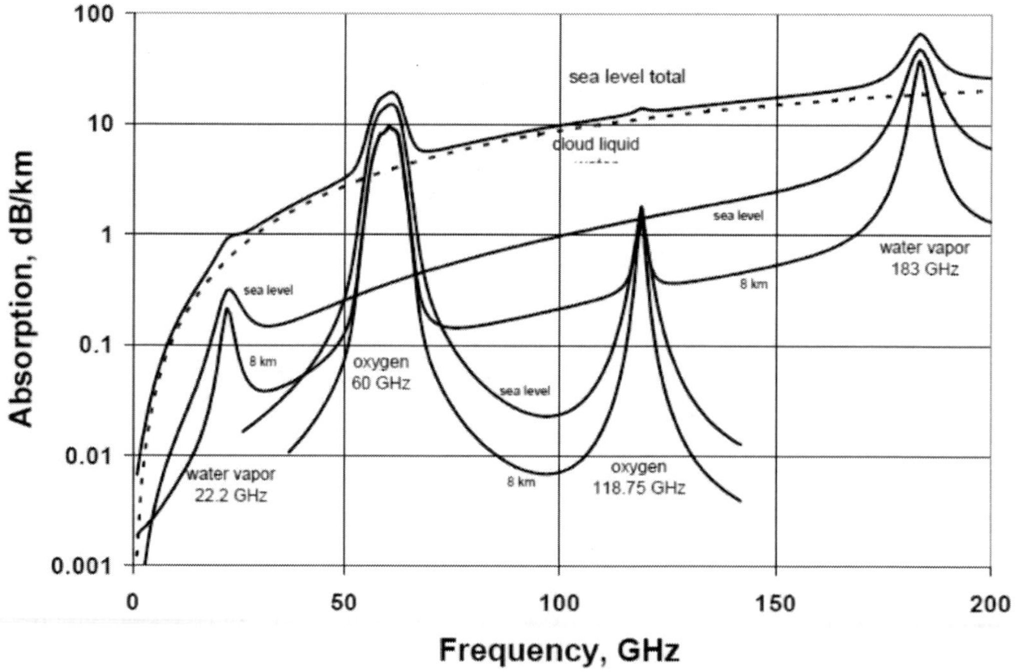

Figure 15. Absorption spectrum of a typical mid-latitude atmosphere for two altitudes and two water vapor densities.

Typically, the atmospheric brightness temperature (T_B) is measured at one frequency on the wing of the water vapor line at 22.235 GHz and at a second frequency in the window region around 30 GHz. The first frequency is chosen so that the water vapor absorption coefficient is nearly independent of altitude. Because the emission of cloud liquid water increases with the frequency squared, the signal at the second frequency is dominated by liquid water contribution. From measurements at both the frequencies, the liquid water path (LWP) and integrated water vapor content (IWV) can be retrieved simultaneously. It has been suggested recently that LWP retrievals can be improved by adding a temperature dependent frequency around 50 GHz [36] or an additional frequency sensitive to cloud liquid water at 85 GHz [37]. The 90 GHz channel is most valuable when deriving LWP due to its high sensitivity to liquid water content. When all four channels are used, the root mean square error between modeled and retrieved LWP is reduced by a factor of two. Although 90 GHz channel is more sensitive to liquid water than the 50 GHz channel (i.e., the same amount of liquid water content increase leads to a larger T_B increase at 90 GHz than at 50 GHz. A physical reason for this is the higher water vapor contribution at 90 GHz which is even higher than at the center of the water vapor absorption line at 22.235 GHz leading to a more indirect relationship between LWP and the 90 GHz channel. Thus, the 90 GHz channel complements the lower two channels better than the 50.8 GHz channel [2].

6. Molecular Spectroscopic Database

The transmission of the atmosphere is modeled by extensive databases. Most prominent of these are the HITRAN and LOWTRAN databases. In particular, the LOWTRAN program calculates the transmission spectrum for molecular absorption, Rayleigh scattering and aerosol extinction at a moderate resolution of 20 cm^{-1}.

The HITRAN compilation consists of information about spectral line parameters of atmospheric gases and pollutants that is essential for solving problems of atmospheric optics and for modeling and visualization of molecular absorption spectra. The standard database HITRAN mainly covers the relatively well-studied infrared spectral region but is extended to wider spectral ranges from microwave to ultraviolet (0.00001-25232.0 cm^{-1}). The line-by-line portion of the database contains spectroscopic parameters for 42 molecules including many of their isotopologues. Computation of spectral line profiles of various gas molecules is based on the *line-by-line* method that is summing of the standard line shapes of single absorption lines. As a matter of rule, lines symbolize rotation-vibrational transitions that are grouped into vibrational bands. Line center positions and intensities are recorded by high resolution spectrometers under controlled laboratory conditions. In case of incomplete experimental data, theoretical calculations of these quantities are used. When modeling high-resolution spectra, data are selected either via the profile Molecule/Isotope/band or within a wavenumber range selected for a gas mixture chosen or composed by the user. This HITRAN database for several components, including aerosol species, serve as input for radiative-transfer codes: individual line parameters for microwave through visible spectra of molecules in the gas phase; absorption cross-sections for molecules having dense spectral features, i.e., spectra in which the individual lines are not resolved; individual line parameters and absorption cross-sections for bands in the ultra-violet; refractive indices of aerosols, tables and files of general properties associated with the database; and database management software. Further updated information can be found at http://hitran.iao.ru/.

LOWTRAN (current version 7) is a low-resolution propagation model and computer code for predicting atmospheric transmission and background radiance from 0 to 50000 cm^{-1}

at a resolution of 20 cm^{-1} (from ground to space as a function of height above sea-level). Multiple scattered radiation is treated in the model along with the new molecular band model parameters and new ozone and molecular oxygen absorption parameters for the ultra-violet. Most of the atmospheric correction models for imaging spectroscopy have included residual or scene average, internal average relative reflectance, flat correction, single spectrum and empirical line or regression methods. Basic assumptions of these methods are a priori knowledge of the location/site, the existence of a spectrally neutral area with no vegetation or sparse vegetation cover, wavelength-independent average scene reflectance or the existence of field measurements taken at the time of image acquisition. LOWTRAN 7 has the potential to overcome most of these limitations imposed in previous methods and is a versatile simulation model that has been parameterized for a variety of conditions. Explicitly, in LOWTRAN model two basic assumptions namely, a horizontal homogeneous atmosphere and a small elevation variation within the scene are considered.

7. Future Scope

With the advanced developments and technological improvements on the laser and spectroscopy, opt-electrical detectors and signal processing, lidar is becoming a sophisticated and industrialized remote sensing instrument without neither additional maintenance compared to neither standard ceilometers nor specific alignment processes. It can provide three-dimensional temporal-spatial resolution of atmospheric composition and has wide applications in the field of weather, climate radiation and atmospheric pollution. For instance, the ground-based lidar network like the European Aerosol Lidar Network, EARLINET, and space-borne lidar like CALIPSO, can make the regional and global observation of aerosols and clouds, their products are being used to validate and verify weather and climate-modeling aspects such as to see cloud climatology and radiative effects, aerosol-cloud interactions, long-distance transport of pollutants, as well as being assimilated into the models. This chapter briefly introduces the fundamental and physical process of lidar monitoring atmosphere. The typical applications and potentials of lidar measuring aerosol plumes, water vapor and clouds, and PBL-height are illustrated. The observation examples demonstrate the advantages of lidar remote sensing such as highly temporal-spatial resolution, range-resolved information, continuous and long-term observations and multiple parameters retrievals.

These few examples of current optical remote sensing technologies underscore their potential applications in future environmental management programs. Further applications of active optical remote sensing instruments include studies related to tropospheric chemistry, global change, regional photochemical measurements and modeling, air pollution exposure, pollution prevention, and general air quality. Measurement challenges posed by these studies include: detection of multiple gases, sufficient instrument sensitivity, short enough temporal resolution, and appropriate spatial coverage. In order to meet these measurement challenges in a cost-effective manner, remote sensing technologies will need to be further developed in order to provide for real-time, multiple-pollutant, extended spatial coverage monitoring capabilities. Further, technological development is required not only by researchers but also by industry that must remain compliance with strict regulations imposed on them to control air pollution emissions.

REFERENCES

[1] G. Warnecke, *Meteorologie und Umwell*, Springer, Heidelberg, p 70 (1991).

[2] U. Löhnert, S. Crewell, *Radio Science,* 38(3) (2003) 8041, doi: 10.1029/2002RS002654.

[3] G. E. Thomas, K. Stamnes, *Radiative Transfer in the Atmosphere and Ocean,* Cambridge University Press, pp. 517 (1999).

[4] J. Lenoble, *Radiative transfer in scattering and absorbing atmospheres: Standard computational procedures*, A. Deepak (Hampton, Va., USA), xv, 300 p., ISBN 0937194050 (1985).

[5] J. Grainger, J. Ring, *Nature* 193 (1962) 762.

[6] A.W. Brewer, C.T. McElroy, J. B. Kerr, *Nature* 246 (1973) 129–133.

[7] J. F. Noxon, *Science* 189 (1975) 547–549.

[8] D. Perner, D. H. Ehhalt, H. W. Pätz, U. Platt, E. P. Röth, A. Volz, *Geophys. Res. Lett.* 3 (1976) 466–468.

[9] U. Platt, D. Perner, H.W. Patz, *J. Geophys. Res*. 84 (1979) 6329–6335.

[10] D. Perner, U. Platt, *Geophys. Res. Lett*. 6 (1979) 917–920.

[11] U. Platt, Differential optical absorption spectroscopy (DOAS), in: Air Monitoring by Spectroscopic Techniques, edited by: M. W. Sigrist, *Chemical Analysis Series*, John Wiley, New York, 127, 27–84 (1994).

[12] U. Platt, L. Marquard, T. Wagner, D. Perner, *Geophys. Res. Lett*. 24 (1997) 1759–1762

[13] J. P. Burrows, M. Weber, M. Buchwitz, V. Rozanov, A. Ladstatter-Weißenmayer, A. Richter, R. DeBeek, R. Hoogen, K. Bramstedt, K.-U. Eichmann, M. Eisinger, and D. Perner, *J. Atmosph. Sci.* 56 (1999) 151–175.

[14] J. P. Burrows, et al*., SCIAMACHY – A European Proposal for Atmospheric Remote Sensing from the ESA Polar Platform, Max-Planck Institute of Chemistry*, Mainz, Germany, (1988).

[15] J. P. Burrows, E. Hölzle, A. P. H. Goede, H. Visser, W. Fricke, *Acta Astr.* 35 (1995) 445–451.

[16] H. Bovensmann, J. P. Burrows, M. Buchwitz, J. Frerick, S. Noel, V.V. Rozanov, K.V. Chance, A. H. P. Goede, *J. Atmosph. Sci.* 56 (1999) 127–150.

[17] Richter, A. and T. Wagner (2001), Diffuser plate spectral structures and their influence on GOME slant columns, Technical note, available at: *http://www.iup.uni-bremen.de/gome/data/diffuser_gome.pdf*

[18] G. M. B. Dobson, D.N. Harrison, *Proceedings of Royal Society, London*, A110 (1926) 660–693.

[19] C. Weitkamp, Ed., Lidar: Range-Resolved Optical Remote Sensing of the Atmosphere, Springer *Series in Optical Sciences,* New York, USA, ISBN 0342-4111 (2005).

[20] J. D. Klett, *Appl. Opt.* 20 (1981) 211-220.

[21] F. G. Fernald, *Appl. Opt.* 23 (1984) 652-653.

[22] J. Pelon, C. Flamant, P. Chazette, J.-F. Leon, D. Tanre, M. Sicard, and S.K. Satheesh (2002), Characterization of aerosol spatial distribution and optical properties over the Indian Ocean from airborne LIDAR and radiometry during INDOEX'99, *Journal of Geophysical Research,* 107, doi: 10.1029/2001JD000402.

[23] J. Goldsmith, F.H. Blair, S.E. Bisson, and D.D. Turner, *Appl. Opt*. 37 (1998) 4979–4990.

[24] V. Simeonov, T. Dinoev, B. Calpini, S. Bobrovnikov, Y. Arshinov, P. Ristori, H. van den Bergh, M. Parlange, *A Raman lidar as operational tool for water vapor profiling in the Swiss meteorological office, ILRC 25*, St. Petersbourgh, 1175–1178 (2010).

[25] K. Sassen, *Bull. Am. Meteorological Soc*. 72 (1991) 1848-1866.

[26] Y. Wu, S. Chaw, B. Gross, F. Moshary, S. Ahmed, *Appl. Opt*. 48 (2009) 1218-1227.

[27] Y. Wu, L. Cordero, B. Gross, F. Moshary, et al., *Atmospheric Environment* 63 (2012) 32-42.

[28] W. B. Grant, R. H. Kagann, W. A. McCleanny, *J. Air Waste Management Association*, 42 (1992) 18-30.

[29] C.E. Kolb, Instrumentation for chemical species measurements in the troposphere and stratosphere, Reviews of Geophysics, Supplement, U.S. *National Report to International Union of Geodesy and Geophysics*, pp. 25-36 (1991).

[30] B. J. F. -Pitts, J. N. Pitts, *Atmospheric Chemistry: Fundamentals and Experimental Techniques,* Wiley, New York, pp. 337-356 (1986).

[31] A. Wahner, R. O. Jakoubek, G.H. Mount, A.R. Ravishankara, and A.L. Schmeltekopf (1989), Remote sensing observations of daytime column NO_2 during the airborne Antarctic ozone experiment, August 22 to October 2, 1987, *Journal of Geophysical Research,* 94, 16619-16632.

[32] J. L. Jimenez, J.T. Jayne, Q. Shi, C.E. Kolb, D.R. Worsnop, I. Yourshaw, J.H. Seinfeld, R.C. Flagan, X. Zhang, K.A. Smith, J.W. Morris, P. Davidovits (2003), Ambient aerosol sampling with an Aerosol Mass Spectrometer, *Journal of Geophysical Research,* 108:8425, doi: 8410:1029/200JD001213.

[33] M. R. Alfarra, *Insights into atmospheric organic aerosols using an aerosol mass spectrometer*, PhD Thesis, University of Manchester (2004).

[34] M. R. Canagaratna, J.T. Jayne, J.L. Jimenez, J.D. Allan, M.R. Alfarra, Q. Zhang, T.B. Onasch, F. Drewnick, H. Coe, A. Middlebrook, A. Delia, L.R. Williams, A.M. Trimborn, M.J. Northway, P.F. DeCarlo, C.E. Kolb, P. Davidovits, D.R. Worsnop, *Mass Spectrometry Rev*. 26 (2007) 185-222.

[35] W. L. Gates et al. *Bull. Am. Meteorological Soc*. 80 (1999) 29-55.

[36] A.V. Bosisio, C. Mallet, *Radio Science* 33 (1998) 929-939.

[37] J. P. Bobak, C. S. Ruf, IEEE Transactions on Geosci. *Remote Sens*. 38 (2000) 214-225.

In: Spectroscopic Techniques for Security, Forensic ... ISBN: 978-1-63117-404-9
Editors: Y. Dwivedi, S. B. Rai and J. P. Singh

Chapter 10

RARE EARTH SPECTROSCOPY FOR CLEAN ENERGY AND SENSOR APPLICATIONS

Yashashchandra Dwivedi[1,2*], Sergio Carlos Zilio[1] and Shyam Bahadur Rai[3]

[1]Instituto de Física de São Carlos, Universidade de São Paulo, São Carlos, SP, Brazil
[2]Department of Physics, National Institute of Technology Kurukshetra, Haryana, India
[3]Physics Department, Banaras Hindu University, U.P., India

ABSTRACT

Now, it is well documented that our planet may face an energy crisis in the near future, as natural resources (fossil fuels like oil, natural gas etc.) would be exhausted by human use. This situation has created much enthusiasm among scientists and researchers to search for alternative energy sources or reduce the load on natural energy resources by sharing various energy sources for different individual requirements. One of the main energy consumption sectors is the household/offices/streets in the form of light, warming the water/houses etc. which is more than 20% of all energy consumption. Rare earth (RE) ion-based materials have been used extensively to solve some of these problems. Alternative energy sources like nuclear reactors, solar cells, and wind turbines utilize rare earth elements extensively. In addition to the needs of eco-friendly energy sources another embarking field in which RE elements are being used is field of sensors for gases, temperature, and magnetic field etc. RE elements are predicted to contribute significantly in these major fields. Such sensors show potential for improving the selectivity and sensitivity. This chapter reviews the recent developments in the field of RE spectroscopy with special emphasis on sensors and production of clean energy.

* Email: yashjidwivedi@gmail.com.

1. INTRODUCTION

Elements having unfilled or partially filled d or *f*-shell electrons are of fundamental importance due to their unique optical and magnetic properties. Spectroscopic research on Rare earth (RE) elements is a hot topic of research as it not only spawns vast commercial applications but also helps to understand the basic mechanism of various emission dynamics. RE elements are wealthy by several characteristic luminescent properties such as sharp emission lines, meta-stable states and high quantum efficiency which make them suitable for fabrication of various optical components [1]. These elements have great technological importance in a variety of devices like superconductors, miniaturized magnets, catalysts, lamps, display devices, lasers, biomedical, memory devices, sensors, optical amplifiers, etc. In particular, on the question of energy utilization and its harmful impact on environment, RE doped materials are considered as an alternative source of clean energy production through its application in Nuclear reactors, solar cell, wind turbine, Light Emitting Diode (LED) etc. RE ions exhibits unique luminescent properties, including the ability to convert long-wavelength near infrared radiation into shorter wavelengths in visible region through a process known as *frequency upconversion*. Several exclusive optical upconversion phenomenons like APTE [Addition de Photon par Transferts d'Energie], cooperative emission, multi-photon absorption, photon avalanche etc. have been discovered in RE elements. With the advent of nanotechnology it is worthwhile to review these processes by confining these ions in nanodimensions as well. Despite the myriad of potential applications of RE doped materials several perplexing problems have yet to be solved which evokes research interest among researchers. Some of the technological challenges for rare earth spectroscopy is to understand optical-magnetic interaction, optimizing black body radiation (heat generation), finding single phase and multifunctional luminescent material compatible for biomedical applications, etc. Another aspect of the growing use of lanthanides is their ecological and biological impact because true toxicity level of these ions is not well known, especially when doped in nanostructures.

It is interesting to note that on one hand these elements are reported to be key elements for next generation technologies but on the other hand this sector is under threat of serious supply cut-off after an announcement was made by the largest supplier of these elements to curtail export by 40% (especially, heavy lanthanide (Ln) elements). The shortage in supply of these elements affects the global market and prices of these elements face unexpected rise. In addition to economic aspects, questions arise in future research on these elements and in turn encourage governments to make policy on regulation of these metals and assure sustainable production. It is interesting to note that worldwide annual consumption of rare earth elements was ~136 million tonnes in 2010 [2] in prime sectors like electric vehicles, wind turbines, polishing, magnet, petroleum product, catalytic application, solid state lasers, optical telecommunications, biological imaging [3], lighting and displays, agricultural fertilizer [4] and hundreds of other products. However, the excess, use of these elements leads to bioaccumulation, which is seen as a threat to the liver in addition to damage cell membranes of the water species with negative effects on the reproduction and the nervous systems [5].

In this chapter, amidst many positive and negative aspects of RE doped materials, techniques and technology, we have reviewed at least two of the major applications of RE elements, namely clean energy production and sensors, followed by a brief discussion on its spectroscopic part.

2. BRIEF ABOUT RE SPECTROSCOPY

Discovery of rare earth elements is believed to have first been reported in the summer of 1787, at Roslaga island near Stockholm (Sweden), when an exceptionally heavy piece of black rock found in the dumps of the Ytterby quarry grabbed the attention of Lieutenant Carl Arrhenius (1757-1824). An analysis of the stone revealed the existence of a new element which later was named Yttrium (atomic no. 39). Thereafter, a whole range of elements with promixate properties was discovered and is now known as the Lanthanide (Ln) elements, atomic no. 57-71. Although Lanthanide elements and Yttriium occupy different positions on the periodic table, due to similar chemical properties, they are collectively referred to as RE elements. Occasionally, Scandium (Sc) and Thorium (Th) are also referred to as RE elements. The name of this unique family is misleading. As a matter of fact, these elements are a rich ingredient in the earth's crust. For example Tm, which is estimated to be the least among Lanthanides, is comparable to silver and cadmium (Cd) in the earth's crust while there is more Ce than Tin (Sn). However, as most of the REs found in same type of ore, to obtain ultrapure lanthanides is still rare.

In the following texts we will discuss only discuss spectroscopy of Ln elements. Ln elements possess electronic configuration [Xe] $4f^n\ 6s^2$ or [Xe] $4f^{n-1}\ 5d\ 6s^2$, where [Xe] represents Xenon electronic structure. When the Ln elements disperse in a host matrix, it may attain either +2, +3 or +4 oxidation state among which +3 oxidation states is most common. When a trivalent Ln ion is formed, the atom give up its outermost 6s electrons and the 5d electron or 4f electron leaving partially filled *4f* shell shielded by the outer 5s and 5p shells. In an appropriate conditions Ln ions also show +2 and +4 oxidation states e.g., Sm^{2+}, Eu^{2+}, Yb^{2+} (Stability order $Eu^{2+} >> Yb^{2+} >> Sm^{2+}$); and Ce^{4+}, Pr^{4+}, Tb^{4+}. Similar to Ln ions, Y also shows a stable +3 ionic state but this ion does not fluoresce. Among Ln ions, La^{3+} and Lu^{3+} have a completely empty and a completely filled *4f* shell, respectively, and therefore have no optical transitions. Ce^{3+} has one electron and one *4f* level just above the ground state. Due to small energy difference between the inner *4f* levels to the outer or valence electrons, it takes only a minimal amount of energy to change the relative occupancy of electronic levels. Hence, Ce possesses dual valence states Ce^{4+} (Ceric) ($4f^0$) and Ce^{3+} (Cerous) ($4f^1$) [6]. Naturally, Ce^{4+} ($4f^0$) ion will have higher stability as a consequence of unfilled 4f orbital. Hence, it is reasonable to assume that Ce^{3+} ion has a lower ionization potential and would be a good 'hole' trap. The tetravalent oxidation state of cerium ions does not contain any electron in the 4f orbital hence no emission from Ce^{4+} ions is reported. Trivalent Yb ion also possesses one excited state in *4f* configuration at ~10,000 cm^{-1}. Another Ln ion, Pm is a non-natural occurring, radioactive material and exists, on earth, only as a temporary intermediate product in the ongoing radioactive decay of other elements.

The *4f* orbital in the Ln^{3+} ion do not participate directly in bonding, being well shielded by the $5s^2$ & $5p^6$ orbital. Thus their spectroscopic and magnetic properties are thus almost uninfluenced by the host. In Ln ions different type of transitions may takes place namely, $4f \rightarrow 4f$, $4f5d \rightarrow 4f$, and Charge transfer band $\rightarrow 4f$. Among these transitions, intrashell $4f \rightarrow 4f$

transition is parity forbidden following the Laporte rule which states that electric dipole transitions can only occur between levels of opposite parity. However, these transitions appear in low intensity as it was suggested that crystal field, apart from splitting the energy levels by removing the 2J+1 degeneracy, could also admix higher states of opposite parity into the 4*f* levels. Odd power terms in the expansion of the crystal field potential can mix states of opposite parity into $4f^N$ from higher configurations, such as $4f^{N-1}\,5d$ that partially release the Laporte rule and small intensity of these transitions appears [7, 8]. A weaker mechanism that can also contribute to the admixing of states is the interaction of the electronic states with lattice vibrations. This mechanism becomes the dominant one when ions are placed in sites with inversion symmetry, since the crystal field contribution becomes zero in this case. Since electric dipole transition is no longer forbidden, these transitions are characterized by low absorption coefficients and relatively long lifetimes (up to several milliseconds). Since 4f electrons do not take part in bond making with host and well shield by s, p shell, resulting narrower line like transitions in absorption and emission spectra. Though, the shielding of the 4f electrons makes them almost insensitive towards the proximate field, a few transitions exaggerate the small changes in the lattice coordination sphere. These transitions are termed as hypersensitive transitions. Their intensity is significantly increased as the symmetry of the lanthanide complex is lowered or the polarizability of the ligands increased. Jørgenson and Judd noted that hypersensitive transitions obey the selection rules $\Delta J \geq 2$, $\Delta L \geq 2$ and $\Delta S = 0$ [9]. In contrast to electric dipole, magnetic dipole transitions do not violate the parity selection rule (e.g., it can occur between states of the same parity), however, strength of such transitions are much weaker than allowed electric dipole transitions. Magnetic dipole contributes to the intra-4*f* electronic transitions of Ln ions such as Er^{3+}, Eu^{3+} and Tb^{3+}. As detailed discussion on rare earth spectroscopy that includes energy level calculation, strength of the transitions and other spectroscopic characterizations techniques are not in the scope of present chapter, readers are referred to ref. [10-15].

As 4f–4f transition probabilities are very weak to achieve efficient light converters, some options to improve emissions are to utilize enhance near-electric field of metal NPs, core-shell structures, sensitization via host, organic, or Ln ions etc. [16, 17]. The lanthanide ions belong to the hard acid class [18] and the tendency to bind with hard bases, primarily those that contain oxygen and nitrogen as donor atoms resulting hydrophilic nature of RE tri-cations. Dense energy levels and their splitting bring several exclusive optical phenomena in Ln ions viz. excited state absorption, cooperative emission, multiphoton absorption, photon avalanche, cross-relaxation etc. These processes produce photons of higher energy on excitation with low energy photons and the phenomenon is known as upconversion [19-22]. Upconversion may also occur through energy transfer via host, hetro RE ion, organics etc; phonon assisted energy transfer; and cooperative processes [23-26]. Upconversion processes can be induced by a low power (1–10^3 Wcm^{-2}) continuous wave laser, as opposed to a costly high-intensity (10^6–10^9 Wcm^{-2}) pulsed laser source for the generation of two-photon process. In addition to upconversion, an interesting process referred to as quantum cutting, which converts a UV-Vis photon to two or more IR photons, is also of equal importance.

3. Clean Energy Application

Growing global concerns of the predicted reduction in fossil energy (natural resources like oil, coal, natural gas etc.) and environmental pollution ignite lots of enthusiasm for clean energy. In addition to that ever increasing demand of electricity consumption for various purposes aggravates the situation. To counter this situation two approaches are being considered. First, to search for an alternate clean (low carbon) or renewable energy source for individual purposes and second, to improve the efficiency of the existing equipment that reduce energy consumption. In recent years, at least three different technologies viz. wind turbine, Nuclear reactors and solar cell boost up clean energy sector and proposed to be the future energy sources. According to the report of Nuclear Energy Institute [27] contribution of nuclear energy is 64%, solar/wind and hydro energy are 13.5% and 22.6 %, respectively, in 2012.

Energy sources that release carbon-intensity less than 0.82 metric tons are considered as clean energy sources [28]. Existing overall natural energy is distributed in the various forms that includes water, wind, Sun light etc. which can be renewed in the form of electricity and considered as clean energy sources. Nuclear energy is also an option though there may be a debate whether it may be considered as renewable or clean energy sources as byproducts of nuclear energy may arose serious threat to the environment. In addition to this, leakage from big nuclear reactors is also a concern and due to this problem, construction of nuclear reactors are opposed by local residents as leak may contaminate water, soil etc. and affect the environment. Nuclear power plants do not emit toxic gases like CO_2, or SO_2. Despite various concerns about nuclear energy production, it is proposed as a clean energy source and is now being used by almost every major country around the globe. As of May 2013, 30 countries worldwide have been operating 436 nuclear reactors for electricity generation and 70 new nuclear plants are under construction in 14 countries [27].

The major contribution of RE (especially Ln) elements for clean energy production is multifaceted. It includes preparing Neutron absorber in Nuclear reactor, efficient magnets for wind turbines, quantum cutting and IR absorbing materials for solar cell application, coating materials for LEDs, phosphor for lamps etc. Use of Ln in nuclear reactors is to control fusion as the Lns have large neutron cross sections and so are highly efficient for capturing neutrons. Lanthanides such as Samarium, Gadolinium, and Europium are being used for this purpose. One of the Samarium isotope, Sm^{149}, is mostly used as nuclear control rods. In addition to nuclear energy source, one other alternate source that is widely discussed is wind energy, which uses kinetic energy in the wind to produce mechanical energy. This mechanical power can be used for specific tasks (such as grinding grain or pumping water) or a generator which can convert the mechanical power into electricity. Wind energy can transform into electrical energy through giant wind turbines which essentially need high performance Ln super-magnets. Wind turbines can produce energies in the range of a few kilowatts to as large as several megawatts. Single small turbines, below 100 kilowatts, are used for domestic purposes, telecommunications dishes, or water pumping. China has the largest wind resources in the world and claimed to meet all of its electricity demands from wind power by 2030 [29].

Wind turbines use lanthanide-flecked super-magnets to generate electricity. Various Ln based magnets such as Neodymium-Iron-Boron and Samarium-Cobolt magnets are powerful, lightweight, strong, and relatively cheap and the best available. Apparently the whole wind energy sector is dependent on neodymium-iron-boron magnets, which are used inside wind turbines. Neodymium magnets are substantially cheaper than samarium-cobalt magnets. World production of Neodymium is about 7,000 tons per year and now its magnet is being used in hundreds of other products like cars, engines, headphone etc. Ln alloy improve efficiency e.g., iron alloy of terbium and dysprosium elements. This blend expands and contracts efficiently in the presence of a magnetic field, helping sensors, actuators and injectors to perform better. Car batteries used in electric-powered vehicles also rely heavily on rare earth elements. As this chapter only discusses the spectroscopy of Ln, we will not discuss other aspects of Ln magnets and neutron absorbing rod. Details about the use of Ln in these two sectors can be found in ref. [30].

Two major sectors looking for some breakthrough using Ln elements is solar cell and the Light Emitting Diode (LED). Similar to wind energy, solar energy is also considered as a clean and renewable energy as it directly converts sunlight into electrical energy while the use of efficient LEDs could reduce the worldwide electricity consumptions by up to 20%.

3.1. Solar Cell

A solar cell or photovoltaic cell is a device which converts energy radiated from the sun into heat or electricity. When a photon hits a photovoltaic cell it absorbs in bandgap and an electron-hole pair is generated and if collected properly, these carriers lead to an electric current. The process of converting solar energy to electricity is called *photovoltaics*. Photovoltaics (PV) includes various processes that include the absorption of light, generating either electron-hole pairs or excitons; separation of charge carriers of opposite types; and separate extraction of those carriers to an external circuit. Though the theoretical limit of efficiency (ratio of the electrical output of a solar cell to the incident energy in the form of sunlight) is estimated to be ~33.7% (in a Single p–n junction crystalline silicon) [31] However, in practice the maximum efficiency so far reported is ~24 % at the cell level and 20 % at the module level as reported by SunPower company [32] while a recent report from the University of New South Wales indicate efficiency achievement of ~25%. Details of efficiency comparisons can found in Ref. [33]. The efficiency of solar cell further increases with multiple-junction solar cell up to ~40.7% and 42.8% with multiple dies assembled into a hybrid package [34]. The efficiency of solar cell is the function of various factors that includes reflectance efficiency, thermodynamic efficiency, charge carrier separation efficiency and conductive efficiency. The overall efficiency is a product of each of these individual efficiencies.

Sun is regarded as a black-body radiator with temperature around 6000 K and emits in very broad spectral regions ranging from 300 - 2200 nm having a peak around 500 nm (2.4 eV) however, the silicon can absorb only a limited spectral region (Figure 1). The main hurdle to improve the efficiency in semiconductors is the low energy photon (sub-bandgap energy) which cannot excite the electrons from the valence to the conduction band and simply pass through while high energy photons (above-bandgap energy) even though they generate a pair of e-hole but lose their energy as heat, limiting the cell performance. Approximately, 47% of the solar energy gets converted to heat and 18% of the photons pass through the solar cell. Another option known as 'third generation solar cell' based on dyes and refer as dye-sensitized solar cell (DSSC) invented by Michael Grätzel and co-workers is still in process of development and so far maximum efficiency achieved is only 15 % (*http://actu.epfl.ch/news/dye-sensitized-solar-cells-rival-conventional-ce-2/*).

These losses are few of the main limitations and due to limited absorption by Si based solar cells, these radiations (UV and IR) are therefore considered as unwanted radiations. Ln elements can assist to absorb these unwanted radiations and convert photons that can be absorbed by silicon and further improve efficiency. Ln doped hosts are useful to increase the efficiency of solar cell through a process of down conversion (DC) (especially quantum cutting by absorbing UV solar radiation to produce two IR photons) [36] and upconversion (UC) by absorbing infrared radiations [37]. Through the process of downconversion and/or quantum cutting 32% (149 Wm^{-2}) and through upconversion 35% (164 Wm^{-2}) of the sunlight intensity is accessible [38]. Hence, it may significantly improve the solar cell efficiency. In typical DC/UC-Si Solar cell, a bifacial solar cell, i.e., one that absorbs light in the front and back is used and DC or UC layer is coated on the back side of the cell (Figure 2). A back reflector reflects all emitted photons back into the solar cell.

A theoretical study on realistically modeled UC solar cell reveals that the cell efficiency can be enhanced till ~44% [39]. Some of the old studies show the upper limit of the photovoltaic conversion efficiency of a single junction solar cell coupled to an ideal UC device is 47.6%, when the sun is modeled as a 6000 K blackbody and 50.7% under the standard air-mass (AM1.5) terrestrial spectrum [37, 40]. Briggs *et al.* investigated the effect of narrow-band upconversion on the efficiency of a solar cell and determined that the addition of an ideal upconverter with absorption bandwidth between 0.1 eV and 0.5 eV should boost the efficiency of a 1.7 eV bandgap cell from 28.2% to between 33.5% and 43.6% [41].

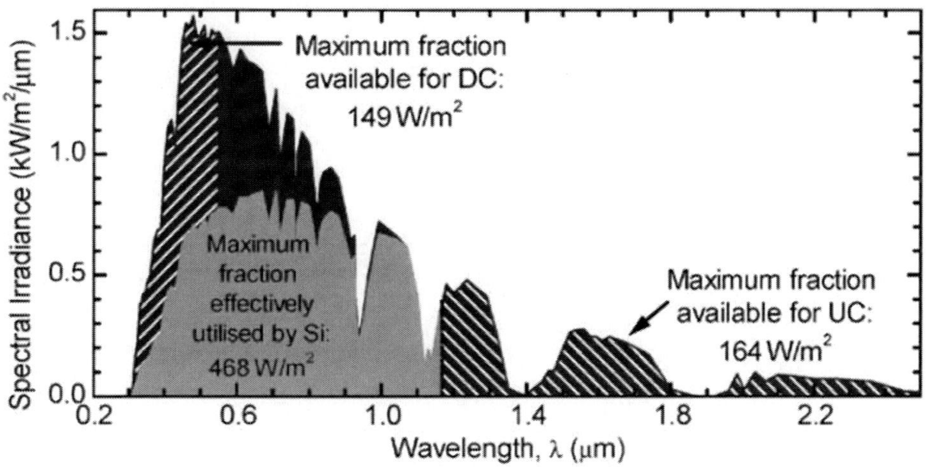

Figure 1. Absorption spectra of Solar radiation and Silicon. Reprinted with permission of ref. [35].

These losses are among few of the main limitations which restrict absorption by Si based solar cells. In fact due to poor absorption by Si, these radiations (UV and IR) are therefore considered as unwanted radiations. Ln elements can assist to absorb these unwanted radiations and convert in photons that can be absorbed by Si and improve efficiency. Ln doped hosts are useful to increase the efficiency of solar cell through a process of down conversion (DC) (especially quantum cutting by absorbing UV solar radiation to produce two

IR photons) [36] and upconversion (UC) by absorbing infrared radiations [37]. Through the process of downconversion and/or quantum cutting 32% (149 Wm^{-2}) and through upconversion 35% (164 Wm^{-2}) of the sunlight intensity is accessible [38]. Hence, it may significantly improve the solar cell efficiency. In typical DC/UC-Si Solar cell, a bifacial solar cell, i.e., one that absorbs light in the front and back is used and DC or UC layer is coated on the front and back side of the cell (Figure 2). A back reflector use to reflect all emitted photons back into the solar cell.

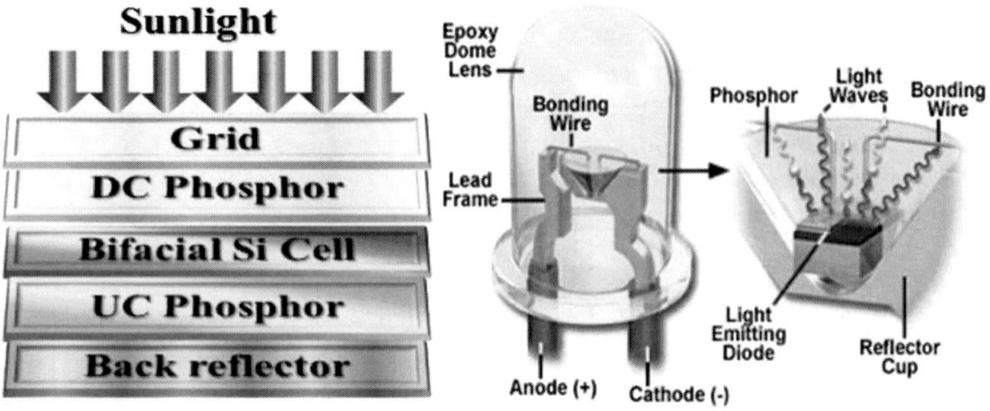

Figure 2. (left) Schematic layer arrangement for upconversion or downconversion phosphors in solar cell; (right) construction of LED and application of phosphor layer.

A theoretical study on realistically modeled UC solar cell reveals that the cell efficiency can be enhanced till ~44% [39]. Some of the old studies show the upper limit of the photovoltaic conversion efficiency of a single junction solar cell coupled to an ideal UC device is 47.6%, when the sun is modeled as a 6000 K blackbody and 50.7% under the standard air-mass (AM1.5) terrestrial spectrum [37, 40]. Briggs *et al.* investigated the effect of narrow-band upconversion on the efficiency of a solar cell and determined that the addition of an ideal upconverter with absorption bandwidth between 0.1 eV and 0.5 eV should boost the efficiency of a 1.7 eV bandgap cell from 28.2% to between 33.5% and 43.6% [41].

Among the Ln, Er and Yb ions are promising candidates for UC solar cells. However, other elements like Tm, Ce and Ho were also reported suitable for solar cell application [42]. Er^{3+} is a favorable ion to study upconversion process with NIR and IR lasers (796, 976 and 1550 nm) as it contains several metastable levels ($^4I_{9/2}$, $^4I_{11/2}$ and $^4I_{13/2}$ near ~12,000, ~10,000 and ~6450 cm^{-1}) which offer interesting possibilities of excited-state absorption and cross-relaxation processes. One inherent problem with Ln is the weak absorption, which can be solve by using sensitizer (with broad emission) or through plasmonic coupling [43]. Absorption cross section of the trivalent erbium ion at 980 nm is of the order of 10^{-21} cm^{-2} [44] while absorption cross-section for the $^2F_{5/2} \leftarrow {}^2F_{7/2}$ transition of Yb is roughly 10 times larger which act as an effective sensitizer for Er ion and improve absorption quite effectively at ~980 nm. However, addition of Yb may induce a possible loss mechanism as well since Si and Yb both absorb ~1000 nm wavelength. Several efforts have been made to improve the efficiency by using most efficient upconversion material which is $NaYF_4$ doped with Er and Yb ions. Shalav *et al.* showed upconversion under lower excitation density of 2.4

W/cm^2 reaching 3.4 % quantum efficiency (QE) at 1523 nm in a crystalline silicon solar cell with $NaYF_4$ doped with Er^{3+} as upconverter [45].

Henke *et al.* reported improvement in external quantum efficiency while using Er:Yb codoped ZBLAN glasses on 1540 nm excitations [46] J. de Wild *et al.* investigated β-$NaYF_4$:Yb^{3+} (18%), Er^{3+} (2%) coated solar cell on excitation with 980 nm diode (28 mW) and noted a maximum current enhancement of 6.2 μA which corresponds to external quantum efficiency of 0.03% of the solar cell [47]. J. de Wild *et al.* also noticed broad-band excitations may be more effective because simultaneously different wavelengths can be absorbed and converted. The response of the solar cells due to upconversion emission from Gd_2O_2S: Er:Yb under broad-band excitation was compared to that under monochromatic laser light excitation. The response due to the broadband was nearly the same as that under laser light excitation, though the power density of the laser was three times higher [48].

Similar to Er^{3+} ions, Ho^{3+} ion also emits Vis-NIR emission on excitation with infrared (~1 micrometer) radiation which is beneficial for solar cell application [49]. Lin *et al.* [50] observed enhancement of NIR emission in Ho:Yb combination doped in $BaGdF_5$ host which was mainly through second- and first-order ET processes: $Ho(^5F_3) \rightarrow 2Yb(^2F_{5/2}) + h\nu$; $Ho(^5S_2) + Yb(^2F_{7/2}) \rightarrow Ho\ (^5I_6) + Yb\ (^2F_{5/2}) + h\nu$ and corresponding QE were calculated to be 192%.

Pillai *et al.* [51] carried out investigations on the absorption characteristics of Si in presence of Ag NPs and reported that surface plasmons can increase the spectral response of thin-film cells over almost the entire solar spectrum. At wavelengths close to the band gap of Si, a significant enhancement of the absorption for both thin-film and wafer-based structures has been observed and further report a sevenfold enhancement for wafer-based cells at λ = 1200 nm and up to 16-fold enhancement at λ = 1050 nm for 1.25 μm thin silicon-on-insulator (SOI) cell, and compare the results with a theoretical dipole-waveguide model. An effort has been made to harvest solar energy using an upconverter-doped plasmonic nanocrescent by Atre *et al.* [52]. Numerical calculation reveals >10-fold absorption enhancement for a broad range of sub-bandgap wavelengths throughout the entire upconverting core and nanocrescent enables a 100-fold increase in above-bandgap power emission towards the solar cell.

In contrast to UC process in which successive absorption of infrared photons results UV-Vis photons, in DC process a UV-blue photon yields low energy photon (down shift) or two photons (quantum cutting (53)). Concept of down shifting is analogues to downconversion which converts high energy solar photon to lower energy photon which can be absorbed by Solar cell. Quantum cutting (QC) is a very important process, in which two or more visible/NIR photons emit on the cost of a single UV photon increasing the efficiency of light-emitting device >100%. Mechanism for QC or cascade emission mainly includes two different processes one: a simultaneous energy transfer from a donor to two acceptors equally shares the energy, and in second: a sequential emission of two photons on absorption of a UV photon in an Ln ion. [54] Examples of first type of QC are Tb:Yb, Bi:Yb, Ce:Yb, Eu^{2+}:Yb, $VO_{3-}{}^{4}$ group:Yb^{3+} [55] while for second type of QC YF_3:Pr^{3+} [54]. One another type of QC is also reported that includes two step energy transfer process in Gd:Eu system [56]. In this process a 202 nm photon absorb in ^6G_J levels of Gd ion and the energy ultimately transferred to Eu ions via two steps and bright visible emission appears. Low energy photon can also obtained through process of phonon assisted energy transfer as reported in case of Eu:Yb codoped system [57].

In case of Pr ion, QC occurs from $^1S_0$ (~47,000 cm^{-1}) state only when it is not buried in 4f5d band. Excited Pr ion relax to ground state through two step process 1)

$^1S_0 \rightarrow {}^1I_6, {}^3P_J$ (~400 nm); 2) $^3P_0 \rightarrow {}^3F_J, {}^3H_J$ (480–700 nm). QC process is observed in various Pr doped fluoride hosts YF_3 (on 185 nm excitation) [54], oxide hosts $SrAl_{12}O_{19}$ [58], $LaMgB_5O_{10}$ [59], LaB_3O_6 [60]. Kuck *et al.* [61] investigated Pr doped different fluoride YF_3, LuF_3, PrF_3, $BaMgF_4$, $LiCaAlF_6$, $LiSrAlF_6$ and $KMgF_3$ which exhibits QC under VUV synchrotron excitation ~200 nm with a theoretical, internal quantum efficiency (QE) higher than unity. However, application of these materials is limited for lamp purpose as emission occurs at ~405 nm in which region human eye is not much sensitive.

Ikeda *et al.* [62] observed QC in $SrAl_{12}O_{19}$:Pr on 172 nm excitation and observed two visible photons, one at 401.8 nm and the other at 486.3, 526.3, 623.7 or 644.6 nm via $^1S_0$ state. The photoluminescence measurements show that the relative quantum efficiency of $SrAl_{12}O_{19}$:Pr is 38.4% of $BaMgAl_{10}O_{17}$: Eu^{2+}. Lee *et al.* [63] observed Pr sensitized QC in K_2GdF_5:Eu^{3+} phosphor through a cross-relaxation and energy transfer process Gd^{3+}–Eu^{3+}, Pr^{3+}. It was suggested that due to the improved absorption in the UV and VUV spectral regions the theoretical QE of K_2GdF_5: Eu^{3+}, Pr^{3+} was increased to 138%. Loef *et al.* [64] discussed a case of visible QC in $LiGdF_4$: Er and Tb. Upon VUV excitation in the $4f^{10}5d$ levels of Er, part of the excitation energy can be transferred to Gd by cross-relaxation, yielding both Er and Gd in an excited state. This is followed by the emission of two visible photons: one from Er and, after energy transfer from Gd to Tb, one from Tb. Recently, Chen *et al.* [65] reported QC in Pr/Yb: YF_3 nanocrystals on excitation with 482 nm which yield 976 nm through an efficient cooperative energy transfer from Pr to Yb. The optimal quantum efficiency was found to be ~ 200%. Dwivedi *et al.* [66] observed bright red emission on 266 nm excitation in Pr:Y_2O_3 nanocrystals with QE >80 %.

Recently, Han *et al.* [67] observed QC in $Na_3Gd_{0.9-x}Y_xSi_3O_9$:$0.1Tb^{3+}$ phosphor on 274 nm exc. QC in Nd:Yb system is also observe by Sun *et al.* [68] and Meijer *et al.* [69] in $CaWO_4$, and YF_3 hosts, respectively. Wegh *et al.* have studied the visible QC of RE ions in fluoride, $LiGdF_4$: Eu [56] and $LiGdF_4$: Er; Tb [70] whose QC efficiency has reached 190% and 130%, respectively, when the absorption at centers other than Gd or Er is neglected. Liu *et al.* [71] reported two step energy transfer QC in BaF_2: Gd; Eu with visible quantum efficiency up to 194%. QC process in Tm-Yb doped in LaF_3 nanocrystals were observed on 468 nm excitation in Tm $^1G_4$ energy level which results emission in 950-1100 nm range corresponding to Yb ions. Maximum QE was reported 162% in 0.5Tm-8Yb composition. Deng *et al.* had demonstrated QC in $NaYF_4$:Ho^{3+}, Yb^{3+} synthesized by hydrothermal method [72]. Upon 359 nm excitation near-infrared QC could occur through a two-step resonance energy transfer from Ho to Yb by cross relaxation, with a maximum QE of 155.2%. Yu *et al.* [73] also observed QC in hexagonal beta- $NaYF_4$ doped with Ho:Yb on excitation with 300–550 nm which yielded two NIR photons. It is found that NIR-QC can occur via two possible energy transfer approaches one, through simultaneously excitation of two Yb ions via cooperative energy transfer and the second, via resonant energy transfer process and then a sequential $^5I_6 \rightarrow {}^5I_8$ transition of Ho^{3+}. Zhou *et al.* [74] observed three photon cutting process in Yb:$CaAl_2O_4$ phosphor via energy transfer from Yb^{2+} ions to Yb^{3+} ions. Chen *et al.* [75] reported two photon QC in Ce^{3+}/Yb^{3+} codoped borate glasses on 330 nm excitation, which excite Ce^{3+}:5d ions and energy transferred to Yb^{3+}: $^2F_{5/2}$ ions. The maximum ET efficiency and the corresponding DC, QE were estimated to be 174%, respectively. QC in Chalcohalide glasses is also observed which occur from Eu^{2+} to Yb^{3+} ion. [76] Detail information on various upconversion and downconversion materials that may be used in Solar cell can found in reviews [77].

3.2. Light Emitting Diode

As stated earlier possible ways to tackle interconnected problem of pollution and clean energy production are 1) to use efficient renewable or green technology sources and 2) to make devices more efficient. It is interesting to note that only for illumination purpose ~20 % of the worldwide electricity consume. The global electricity requirement has increasing faster than the overall energy production because of the versatile nature of the electricity production, as well as its consumption. Worldwide electricity consumption in 2006 was 16378 TWh, which was 11.8% of the total primary energy consumption. According to the International Energy Outlook 2009, the world's total net electricity generation in 2030 is expected to be increased by 77% from the 2006 level [78].

Another concern is the bulky, expensive and less efficient conventional illumination sources (lamp, bulb, tube light etc.). They are also one of the primary domestic pollutant sources as they emit toxic materials like mercury (Hg), sulfur, carbon dioxide (CO_2), uranium etc. Necessity of efficient, economical and eco-friendly sources in future, a surge of research works has been done and a most suitable replacement of conventional illumination sources suggested is Light Emitting Device (LED) which is supposed to reduce electric consumption up to 70-80 % compared to the conventional lamps. According to researchers from Strategies Unlimited LED sales reached to $9.4 billion in 2012 from $2.7 billion in 2008. And the demand grows at a breathtaking compound annual rate of ~20%. The research groups noted that more LED lighting products are becoming popular in various markets. Frost & Sullivan, a market trends research group, anticipates revenues in the efficient lighting market in Europe to hit just under $2 bn by 2018, double of last year's figure, driven largely by legislation by the European Commission to phase out various traditional light bulbs [79]. The concept of LED [80] is attractive for general illumination purpose due to the reliability of solid-state devices, potential for delivering very high luminous efficiency compared to conventional incandescent and fluorescent sources, long life, vivid color emission, possibility of variable and multi-hued devices, its compact, environmentally friendly (mercury free) and various other desirable features.

Broadly, LEDs are categorized in two namely inorganic; and organic LEDs. Recently, Reineke *et al.* have reported excellent white organic LED more than fluorescent tube light [81]. However, the performance of organic LEDs is deleteriously affected by high temperature and additionally, the materials in an Organic LED are sensitive to air and moisture which laid to degradation of the material themselves as well as quenching of excited states within the molecule [82]. As most of the highly fluorescent materials yields poor electroluminescence compare to optical excitation, a concept of hybrid LED coated with yellow organic/inorganic phosphor materials is also mostly acceptable [83].

Electroluminescent (EL) devices based on Ln-complex thin layers are of great interest because of their possible application in large-area light-emitting displays operating at the low voltage. In general a thin layer of Ln-complex (amorphous or crystalline) is sandwiched between a hole transport layer and electron transport layer. When an electric field applies, generation of light results due to recombination of holes and electron carriers injected from the electrodes. Such recombination in the emitter layer excites the emitter material. Efficiency can be further improved by blocking electrons and holes from moving into the electrode without recombination.

Problems related to organic LED would be solved up to some extent by using Ln doped inorganic phosphors which provide narrower and efficient emission. For single color LED mostly Eu^{3+} (red), Eu^{2+} (blue, yellow), Tb (green), Tm (blue), Dy (yellow), and Ce (blue, yellow) ions complexes are used.

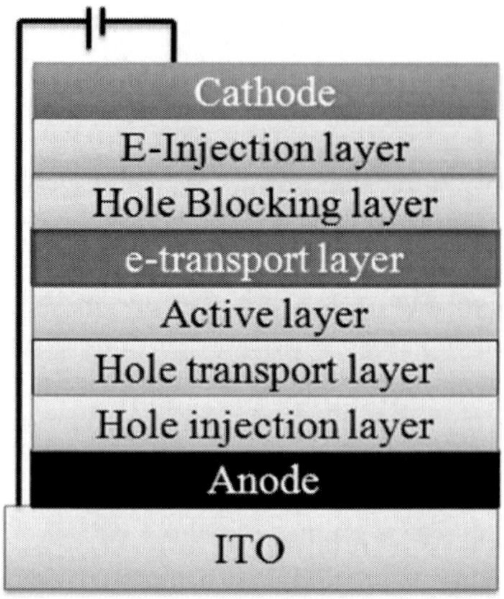

Figure 3. General configuration of electroluminescence cell.

Organic molecules that can be act as sensitizer for Ln ions are beta-diketone functional groups such as acetylacetone, dibenzoylmethane, 2-Phenoyltrifluoroacetone; carboxylic groups such as o-benzoyl benzoic acid, salicylic acid, o-phthalic acid; organic compounds which have ketone or aldehyde groups adjacent to hydroxy groups, such as salicyl aldehyde, o-hydroxyacetonephenone or o-hydroxybenzophenone; oxine analogues such as 8-hydroxyquinoline or 5,7-dibromo oxine; pyridine analogues such as 2,2'-dipyridine, 2,2'2"-tripyridine, 1,10-phenanthroline and cryptates and crown ethers [84]. These ligands can also be combined, using more than one ligand in the same device. Additionally, various semiconductor hosts and inorganic nanophosphors are also predicted for LED applications. Recently, GaN based devices, codoped with Eu and Mg, shows luminescence efficiency as high as 77 percent. Mg doping selectively enhance a specific emission site and increase emission intensity at least one order suggesting codoping of Eu and Mg in GaN may be utilized for new devices [85]. However for general household lighting purpose mostly white light is desired. It is interesting to note that the human eye perceives light as being white if the three types of photosensory cone cells, located in the retina, are stimulated in particular ratios.

In practice, blue emitting GaInN LED chip is coat with Cerium doped Yttrium Aluminium Garnet (YAG). Ce:YAG absorb blue emission from LED and yield yellow emission which apparently looks white to the eye [86]. However, concept of yellow and blue mixing may not be good as it has many disadvantages, such as poor color rendering index (CRI) property, blue/yellow color separation and lacking of the red component in the spectrum of this kind of LED. This problem can be solved up to some extent by codoping with Gd ion along with Ce, which causes a shift in yellow-orange range [87].

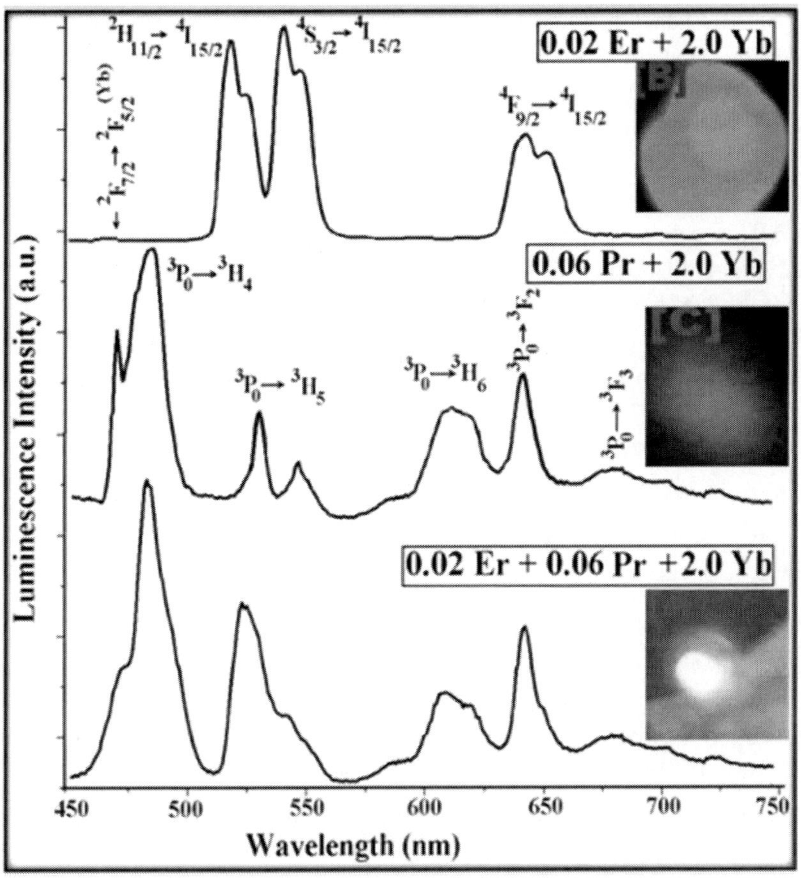

Figure 4. Luminescence spectra of (A) Er:Pr:Yb, (B) Er:Yb (C) Pr:Yb co-doped tellurite glasses. Reprinted with permission of ref. [101].

Divalent Eu ion [88] usually yields blue emission but it can also be tune in wide range of green, yellow and red regions as well. Park *et al.* [89] report bright green emission in $(Ba_{1.2}Ca_{0.8-x}Eu_x)SiO_4$ phosphor. Xiea *et al.* [90] have reported bright yellow emission in Eu doped oxynitride phosphor a-SiAlON. Due to formation of Eu^{2+} ion it absorbs light efficiently in the UV–visible spectral region, and shows a single intense broadband emission in 583–603 nm regions which further use to create white light on 450 nm LED excitation. Kuo *et al.* [91] reported a red phosphor $CaZnOS:Eu^{2+}$ which has broad absorption band in 450-550 nm and emit in ~600-700 nm region. This phosphor was combined with green ($(Ba,Sr)_2SiO_4:Eu^{2+}$) and yellow (YAG:Ce) emitting phosphors to get white light. Similarly, broad red emission is also observed in $Sr_2Si_5N_8:Eu^{2+}$ [92], $CaAlSiN_3:Eu^{2+}$ [93] white light emission was observed in single crystalline $Ca_9Gd(PO_4)_7:Eu^{2+},Mn^{2+}$ phosphor which exhibit a broad blue-green emission band at 490 nm and a red emission band at 645 nm, which originates from the Eu^{2+} and Mn^{2+} ions, respectively [94]. When $Sr_3SiO_5:Ce^{3+},Li^{+}$ and high quality TOP/TOPO/HDA-capped CdSe QDs were coated onto a blue LED chip, white LEDs with an excellent color rendering property were obtained. The white LEDs exhibited a luminous efficiency of 14.0 lmW^{-1}, and R_a of 90.1, and a T_c of 8864 K, which is acceptable for general lighting, indicating that the combination of phosphor and QDs in LEDs can be a good solution to obtain white light with high color rendering properties [95]. Kim *et al.* [96]

fabricated GaN-LED using mixture of $Ba_3MgSi_2O_8$: Eu^{2+} and Sr_2SiO_4:Eu^{2+} phosphors which yield a broad-band spectrum, higher color rendering index and higher color stability against forward bias currents than $Y_3Al_5O_{12}$:Ce^{3+}-based white-LEDs. Li *et al.* [97] synthesize a new material $MSi_2O_{2-x}N_{2+2/3x}$:Eu^{2+} (M= Ca, Sr, Ba) which yield blue-green emission in 490-500 nm region is observed for $BaSi_2$ O_2-$xN_{2+2/3x}$:Eu^{2+}, yellow emission at 560 nm for $CaSi_2O_2$-$xN_{2+2/3x}$:Eu^{2+} (x~0), and a green-yellow emission peaking from 530 to 570 nm for $SrSi_2O_{2-x}N_{2+2/3x}$:Eu^{2+}(x~1), the position depending on the exact value of x. $BaSi_2O_2N_2$:Eu^{2+} is the most promising conversion phosphor for white-light LEDs due to its high conversion efficiency for blue light from InGaN-based LEDs related to its very small Stokes shift. White light also reported by combination of complementary colors in Dy doped $SrSiO_3$ long-lasting phosphor which yields two emissions at 480 nm and 572 nm [98]. White light also stimulates by RGB color mixing in TmErYb [99], TmHoYb [100], PrErYb (Figure 4) [101] however, suitability of these materials still need to test in actual device.

SENSORS

It has been previously recognized that Ln complexes exhibit characteristically narrow absorption/emission bands, intense fluorescence and long excited-state lifetimes. These properties are most suitable to be used as chemical sensors [102]. Ln exhibits intra-configuration transitions and just a few of the optical transitions, known as hypersensitive, are susceptible enough for environment sensing. However, in addition to them, paramagnetic lanthanide elements have a large number of degenerate energy levels and are sensitive to a variety of metal ions, anions, humidity, hydrogen peroxide, magnetic field, pH and various gases like NH_3 and O_2. Readers are referring to Ref. [103] for more detailed information. In the following section we will discuss about variety of Ln based sensor except temperature as it is already discussed in one of the chapters of this book by V. K. Rai *et al.*

As mentioned earlier, Ln elements possess small absorption cross-section and then, to improve their sensitivities sensitizer ligand molecules are usually employed. It is interesting to note that the coordination number of Ln is >6, providing enough sites to accommodate both the sensitizer ligand and the analyte in their inner coordination sphere. Two type of coordination ligand is in practices 1) saturated (e.g., Caged complex, cyclen complex, helicate etc.) and 2) unsaturated ligands (e.g., Beta-Diketonate complex, tripode complex, porphyrinate complex, oligopyridine– polycarboxylate complex, dendrimer complex etc). Saturated sensitizer ligand also provides additional advantage of shielding of Ln ion from ambient hydroxylic like solvent molecules that would otherwise quench the Ln luminescence; and used for the labeling of biological proteins and DNA molecules. However, for sensing purposes, the ligand should be sufficiently close to the emissive site to permit efficient energy transfer, besides containing vacant sites for the guest ion. The complexes of Eu^{3+} and Tb^{3+} ions exhibit red and green luminescence, respectively, while Nd^{3+} and Yb^{3+} complexes yield NIR emission. These complexes are highly luminescent and desirable for in vitro and in vivo analyses. The long-lived energy levels of Ln based sensors also offer advantage over shorter lived background emission (auto-fluorescence) and light scattering from biological environments [104].

Ln ions probe some of the toxic war gases as soman, which was reported to be used in wars [105]. It should be noted that one of the serious threat of chemical warfare is the use of nerve agents that causes rapid and severe effects on human health. Basically, these nerve agents are made of organo-phosphate that stops the function of the acetylcholinesterase enzyme, leading to various organ failures. Ln has the property of linking to organo phosphate and helps to detect the nerve agent [106]. Detection of nerve agents is based on variations of the fluorescence intensity upon complexation to lanthanide (Eu and Tb) ions (Figure 5). Comparing the fluorescence intensity of free ligand and complex, chemical agent can be detected. By tailoring the nature of the metal-ligand interactions, it is possible to enhance the sensitivity of such systems. Another advantage of Ln sensors is its exceptional selectivity to other chemical compounds like ester, ketones, etc.

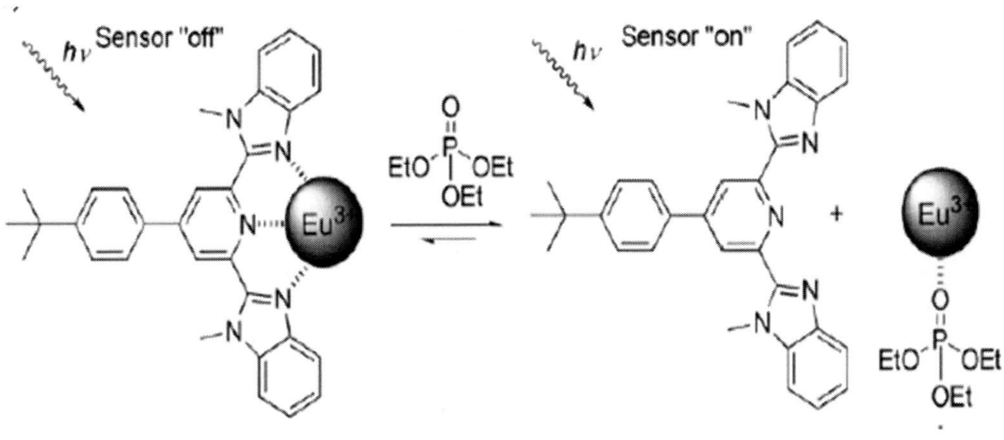

Figure 5. Mechanism for supramolecular sensors. Reprinted with permission from ref. [107].

Knapton *et al.* [108] reported Mebip-functionalized ethynylene derivatives for the detection of organophosphate and demonstrated the ability of readily controlling the extent to which the fluorescence of the Mebip ligands changes upon metal binding. It was observed that on binding with La^{3+}, the emission spectrum of the ligand experiences a significant bathochromic shift (Δ_{max}=73 nm, emission color change from blue to green and the fluorescence intensity is significantly decreased).

Tremblay *et al.* [109] proposed a general strategy for the rational design of ratiometric, luminescent probes based on a heterometallic bislanthanide (Eu^{3+} and Tb^{3+} (or Dy^{3+})) platform. The relative emission intensities can be tuned by adjusting the stoichiometry and since the luminescence is sensitized by a single antenna, changes in its electronic structure can manifest themselves as changes in the relative ratio of luminescence from each metal. Jenkins *et al.* [110] reported a fluorescence based fiber optic probe for organo phosphates, using a supramolecular complex up to the detection limit of 660 ppq. They used a molecularly imprinted polymer sensor of [Eu-(dvmb)$_3$(pmp)(NO$_3$)$_2$], a Eu^{3+} complex with three different ligands. Although the response of this sensor is limited by analyte diffusion, the sensitivity is high enough for precise detection even in presence of chemically similar organo phosphates.

Silica particle coated with Eu complex shows instantaneous and significant color change, from pink to blue, on exposure to liquid and vapor (for 2 h at 60 ^{0}C) triethyl phosphate. Another approach for detection is using an enzyme as the binding motif in presence of metal nanoparticles like Au-NPs. This concept is based on the variation of the fluorescence intensity of lanthanide ions with the relative distance to the metal NPs, so as the field near lanthanide ion. Detachment of NPs from lanthanide complex occurs in the presence of a weakly bound phosphate derivative, fluorophore, to the enzyme which is in turn covalently attached to gold nanoparticles. Separation of the metal NP from the complex reduces the emission intensity of the Ln complex. This displacement occurs through the competitive binding with an analyte that has a higher binding affinity to the enzyme than the phosphate derivative. Simonian *et al.* [111] reported organophosphorus hydrolase as the receptor for organophophorus analytes.

Koshizaki and Oyama [112] investigated the selective enhancement of NO_2 against NO sensing by rare earth element doping. Er-doped ZnO film exhibited good sensitivity in the sub-ppm concentration range, good selectivity and fast response time to NO above 200 °C. H_2S gas over >250 ppm is known to be a potentially hazardous gas for the health. Hence, it is necessary to determine H_2S quantitatively, especially at the coal, oil or natural gas manufacturing sites or laboratories. Zhong *et al.* [113] reported a novel Ln (Eu,Gd,Ho):In_2O_3-based H_2S gas sensor and found that 5 wt% Ho^{3+}- In_2O_3 sensor exhibits the highest response value, excellent selectivity and quick responsive in the range of 1-100 ppm.

Li et al. [114] studied a simple and effective solvent swelling method to prepare PMMA/Eu complex hybrid nanospheres (100-200 nm). These luminescent nanospheres show high sensitivity in the luminescence sensing of hydrogen peroxide in a broad linear concentration range.

Tsang and Bulpitt [115] reported superior performance of Ln oxides over tin oxide-based ethanol sensors. They proposed that adsorbed oxygen of strong basicity can facilitate the oxidative dehydrogenation of ethanol, with the intermediate acetaldehyde giving a high molecular response. They further investigated a different sensing mechanism in case of Pr oxide-based sensor, where their labile lattice oxygen is active for ethanol oxidation even at low temperatures (≥150°C). The electron hopping sites, associated to the mixed valence cations, are partially destroyed in a stream of ethanol vapor causing an increase in the electrical resistance. Upon removal of ethanol, molecular oxygen in air replenishes the vacant oxygen sites reverting sensor to its initial resistance value. The lattice oxygen can be tuned to react only with ethanol but not with n-butane at or below 300°C, achieving 100% selectivity. This mechanism can selectively transduce ethanol concentration into an electrical signal. Niu *et al.* [116] reported nanometer materials of $REFeO_3$ (RE=Eu, Gd) which were sensitive to C_2H_5OH, H_2S, gasoline and H_2 gas. The studies of gas-sensing characteristics show the sensitivities of $EuFeO_3$ and $GdFeO_3$ nanometer materials to gasoline are 39 and 45 at 340 °C, respectively, and both of them have good selectivity too. Aono *et al.* [117] accounted for NO_2 sensing using $SmFeO_3$ and $REFeO_3$ (Re=La, Nd, Sm, Gd, and Dy).

Ln based sensors were studied for the detection of metal cation, organic and inorganic anions, small neutral molecules in solution and various volatile organic chemicals detection. The use of encapsulated Ln ions as sensitized emission centres capable of multiple bindings has emerged as a compelling strategy for luminescent supramolecular sensor development [118]. They utilize the large absorption cross-sections of selected organic molecules and the long lifetime and sharp emission characteristics of Ln ions to generate very efficient molecular devices. One of the major constituents of our environmental system, Nitrogen, is generally not consider toxic for living bodies though its metabolites may be carcinogens [119]. However, the use in large scale of nitrogen- and phosphorus-based fertilizers in the agriculture sector may pollute the soil, wastewater and ultimately food, via plants. The detection of excess amount of nitrates generally is based on the ion exchange principle. It is known that counter ions can coordinate to Ln ions in unsaturated ligands to satisfy the coordination requirements of the Ln ions. Generally, there are two ways of modulating the emission property of Ln complexes by anions. First, water molecules bound to the Ln centre could displace by anions, leading to an enhanced luminescent intensity and second is the displacement of ligands by anions resulting in the increase or decrease of the luminescent intensity [120]. Ln based anions detection can be possible by the careful design of antenna ligands that leaves some coordinate vacancy for anion binding to Ln.

Douce *et al.* [121] synthesized the Eu complex of bis-bipyridinephenylphosphine oxide ligand and reported an increase in luminescence yield and lifetime due to the anion binding. The authors concluded that this new system is applicable to other molecules because it also detects traces of halides and nitrates by monitoring luminescence of Eu, whose long excited-state lifetime fasilitate the enhancement of the signal to noise ratio with the use of time-resolved spectroscopy. Santos *et al.* [122] investigated interesting anion-sensing with cyclen Ln complexes. Zhang *et al.* [123] synthesed an organoeuropium complex, $Eu(DBM)_3DPPZ$ (DBM = dibenzoylmethane, DPPZ = dipyrido [3,2-*x*2′,3′-c]phenazine), which was found to be efficiently to recognize AcO^- ions over other anions due to the replacement of DPPZ by AcO^- ion. Based on this characteristic, this “off–on”-type luminescent sensor could successfully mimic an IMP (IMPLICATION) gate at molecular level with AcO^-ion and oxygen as inputs. The output of a sharp characteristic Eu^{3+} emission is read as “***0***” only in the absence of AcO^- ion under oxygen atmosphere. Optical sensing of Cl^- and NO^{3-} anions were reported in Eu and Tb complexes with tetradentate tripodes [124]. Parker *et al.* [125] reported the emission intensity variation of Eu ions as a function of halide ions (Cl, Br, I) in aqueous solution of Eu-octadentate monoamindetriphosphinate or chiral tetraamide complex, which was attributed to the competition between the excited-state quenching rate of the chromophore and the rate of intersystem crossing from triplet state of the chromophore to the 4*f* level of Eu ion. Faulkner *et al.* [126] showed the selective stabilization of a more luminescent bound form to give an intense near-infrared Yb^{3+} luminescence on addition of SCN^- ion in a dendrimer complex. Parker [127] reported on heptadentate cyclen ligands for Ln ions that selectively bind bidentates such as HCO^{3-} and carboxylate anions and found greater dynamic displacement of the coordinated water molecules by the guest anions than octadentate cyclen complexes. Distinctive anion-selective (NO_3^- and Cl^-) luminescence enhancement was demonstrated by Kataoka *et al.* [128] in different Eu- tripodes complexs bearing quinoline chromophores.

A Eu:Tb ion dispersed in a polymer hybrid doped with a proton-sensitive amide-type beta-diketonated photosensitizer, N-(2-pyridinyl)benzoylacetamide was shown to be highly luminescence sensitive to OH^- concentrations, thus representing a new paradigm for fullcolor displays[129]. At the same time, base–acid vapor-responsive luminescent films with good stability, sensitivity, reversibility and quick response might provide access to sensor materials.

Various cation sensors were developed for selective detection of alkali and alkaline earth metals (Li, Na, K, Zn, Ca, Mg) that are necessary to monitor and regulate a number of cellular functions and heavy metal cation (Pb, Cu, Hg) for biomedical purposes. It is interesting to know that bound Cu^{2+} plays a vital role in many enzymatic processes, while free Cu^{2+} can disrupt and damage various biological processes [130]. Similarly, Hg^{2+} is also toxic to humans and can accumulate in the blood-brain barrier, resulting in severe neurological disorders [131]. Similarly, lead is a cumulative poison, with foetuses, infants, and pregnant women being the most susceptible to adverse health effects. Lead can severely affect the central nervous system. Evident signs of acute intoxication include dullness, restlessness, irritability, poor attention, headaches, muscle tremor, hallucinations and loss of memory [132]. In addition, there are several other environmental concerns regarding cations. The emission intensity of Eu ions is enhanced after the addition of alkali metal ions (Na^+, K^+) by at least one order of magnitude in aza crown molecular complex [133] due to the suppression of electron transfer from the nitrogen atom of aza crown to excited Eu ion. Nevertheless, this result was analogous to the case of Tb-complex, which is thought to be due to the existence of additional relaxation nonradiative pathways for $^5D_4$:Tb ion.

A Tb^{3+}-cyclen-based sensor was proposed for the pH independent, selective detection of Cu^{2+} and Hg^{2+} ions in water. The changes in the Tb^{3+} emission were monitored by carrying out metal titrations which quenches Tb emission by *ca.* 65% (Cu) and 40% (Hg) [134]. Plush and Gunnlaugsson [135] design a novel stable dinuclear Eu conjugate by tethering a mono-aza-18-crown-6-ether moiety to a cyclen macrocyle. The analysis of this conjugate showed that anions such as acetate and L-aspartate quench Eu^{3+} emission. Malonate gave rise to large luminescent enhancements in the hypersensitive transition.

Hanaoka *et al.* [136] reported a novel Ln molecule, using the quinolyl ligand, to detect Zn^{2+} ions. It was observed that the emission intensity of Eu^{3+} considerably enhanced on addition of Zn^{2+}. One of the important advantages of this complex is that it can be excited with comparatively longer excitation wavelengths (~340 nm) than previously reported Ln based Zn^{2+}-sensor, whose excitation wavelength is at a too high energy level for biological applications. This complex was claimed as the first Zn^{2+} sensitive Ln chemosensor which can be used for the study of the biological functions of Zn^{2+}. The design strategy should yield a range of long-lived Ln probes for sensing Zn^{2+} or, after appropriate modification of the acceptor moiety, other molecules of interest in biological applications.

Yu *et al.* and Pikramenou *et al.* reported Ln–cryptate assemblies, which react to the addition of light harvesting molecules like picolinic or benzoic acid, resulting in the binding to Eu ion, energy transfer and sensitized Ln emission [137]. Charbonniere *et al.* [138] observed the increase of the Eu complex luminescence lifetime (combining a hard donor with two soft pyridine donors) upon the addition of Adenosine triphasphate (ATP), whereas Adenosine diphosphate (ADP) and Adenosine monophosphate (AMP) did not induce significant changes. The application of Ln ions in liquid chromatography [139] extend the possibility of detection of various other species including single stranded nucleic acids [140], hormones [141], amino compounds [142] and labeled thiols [143].

Lanthanide ions are also use for protein labeling, specifically metalloproteins which are reported to form stable lanthanide complexes. Two of the natural occurring proteins, transferrin and lactoferrin, contain a Fe^{3+} metal center. It is interesting to note that the two bound Fe^{3+} cations released from lactoferrin at pH 2.5-4.0 and the transferrin loses its Fe^{3+} cations at pH 4.0–6.0 which is reported to be substitute by Ln ions and thus form emissive complex [144]. Another protein, apoferritin, is reported to form complex with Tb ions [145] and lengthen Tb luminescence. Kataoka *et al.* [146] reported pH-dependent Tb luminescence in proteins apolactoferrin and apotransferrin.

Since hydrocarbons do not coordinate to Ln ions, a rather different technique is needed to detect them. In this case, two different ligands are used, one for the Ln ion (aza crown) and the other for the hydrocarbon (functionalized cyclodextrin) [147]. Mortekkari *et al.* [148]

introduced a tri-anionic diethylenetriaminepentaacetate to the Tb- cyclodextrin complex to improve the strength of aromatic hydrocarbon binding and reported effective sensitivity of such complex towards hydrocarbon. The detection of protease inhibitors is very important due to its application in curing different diseases (cancer, Acquired Immunodeficiency Syndrome (AIDS), inflammatory disorders, etc.). Karvinen *et al.* [149,150] reported a protease assay based on a complicated time-resolved fluorescence resonance energy transfer (TR-FRET). Mizukami *et al.* [151] developed a novel luminogenic Tb probes for detecting protease activities.

Magnetic field sensing is one of the requirements in the fields of navigation, vehicle detection, current sensing, and spatial and geophysical research. Optical magnetic field sensors are desirable because of their immunity to electromagnetic interference, low weight, small size and long-distance signal transmission for remote operation. The well-known Faraday Effect rotates the plane of polarization of an optical wave in the presence of an external magnetic field and hence it can be used for magnetic field sensing [152]. Optical current/magnetic-field sensors based on the Faraday Effect offer the advantages of simplicity and high electrical insulation. Optical magnetic-field sensors based on rare earth iron garnets, such as $Y_3Fe_5O_{12}$ and $Gd_2BiFe_3O_{12}$, have extremely large Verdet constant, which characterizes the Faraday Effect, yielding higher sensitivity, frequency response, temperature characteristics and domain effects than sensors based on diamagnetic and paramagnetic materials [153]. The Faraday Effect of rare-earth-containing oxide, fluoride and oxyfluoride glasses has been well discussed [154]. It is known that trivalent Ce, Pr, Tb, Dy and divalent Eu bring about large Verdet constants in the visible range. Tanaka *et al.* [155] measured the Faraday Effect and optical absorption of Pr-containing lithium borate glasses and reported large magneto-optical figures of merit in the visible range. In particular, $25Pr_2O_3$ -$75B_2O_3$ and $20Pr_2O_3$ - $5Li_2O$ -$75B_2O_3$ glasses exhibit figures of merit larger than 250 deg/μm at 400 nm under an external magnetic field of 15 kOe. Therefore, these glasses are promising materials in optical isolator for the (Ga,In)N laser diode with the emission wavelength of 400 nm. Itoh *et al.* [156] developed a 5 mm wide optical magnetic-field sensor based on the novel $Bi_{0.98}Gd_{0.92}La_{0.03}Y_{1.07}Fe_{4.72}Ga_{0.28}O_{12}$ garnet, that demonstrated high accuracy with ±1.0% error and shows high temperature stability for sensitivity of less than 2.0% from -20 to +80 °C. Potter and Sinclair [157] review two inorganic materials classes used for optical sensors: photosensitive glasses and rare-earth doped materials. The discussion on RE doped optical sensor materials emphasizes the important contribution of the host material to observed optical behavior of RE ions. The use of RE doped materials in applications ranging from temperature sensing to fiber-optic gyroscopes is also described. Sun *et al.* [158] accounted for a 2 cm long of 56 wt% terbium doped silicate fiber as Faraday rotator, with a Verdet constant of –24.5 rad/(Tm) at 1053 nm. The sensor had a sensitivity of 0.49 rad/T and can measure magnetic fields from 0.02 to 3.2 T. The highest Faraday rotation in the visible spectral regimes was obtained in the composite film $(BiDy)_3(FeGa)_5O_{12}$:Bi_2O_3 with values 10°/μm, 2.6°/μm, and 1.9°/μm for wavelengths 532, 635 and 670 nm, respectively. In this composite film, it was demonstrated that as the wavelength of the incident light increases, the Faraday rotation decreases [159]. Similarly, at the telecommunication wavelength 1.55 μm, the Faraday rotation measured in CeYIG,Co-doped $CeO_{2\delta}$ films, $CeFeO_3$ films, YbIG, Sc doped EuIG, scandium substituted EuIG films were found to be 0.44, 0.60, 0.60 −0.0420, −0.0450 and −0.0500 deg/μm respectively [160]. Sekhar *et al.* [161] measured Faraday rotation in

$Bi_xCe_{3-x}Fe_5O_{12}$ films prepared by pulsed laser deposition and found 1.6 and 50 times larger than that of CeYIG and YIG, respectively.

ACKNOWLEDGEMENT

Authors SCZ and YD are thankful to Fundação de Amparo à Pesquisa do Estado de São Paulo (FAPESP) and Conselho Nacional de Desenvolvimento Científico e Tecnológico (CNPq), from Brazil, for financial support. YD is acknowledging the Post-doctorale Fellowship from FAPESP. YD would like to acknowledge the assistance from Dr. Jonathan and Kristen Tallant for proof correction and their support. SBR acknowledge the financial support from Department of Science and Technology, New Delhi, India.

REFERENCES

[1] S. B. Rai, Y. Dwivedi (Eds.) '*Synthesis, Characterization and Application of Multifunctional Materials'*, Nova Publishing Pvt. Ltd., USA (2012).

[2] *"Lynas Says Rare Earths Demand to Grow at 9% a Year,"* bloomberg.com/news, October 25 (2010); U.S. Geological Survey (USGS), Mineral Commodity Summaries, (2011).

[3] Jacquier, M. Wendland, L. Do, P. Robert, C. Corot, C. B. Higgins, M. Saeed, *Contrast Media Mol. Imaging* 3 (2008) 112.

[4] X. K. Xu, W. Z. Zhu, Z. J. Wang, G. Witkamp, *J. Sci. Total Environ.* 293 (2002) 97.

[5] M. R. Ganjali, R. Nemati, F. Faridbod, P. Norouzi, F. Darviche, *Int. J. Electrochem. Sci.,* 3 (2008) 1288.

[6] *Fluoride Glass Optical Fibres,* P. W. France, Blakie and Son, Glasgow, UK (1989).

[7] J.H. Van Vleck, *J. Phys. Chem.* 41 (1937) 67.

[8] R. Reisfeld, C.K. Jorgensen, *Lasers and Excited States of Rare Earths,* Springer-Verlag, Berlin (1977).

[9] B. R. *Judd in book Lanthanide and Actinide Chemistry and Spectroscopy,* Chapter 13, pp (1980) 267.

[10] C. G. -Walrand, K. Binnemans, *Handbook on the Physics and Chemistry of Rare Earths,* eds. K. A. Gschneidner Jr., L. Eyring, Vol. 23, Chapter 155, Elsevier Science B.V., Amsterdam (1996).

[11] C. G. –Walrand, K. Binnemans, *Handbook on the Physics and Chemistry of Rare Earths,* eds. K. A. Gschneidner Jr., L. Eyring, Vol. 25, Chapter 167, Elsevier Science B.V., Amsterdam (1998).

[12] *Spectroscopic Properties of Rare Earths in Optical Materials*, Tsinghua University Press & Springer Verlag, Beijing & Heidelberg (2005).

[13] B. M. Walsh, in *Advances in Spectroscopy for Lasers and Sensing,* ed. B. Di Bartolo, O. Forte, Springer Verlag, Berlin (2006) 403-33.

[14] W. T. Carnall, P. R. Fields, K. Rajnak, *J. Chem. Phys.* 49 (1968) 4424; W.T. Carnall, P.R. Fields, K. Rajnak, *J. Chem. Phys.* 49 (1968) 4412.

[15] G. H. Dieke, *Spectra and energy levels of rare earth ions in crystals,* Interscience Publishers, New York (1968).

[16] S. V. Eliseeva, J. –C. G. Bunzli, *Chem. Soc. Rev.* 39 (2010) 189.

[17] R. G. Chaudhuri, S. Paria, *Chem. Rev.,* 112 (2012) 2373.

[18] R. G. Pearson, J. Songstad, *J. Am. Chem. Soc.* 89 (1967) 1827.

[19] F. Auzel, *Chem.Rev.* 304 (2004) 139.

[20] D. Vennerberg, Z. Lin, *Sci. Adv. Mat.* 3 (2011) 26.

[21] M. Wang, G. Abbineni, A. Clevenger, C. Mao, S. Xu, Nanomed.: *Nanotech. Bio. Med.* 7 (2011) 710.

[22] H. Dong, L. D. Sun, C. H. Yan, *Nanoscale* 5 (2013) 5703.

[23] T. Kushida, *J. Phys. Soc. Japan*, 34 (1973) 1318.

[24] D. L. Dexter *J. Chem. Phys.* 21 (1953) 836.

[25] M. Yokota, O. Tanimoto*, J. Phys. Soc. Japan,* 22 (1967) 779.

[26] M. Inokuti, F. Hirayama, *J. Chem. Phys.* 43 (1965) 1978.

[27] http://www.nei.org/.

[28] http://www.energy.senate.gov/public/index.cfm/files/serve?File_id=6d653f29-1df1-45e7-948e-0aa9837e2e59.

[29] http://en.wikipedia.org/wiki/Renewable_energy_in_China; Caprotti F (2009) China's Cleantech Landscape: The Renewable Energy Technology Paradox *Sustainable Development Law & Policy* Spring 2009: 6-10.

[30] P. Gipe, Wind Power: Renewable Energy for Home, Farm, and Business, USA; M. Saito, T. Sawada, *Advanced Nuclear Energy Systems Toward Zero Release of Radioactive Wastes,* vol. 40, Elsevier Science Ltd. UK. (2002).

[31] W. Shockley, H. J. Queisser, *J. App. Phys.* 32 (1961) 510.

[32] P. J. Cousins, D. D. Smith, H. C. Luan, J. Manning, T. D. Dennis, A. Waldhauer, K. E. Wilson, G. Harley, G. P. Mulligan. *Gen III: improved performance at lower cost.* 35th IEEE PVSC, Honolulu, HI, (2010).

[33] M. A. Green, K. Emery, Y. Hishikawa, W. Warta, Prog. Photovolt: Res. Appl. 19 (2011) 84; M. A. Green, K. Emery, D. L. King, S. Igari, Prog. Photovolt. Res. Appl. 8 (2000) 187; M. A. Green, K. Emery, Y. Hishikawa, W. Warta. Prog. Photovolt: Res. Appl. 17 (2009) 85; M. A. Green, K. Emery, Y. Hishikawa, W. Warta. *Prog. Photovolt: Res. Appl.* 18 (2010) 346.

[34] *UD-led team sets solar cell record, joins DuPont on $100 million project.*

[35] B. S. Richards, *Sol. Energy Mater. Sol. Cells* 90 (2006) 2329.

[36] H. J. Kim, J. S. Song, *J. Korean Phys. Soc.* 45(2004) 609.

[37] T. Trupke, M. A. Green, P. Würfel*, J. Appl. Phys.* 92(2002) 4117.

[38] B. M. V. Ende, L. Aartsa, A. Meijerink, *Phys. Chem. Chem. Phys.* 11 (2009) 11081.

[39] C. Atre, J. A. Dionne, *J. Appl. Phys.* 110 (2011) 034505.

[40] T. Trupke, A. Shalav, B. S. Richards, P. Würfel, M. A. Green, *14th Photovoltaic Science and Engineering Conference,* Bangkok, January 2004.

[41] J. A. Briggs, A. C. Atre, J. A. Dionne, *J. App. Phys.* 113 (2013) 124509.

[42] O. Wenger, C. Wickleder, K. Kramer, H. Gudel, J. Lumin. 94-95 (2001) 101; P. Muller, M. Wermuth, H. Gudel, *Chem. Phys. Lett.* 290 (1998) 105.

[43] S. Baluschev, F. Yu, T. Miteva, S. Ahl, A. Yasuda, G. Nelles, W. Knoll, G. Wegner, *Nano Lett.* 12 (2005) 2482.

[44] B. –C. Hwang, S. Jiang, T. Luo, J. Watson, G. Sorbello, and N. Peyghambarian. *J. Opt. Soc. Am.* B 17 (2000) 833.
[45] Shalav, B. S. Richards, T. Trupke, K. W. Kramer, H. U. Gudel, *Appl. Phys. Lett.* 86 (2005) 013505.
[46] Henke, B. Ahrens, J. A. Johnson, P. –T. Miclea, S. Schweizer, *SPIE Newsroom.* DOI: 10.1117/2.1200910.1811.
[47] J. de. Wild, J.K. Rath, A. Meijerink, W.G.J.H.M. van Sark, R.E.I. Schropp, *Solar Energy Mat. Solar Cells* 94 (2010) 2395.
[48] J. de Wild, T. F. Duindam, J. K. Rath, A. Meijerink, W. G. J. H. M. van Sark, R. E. I. Schropp, *IEEE J. Photovolt.* 3 (2013) 17.
[49] F. Lahoz, Opt. Lett. 33 (2008) 2982; Y. Dwivedi, A. Bahadur, S. B. Rai, *J. Appl. Phys.* 110 (2011) 043103.
[50] H. Lin, D. Chen, Y. Yu, A. Yang, Y. Wang, *Opt. Lett*. 36 (2011) 876.
[51] S. Pillai, K. R. Catchpole, T. Trupke, M. A. Green, *J. Appl. Phys.* **101** (2007) 093105.
[52] C. Atre, A. G. Etxarri, H. Alaeian, J. A. Dionne, *J. Opt.* 14 (2012) 024008.
[53] D. L. Dexter, *Phys. Rev.* 108 (1957) 630.
[54] J. L. Sommerdijk, A. Bril, A. W. de Jager, J. Lumin. 8 (1974) 341; W. W. Piper, J. A. DeLuca, F. S. Ham, *J. Lumin.* 8 (1974) 344.
[55] Q. Y. Zhang, C. H. Yang, Y. X. Pan, Appl. Phys. Lett. 90 (2007) 021107; X. Y. Huang, and Q. Y. Zhang, J. Appl. Phys. 107 (2010) 063505; J. Zhou, Y. Zhuang, S. Ye, Y. Teng, G. Lin, B. Zhu, and J. Xie, J. Qiu, Appl. Phys. Lett. 95 (2009) 141101; X.T. Wei, G. Jiang, K. Deng, C. Duan, Y. Chen, M. Yin, Appl Phys B 108 (2012) 463;C. H. Yang, Y. X. Pan, Q. Y. Zhang, and Z. H. Jiang, *J. Fluor.* 17 (2007) 500.
[56] R.T. Wegh, H. Donker, K.D. Oskam, A. Meijerink, *Science* 283 (1999) 663.
[57] V. Jubera, A. Garcia, J.P. Chaminade, F. Guillen, J. Sablayrolles, C. Fouassier, J. Lumin. 124 (2007) 10; Y. Dwivedi, S. C. Zilio, *Opt. Exp.* 21 (2013) 4717.
[58] A.M. Srivastava, W.W. Beers, *J. Lumin,* 71 (1997), pp. 285–290.
[59] A.M. Srivastava, D.A. Doughty, W.W. Beers, *J. Electrochem Soc.* 143 (1996) 4113.
[60] A.M. Srivastava, D.A. Doughty, W.W. Beers, *J. Electrochem Soc.* 144 (1997) L190.
[61] S. Kuck, I. Sokolska, M. Henkec, M. Doring, T. Schefflerc, *J. Lumi.* 102/103 (2003) 176.
[62] Y. Ikeda, K. Masada, H. Kurokawa, H. Motomura, M. Jinno, K. Tachibana, *J. Phys. D: Appl. Phys.* 46 (2013) 065305.
[63] T. -J. Lee, L. –Y. Luo, B.-M. Cheng, W. –G. Diau, T. –M. *Chen Appl. Phys. Lett.* 92 (2008) 081106.
[64] R.T. Wegh, E.V.D. van Loef, A. Meijerink, *J. Lumin*. 90 (2000) 111.
[65] D. Chen, Y. Wang, Y. Yu, P. Huang, F. Weng, *Opt. Lett.* 33 (2008) 1884.
[66] K. Mishra, Y. Dwivedi, A. Rai and S.B. Rai, *Appl. Phys.* B 109 (2012) 663.
[67] L. Han, Y. Wang, L. Zhao, J. Zhang, Y. Wang, Y. Tao, *Mat. Res. Bull.* 48 (2013) 2139.
[68] J. Sun, Y. Sun, C. Cao, Z. Xia, H. Du, *Appl. Phys.* B 111 (2013) 367.
[69] J. M. Meijer, L. Aarts, B. M. van der Ende, T. J. H. Vlugt, and A. Meijerink, *Phys. Rev.* B 81 (2010) 035107.
[70] K. D. Oskam, R. T. Wegh, H. Donker, E.V.D. van Loef, A. Meijerink, *J. Alloys Compd.* 300–301 (2000) 421.
[71] B. Liu, Y. Chen, C. Shi, H. Tang, Y. Tao, *J. Lumi.* 101 (2003) 155.
[72] K. Deng, T. Gong, L. Hu, X. Wei, Y. Chen, M. Yin, *Opt. Exp.* 19 (2011) 1749.

[73] D. C. Yu, X. Y. Huang, S. Ye, Q. Y. Zhang, *J. Alloys Compd.* 509 (2011) 9919.
[74] Y. Teng, J. Zhou, X. Liu, S. Ye, *J. Qiu, Opt. Exp.* 18 (2010) 9671.
[75] D. Chen, Y. Wang, Y. Yu, P. Huang, F. Weng, *J. Appl. Phys.* 104 (2008) 116105.
[76] Q. Yan, J. Ren, Y. Tong, G. Chen, *J. Am. Ceram. Soc.,* 96 (2013) 1349.
[77] B. S. Richards, Sol. Energy Mater Sol Cells 90 (2006) 1189; A. Shalav, B. S. Richards, M.A. Green, Sol. Ener. Mat. Solar Cells 91 (2007) 829; B. M. V. Ende, L. Aartsa, A. Meijerink, Phys. Chem. Chem. Phys. 11 (2009) 11081; W. G. van Sark, J. de Wild, J. K. Rath, A. Meijerink, R. EI Schropp, Nanoscale Res. Lett. 8 (2013) 81; C. Strumpela, M. McCanna, G. Beaucarneb, V. Arkhipovb, A. Slaouic, V. Svrcekc,C. del Canizod, I. Tobias, Sol. Ener. Mate. & Solar Cells 91 (2007) 238; B. M. van der Ende, L. Aarts, A. Meijerink, *Adv. Mater.* 21 (2009) 3073.
[78] International Energy Agency http://iea.org/ (IEA 2006) (EIA 2006) (EIA 2009); http://www.lightinglab.fi/IEAAnnex45/guidebook/2_lighting%20energy%20in%20buildings.pdf
[79] http://blog.metal-pages.com/2012/04/12/led-lighting-a-rapidly-growing-application-for-rare-earths/.
[80] http://www1.eere.energy.gov/buildings/ssl/sslbasics_ledbasics.html.
[81] S. Reineke, F. Lindner, G. Schwartz, N. Seidler, K. Walzer, B. Lussem, K. Leo, *Nature* 459 (2009) 234.
[82] B. Geffroy, P. le Roy, C. Prat, *Polym. Int.* 55 (2006) 572.
[83] J. K. Park, C. H. Kim, S. H. Park, H. D. Park, S. Y. Choi, Appl. Phys. Lett. 84 (2004), 1647; P. Uthirakumar, C. H. Hong, E. K. Suh, Y. S. Lee, React. Funct. Polym. 67 (2007) 341; A.R. Duggal, J.J. Shiang, C.M. Heller, D.F. Foust, *Appl. Phys. Lett.* 80 (2002) 3470.
[84] J. Kido, H. Hayase, K. Hongawa, K. Nagai, K. Okuyamaa, *Appl. Phys. Lett.* 65 (1994) 2124.
[85] H. Sekiguchi, Y. Takagi, T. Otani, R. Matsumura, R. H. Okada, A. Wakahara, *J. Appl. Phys.* 52 (2013) 08JH01.
[86] S. Nakamura, G. *Fasol Springer, Berlin* (1997) pp. 216–221.
[87] K. Li, C. Shen, 5th International Symposium on Advanced Optical Manufacturing and Testing Technologies: Optoelectronic Materials and Devices for Detector, Imager, Display, and Energy Conversion Technology, edited by Y. Jiang, B. Kippelen, J. Yu, *Proc. of SPIE* Vol. 7658, 76583K.
[88] Y. Dwivedi, S. B. Rai, *J. Am. Ceram. Soc.* 93 (2010) 727.
[89] K. Park, J. Kim, P. Kung, S. M. Kim, *Jap. J. Appl. Phys.* 49 (2010) 020214.
[90] R. –J. Xiea, N. Hirosak, K. Sakuma, Y. Yamamoto, M. Mitomo, *Appl. Phys. Lett.* 84 (2004) 5404.
[91] T.-W. Kuo, W. –R. Liu, T. –M. Chen, *Opt. Exp.* 18 (2010) 187.
[92] R. J. Xie, N. Hirosaki, T. Suehiro, F. –F. Xu, M. Mitomo, *Chem. Mater.* 18 (2006) 5578.
[93] K. Uheda, N. Hirosaki, Y. Yamamoto, A. Naito, T. Nakajima, H. Yamamoto Electrochem. *Solid-State Lett.* 9 (2006) H22.
[94] N. Guo, H. You, Y. Song, M. Yang, K. Liu, Y. Zheng, Y. Huang, H. Zhang, *J. Mater. Chem.* 20 (2010) 9061.
[95] H. S. Jang, H. Yang, S. W. Kim, J. Y. Han, S. –G. Lee, D. Y. Jeon, *Adv. Mater.* 20 (2008) 2696.

[96] J.S. Kim, J. Y. Kang, P. E. Jeon, J. C. Choi, H. L. Park, T. W. Kim, *Jap. J. App. Phys.* 43 (2004) 989.

[97] Y. Q. Li, A. C. A. Delsing, G. de With, H. T. Hintzen, *Chem. Mater.* 17 (2005) 3242.

[98] J. Kuang, Y. Liu, J. Zhang *J. Solid State Chem.* 179 (2006) 266.

[99] D. Chen, Y. Wang, K. Zheng, T. Guo, Y. Yu, P. Huang, *Appl. Phys. Lett.* 91 (2007) 251903; G. Glaspell, J. Anderson, J. R. Wilkins, M. S. El-Shall, *Phys. Chem. C,* 112 (2008) 11527.

[100] S. G. -Neto, L. A. Bueno, R. F. do Nascimento, E. A. da Silva, Jr, E. B. da Costa, V. B. do Nascimento, *Appl. Phys. Lett.* 91 (2007) 091114.

[101] Y. Dwivedi, A. Rai, S. B. Rai, *J. Appl Phys.* 104 (2008) 043509.

[102] J. -C. G. Bunzli, C. Piguet, Chem. Soc. Rev. 34 (2005) 1048; B. Zhao, X. Y. Chen, P. Cheng, D.-Z. Liao, S. -P. Yan, Z. H. Jiang, *J. Am. Chem. Soc.* 126 (2004) 15394; D. Parker, Coord. Chem. Rev. 205 (2000) 109.

[103] 'Lanthanide Luminescence: Photophysical, Analytical and Biological Aspects' edited by P. Hnninen, H. Hèarmè, Springer; M. H. Keefe, K. D. Benkstein, J. T. Hupp, Coord. Chem. Rev. 205 (2000) 201; S. Shinoda and H. Tsukube, *Analyst* 136 (2011) 431.

[104] M. Cantuel, C. Lincheneau, T. Buffeteau, L. Jonusauskaite, T. Gunnlaugsson, G. Jonusauskas, N. D. McClenaghan, *Chem. Commun.* 46 (2010) 2486; C. M. G. dos Santos, A. J. Harte, S. Quinn, T. Gunnlaugsson, Coord. Chem. Rev. 252 (2008) 2512.

[105] '*Decaying Sarin-filled Rockets Spark Fears' Jane's Defense Weekly* 25(1996) 3.

[106] M. Baaden, F. Berny, C. Boehme, N. Muzet, R. Schurhammer, G. Wipff, *J. Alloys Compd.* 303 (2000) 104.

[107] M. Burnworth, S. J. Rowan, and C. Weder, *Chem. Eur. J.* 7828 (2007) 129.

[108] D. Knapton, M. Burnworth, S. J. Rowan, C. Weder, *Angew. Chem.* 118 (2006) 5957.

[109] M. S. Tremblay, M. Halim, D. Sames, *J. Am. Chem. Soc.* 129 (2007) 7570.

[110] L. Jenkins, O. M. Uy, G. M. Murray, *Anal. Commun.* 34 (1997) 221; A. L. Jenkins, O. M. Uy, G. M. Murray, *Anal. Chem.* 71 (1999) 373.

[111] L. Simonian, T. A. Good, S.-S. Wang, J. R. Wild, *Anal. Chim. Acta* 534 (2005) 69.

[112] N. Koshizaki, T. Oyama, *Sens. Actu.* B 66 (2000) 119.

[113] X. Niu, H. Zhong, X. Wang, K. Jiang, *Sens. Actuators* B 115 (2006) 434.

[114] Z. J. Li, H. W. Zhang, J. S. Shen, J. *Appl. Poly. Sci.* 128 (2013) 845.

[115] S. C. Tsang, C. Bulpitt, *Sens. Actuators* B 52 (1998) 226.

[116] X.S. Niu, W.M. Du, W.P. Du, *Sens. Actuators* B 99 (2004) 399.

[117] H. Aono, M. Sato, E. Traversa, M. Sakamoto, Y. Sadaoka, J. Am. Ceram. Soc, 84 (2001) 341; H. Aono, E. Traversa, M. Sakamoto, Y. Sadaoka, *Sens. Actuators* B, 94 (2003) 132.

[118] L. Fabbrizzi, G. Francese, M. Licchelli, A. Perotti, A. Tagliette, Chem. Commun. (1997) 581; N.H. Damraver, T.R. Boussie, M. Devenney, J.K. McCusker, J. Am. Chem. Soc. 119 (1997) 8253; F. Barigelletti, L. Flamigni, G. Calogero, L. Hammarstrom, J.-P. Sauvage, J.-P. Collin, *Chem. Commun.* (1998) 2333.

[119] G. Ellis, I. Adatia, M. Yazdanpanah, S. K. Makela, *Clin. Biochem.* 31 (1998) 195.

[120] R. M. -Máñez, F. Sancenón, *Chem Rev,* 103 (2003) 4419.

[121] L. Douce, L. Charbonniere, M. Cesario, R. Ziessel, *New J. Chem.* 25 (2001) 1024.

[122] M. G. dos Santos, A. J. Harte, S. J. Quinn, T. Gunnlaugsson, *Coord. Chem. Rev.* 252 (2008) 2512.

[123] L. Wang, B. Li, L. Zhang, P. Li, H. Jiang, *Dyes and Pigments* 97 (2013) 26.

[124] T. Yamada, S. Shinoda, H. Tsukube, Chem. Commun. (2002) 1218; T. Yamada, S. Shinoda, H. Sugimoto, J. Uenishi and H. Tsukube, *Inorg. Chem.* 42 (2003) 7932.
[125] Parker, P.K. Senanayake, J.A.G. Willliams, *J. Chem. Soc. Perkin Trans.* 2 (1998) 2129.
[126] S. Faulkner, L. S. Natrajan,W. S. Perry, D. Sykes, *Dalton Trans.* (2009) 3890.
[127] Parker, *Chem. Soc. Rev.* 33 (2004) 156.
[128] Y. Kataoka, D. Paul, H. Miyake, S. Shinoda, H. Tsukube, Dalton Trans. (2007) 2784; Y. Kataoka, D. Paul, H. Miyake, T. Yaita, E. Miyoshi, H. Mori, S. Tsukamoto, H. Tatewaki, S. Shinoda, H. Tsukube, *Chem.–Eur. J.* 14 (2008) 5258.
[129] J. Xu, L. Jia, N. Jin, Y. Ma, X. Liu, W. Wu, W. Liu, Y. Tang, F. Zhou, *Chem. Eur. J.*, 19 (2013) 4556.
[130] B. Sarker, *Handbook on Metals in Clinical and Analytical Chemistry,* Marcel Dekker Inc. New York, 1994.
[131] M. Tian, H. Ihmels, *Chem. Commun.* (2009) 3175.
[132] Syracuse Research Corporation Agency for Toxic Substances and Disease Registry (ATSDR). *Toxicological profile for lead. Public Health Service*/U.S. Environmental Protection Agency, 1990.
[133] A.P. de Silva, J.Q.N. Gunaratne, T.E. Rice, S. Stewart, *Chem. Commun.* (1997) 1891.
[134] B. K. McMahon, T. Gunnlaugsson, *Tetrahedron Lett.* 51 (2010) 5406.
[135] S. E. Plush and T. Gunnlaugsson, *Org. Lett.* 9 (2007) 1919.
[136] K. Hanaoka, K. Kikuchi, H. Kojima, Y. Urano, T. Nagano, *J. Am. Chem. Soc.* 126 (2004) 12470.
[137] J. -A. Yu, R. B. Lessard, D. G. Nocera, *Chem. Phys. Lett.* 187 (1991) 263; Z. Pikramenou, J. –A. Yu, R.B. Lessard, A. Ponce, P. Wong, D. G. Nocera, *Coord. Chem. Rev.* 132(1994) 181.
[138] L. J. Charbonniere, R. Ziessel, M. Montalti, L. Prodi, N. Zaccheroni, C. Boehme, G. Wipff, *J. Am. Chem. Soc.* 124 (2002) 7779.
[139] E. DiBella, J. B. Weisman, M. J. Joseph, J. R. Schultz, T. J. Wenzel, *J. Chromatogr.* 328 (1985) 101.
[140] T.J. Wenzel, L.M. Collette, J. *Chromatogr.* 436 (1988) 299.
[141] M. Amin, K Harrington, R. von Wandruszka, *Anal. Chem.* 65 (1993) 2346; R.E. Milofsky, M.G. Mahlberg, J.M. Smith, *J. High Resolut. Chromatogr.* 17 (1994) 731.
[142] T. Iwata, M. Senda, Y. Kurosu, A. Tsuji, M. Maeda, *Anal. Chem.* 69 (1997) 1861.
[143] M. Schreurs, C. Gooijer, N.H. Velthorst, Anal. Chem. 62 (1990) 2051; M. Schreurs, L. Hellendoorn, C. Gooijer, N.H. Velthorst, *J. Chromatogr.* 552 (1991) 625.
[144] V. L. Pecoraro, W. R. Harris, C. J. Carrano, K. N. Raymond, Biochemistry 20 (1981) 7033; O. Zak and P. Aisen, Biochemistry 27 (1988) 1075; K. Ichimura, H. Kihara and S. Matsuzaki, Biochem. *Cell Biol.* 80 (2002) 161.
[145] J. G. Wardeska, B. Viglione, N. D. Chasteen, J. Biol. Chem. 261 (1986) 6677; C. M. Barnes, S. Petoud, S. M. Cohen, K. N. Raymond, *J. Biol. Inorg. Chem.* 8 (2003) 195.
[146] Y. Kataoka, S. Shinoda, H. Tsukube, *J. Nanosci. Nanotechnol.* 9 (2009) 655.
[147] Z. Pikramenou, J.-a Yu, R.B. Lessard, A. Ponce, P. Wong, D.G. Nocera, *Coord. Chem. Rev.* 132(1994) 181.
[148] M. Mortekkari, D.G. Nocera, *J. Am. Chem. Soc.* 118 (1996) 7414.
[149] J. Karvinen, V. Laitala, M. –L. Makinen, O. Mulari, J. Tamminen, J. Hermonen, P. Hurskainen, I. Hammila, Anal. Chem. 76 (2004) 1429; J. Karvinen, A. Elomaa, M. –L.

Makinen, H. Hakala, V. –M. Mukkala, J. Peuralahti, P. Hurskainen, J. Hovinen, I. Hemmila, *Anal. Biochem.* 325 (2004) 317.

[150] L. Stryer, D. D. Thomas, C. F. Meares, *Annu. ReV. Biophys. Bioeng.* 11 (1982) 203.

[151] S. Mizukami, K. Tonai, M. Kaneko, K. Kikuchi, *J. Am. Chem. Soc.* 130 (2008) 14376.

[152] J. Ballato, E. Snitzer, Appl. Optics 34 (1995) 6848; N. F. Borelli, J. Chem. Phys. 41 (1964) 3289; C. C. Robinson, Appl. Opt. 3 (1964) 1163; M. Daybell, W. C. Overton, Jr., H. L. Laquer, *Appl. Phys. Lett.* 11 (1967) 79.

[153] N. Koshizuka, Y. Yokoyama, K. Ando, T. Okuda, A. Murata, IEEE Transl. J. Magn. Japan 1 (1985) 102; O. Kamada, H. Minemoto, N. Itoh, J. Appl. Phys. 75 (1994) 6801; K. B. Rochford, A. H. Rose, G. W. Day, *IEEE Trans. Magn.* 32 (1996) 4113.

[154] S. B. Berger, C. B. Rubinstein, C. R. Kurkjian, A. W. Treptow, Phys. Rev. 133 (1964) A723; G. T. Petrovskii, I. S. Edelman, T. V. Zarubina, A. V. Malakhovskii, V. N. Zabluda, M. Yu Ivanov, J. Non-Cryst. Solids 130 (1991) 35; J. T. Kohli, J. E. Shelby, Phys. Chem. Glasses 32 (1991) 109; J. T. Kohli, Key Eng. Mater. 94–95 (1994) 125; J. Qiu, K. Tanaka, N. Sugimoto, K. Hirao, *J. Non-Cryst. Solids* 213-214 (1997) 193; J. Qiu, K. Tanaka, K. Hirao, *J. Am. Ceram. Soc.* 80 (1997) 2696.

[155] K. Tanaka, N. Tatehata, K. Fujita, K. Hirao and N. Soga, *J. Phys. D: Appl. Phys.* 31 (1998) 2622.

[156] N. Itoh, H. Minemoto, D. Ishiko, S. Ishizuka, *Appl. Optics* 38 (1999) 2047.

[157] B.G. Potter Jr., M.B. Sinclair, *J. Electroceram.* 2 (1998) 295.

[158] L. Sun, S. Jiang, J. R. Marciante, *Opt. Exp.* 18 (2010) 5407.

[159] M. Vasiliev, K. E. Alameh, V. A. Kotov, Y. T. Lee, "Nanostructured engineered materials with high magneto-optic performance for integrated photonics applications," in Proceedings. *IEEE Photonics* Global @Singapore, (IPGC 2008).

[160] T. Shintaku, A. Tate, and S. Mino, Appl. Phys. Lett. 71(1997) 1640; M. Bolduc, A. R. Taussig, A. Rajamani, G. F. Dionne, C. A. Ross, *IEEE. Trans. Mag.* 42 (2006) 3093; M. C. Sekhar, J. Y. Hwang, M. Ferrera, Y. Linzon, L. Razzari, C. Harnagea, M. Zaezjev, A. Pignolet, R. Morandotti, *Appl. Phys. Lett.* 94 (2009) 181916.

[161] M. C. Sekhar, M. R. Singh, S. Basu, S. Pinnepalli, *Opt. Exp.* 20 (2012) 9624.

In: Spectroscopic Techniques for Security, Forensic ... ISBN: 978-1-63117-404-9
Editors: Y. Dwivedi, S. B. Rai and J. P. Singh

Chapter 11

RARE EARTH DOPED MATERIALS FOR TEMPERATURE SENSORS

Anurag Pandey and Vineet Kumar Rai*
Laser and Spectroscopy Laboratory, Department of Applied Physics,
Indian School of Mines, Dhanbad, Jharkhand, India

There has been an increased curiosity on rare earth doped solid materials due to their wide range of potential applications in photonics devices. These materials may also serve as temperature sensor where conventional temperature sensors are not used. These sensors are known as optical sensor. The temperature sensor is based on monitoring the variation in fluorescence intensity from two close lying levels to a common lower level or variation in lifetime of the excited levels with temperature. In the triply ionized rare earths various pairs of energy levels with energy separation of the order of thermal energy or even slightly greater than that are known. The temperature dependent ratio of the fluorescence intensity or lifetime ratio is independent of the source intensity since the emitted intensity is proportional to the populations of each level involved. Therefore, by using the fluorescence intensity ratio or fluorescence lifetime technique the temperature could be determined. The significant development of temperature sensor came in motion after the discovery of lasers. For temperature sensors based on the lanthanides doped materials, generally, the materials with low phonon energies are preferred, because in hosts with low cut-off phonon energies, the emission intensity of rare-earth ions introduced as doping is enhanced by several times. The present chapter describes the detail report on the rare earth doped materials for temperature sensors by utilizing the fluorescence intensity ratio or fluorescence lifetime technique and their experimental validation in terms of sensitivity and temperature range. The idea of optical heating produced due to increase in the pump energy of the excitation source has also been detected using fluorescence intensity ratio variation of two thermally coupled levels.

---

* Corresponding author: Laser and Spectroscopy Laboratory, Department of Applied Physics, Indian School of Mines, Dhanbad-826004, Jharkhand, India, vineetkrrai@yahoo.co.in; rai.vk.ap@ismdhanbad.ac.in.

1. INTRODUCTION

Lanthanides are the elements of lanthanum series in periodic table from lanthanum to lutetium and stable mostly in triply ionized state. Luminescence from triply ionized lanthanides are mainly ascribed due to their intra 4f-4f or 5d-4f transitions, characterized by narrow emission lines with high colour purity [1-5]. Rare earths doped materials have significant applications in diverse fields for example, Plasma display panels, high definition televisions (HDTVs), light emitting diodes, cathode ray tubes (CRTs), finger print detection, sensors, upconverters, biological and medical applications, etc [6-11]. A sensor is a device that is used to measure a physical quantity and convert it into a form that can be read by the observer. There are several types of sensors used for different tasks e.g., biological sensor, chemical sensor, electric and magnetic field based sensors, mechanical sensors, temperature and radiation sensors, etc. [12]. Optical temperature sensors are devices which mainly detect the light signals and used for measuring the physical quantities such as temperature. Though, there are various temperature sensing tools available now days but some limited application of those devices such as inability to measure temperature from distant object upto higher temperature, in corrosive environments etc., encourage to develop a temperature sensing device that works in all types of environments for high temperature sensing purpose [12, 13].

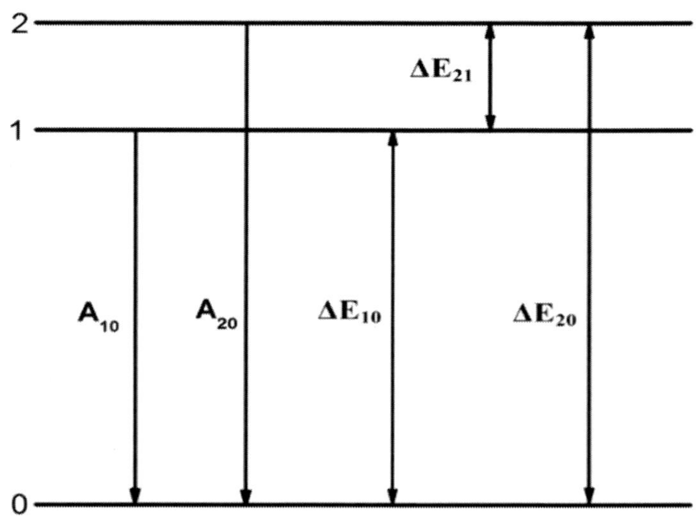

Figure 1. Energy level structure of two thermally coupled levels to observe the mechanism of temperature sensing.

The temperature of a distant object is measured by using the optical temperature sensors which basically depend on intensity change of emergent light or radiations coming from the object. The significant development of optical sensor came in motion after the discovery of lasers. Several methods have been adopted for both point and distributed temperature detecting purposes as Raman scattering, Braggs grating, Lifetime change, Fluorescence intensity variation, Interferometric measurements, etc. [10, 14-18]. Mostly, fluorescence intensity ratio (FIR) and fluorescence lifetime (FL) based temperature monitoring are very popular and useful for temperature sensing purpose. The FIR technique is valuable and noncontact method, used to improve the sensitivity of measurement and minimise the influence of measurement conditions [19-21]. In FIR technique, the ratio of emission intensity arising from two closely spaced levels of one or two rare earth ions are used while in FL technique the lifetimes of emitting states are utilised for thermometric investigation.

Figure 1 represents an energy level structure of two closely spaced upper levels **1** and **2** having a very small separation, populated via a suitable excitation source. After the excitation the population of upper levels decays radiatively to the ground state **0**. The fluorescence decay time can be represented by the formula [12, 22]

$$\tau = \frac{1+b_{21}}{A_{10}+b_{21}A_{20}} \tag{1}$$

where A_{20} and A_{10} are the spontaneous transition rates corresponding to the two levels **2** and **1**. $b_{21} = (g_2/g_1)\exp(-\Delta E_{21}/kT)$ is the ratio of the population in the levels **2** and **1** at equilibrium, g_2 and g_1 are their respective degeneracies, k is the Boltzmann's constant and T is the absolute temperature.

The rate by which the decay time ratio changes with temperature known as the sensitivity of the FL method is given by the following equation

$$S_s = \frac{1}{\tau}\frac{\partial\tau}{\partial T} = \left(1 - \frac{1}{1+b_{21}}\right)\left(\frac{1-\left(\frac{A_{20}}{A_{10}}\right)}{1+\left(\frac{A_{20}}{A_{10}}\right)b_{21}}\right)\frac{\Delta E_{21}}{kT^2} \tag{2}$$

where all the terms have their usual meanings [23].

It is worthwhile noted that the pulse excitation is needed for monitoring the temperature sensing behaviour based on FL technique. The lifetime measurements can be determined by taking a single wavelength or a band including all decays of fluorescence emission. For more than three level schemes the above form can be modified accordingly.

The intensity emitted at a particular wavelength due to i→j level transition is given by

$$I_{ij} = N_i\,\omega_{ij}\,A_{ij} \tag{3}$$

where ω_{ij} is the angular frequency of radiation, A_{ij} is the spontaneous emission rates, N_i is the population of ith state.

So, the intensity of fluorescence from specific energy level depends upon the population of that particular level and transition rates. For a given material, the nonradiative relaxation rate depends upon the energy gap between the two levels and cutoff phonon frequency of host material. If the energy gap between the two levels is very small then their individual population is directly proportional to total population and the relative change in their population can be easily determined by change in their fluorescence intensity.

Therefore, the intensity ratio for transitions from levels **2** and **1** to level **0** can be given as

$$R = I_{20}/I_{10} = N_2\,\omega_{20}\,A_{20} / N_1\,\omega_{10}\,A_{10} \tag{4}$$

N_i (i=1, 2) is the population in the ith state, ω_{i0} (i=1, 2) is the angular frequency of radiation and other terms have their usual meanings.

As the relative population of thermally coupled levels follows a Boltzmann type population distribution [24, 25] and related to the temperature in following manner

$$N_i = g_i \exp(-E_i/kT) \tag{5}$$

where g_i is the degeneracy of the ith state. Therefore, the ratio of the population of two states **2** and **1** may be given as,

$$N_2/N_1 = g_2 \exp(-E_2/kT) / g_1 \exp(-E_1/kT) \text{ or}$$

$$N_2/N_1 = [g_2 / g_1] [\exp(-\Delta E_{21}/kT)] \tag{6}$$

where $\Delta E_{21} = E_2 - E_1$, energy difference of the two states.

Using equation (4) and (6) we get

$$R = I_{20}/I_{10} = B \exp(-\Delta E_{21}/kT) \tag{7}$$

where $B = \omega_{20} A_{20} g_2 / \omega_{10} A_{10} g_1$, is a proportionality constant and other terms have their usual meanings.

For temperature sensing purpose, it is important to know the rate of change of FIR with change in temperature, known as sensitivity 'S' (or absolute sensitivity) is defined by the relation

$$S = (1/R)\, dR/dT = \Delta E_{21}/kT^2 \tag{8}$$

For optical thermometry, it is also required to determine the sensor sensitivity (or relative sensitivity) that can be written as follows

$$S_S = S_R = dR/dT = R\, \Delta E_{21}/kT^2 \tag{10}$$

where $S_S = S_R$ is the sensor or relative sensitivity and other parameters have their usual meanings.

The suitability of rare earth doped materials for temperature sensing purpose is dominant due to the presence of large number of thermally coupled levels and high transition probabilities of nonradiative relaxations [10, 12]. The erbium, europium, holmium, thulium, samarium, dysprosium, neodymium, praseodymium and ytterbium ions have been assigned mainly as dopants for temperature sensing applications due to availability of a number of thermally coupled levels. Not only the two close lying levels of single ions but also that of two different ions as well as the stark components of the same level could be used for thermometry purposes [24, 36, 51].

Numerous studies on temperature sensing performance have been done on the rare earth (RE) doped/codoped materials viz. glass, phosphor, ceramic and crystals using different techniques till now and the investigations for searching better sensing material is still continued [18, 21-54]. These investigations used both FIR and FL techniques and based on the experimental observations the FIR method has been reported to be the superior one for high temperature sensing purposes [25]. A lot of spectroscopic studies have been performed by taking the two thermally coupled levels of green bands of Er^{3+} ions corresponding to the $^2H_{11/2}\rightarrow{}^4I_{15/2}$ and $^4S_{3/2}\rightarrow{}^4I_{15/2}$ transitions while the investigations on other rare earth ions are less. The concept of optical heating has also been detected using FIR variation of two thermally coupled levels [38, 41, 48, 49]. Here we have described the temperature sensing performance of a variety of rare earth ions doped into different host materials in detail.

2. Experimental Setup

The experimental arrangement for measurement of temperature along with fluorescence emission/decay coming out from the sample is shown in Figure 2. The CW/pulsed laser beam enters the sample chamber via reflection from a front coated mirror and conversing through a convex lens. The sample was situated inside a homemade heater enclosed by insulating chamber. The thermocouple was located near the sample and temperature measurement has been performed with the help of multimeter. The temperature of the heater was controlled by a Varivolt. The fluorescence emitted from the sample is carried to the monochromator through optical fiber. A photomultiplier tube (PMT) is mounted on the exit slit of the monochromator for the detection of light. The PMT output was received by a computer for spectral recording. For lifetime measurement the output from PMT was recorded with the help of digital oscilloscope.

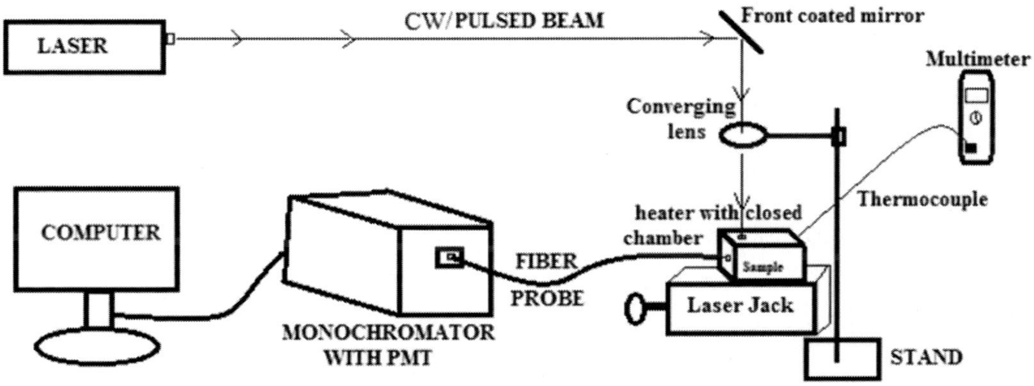

Figure 2. Block diagram of the experimental setup for temperature sensing.

3. Temperature Sensing Performance of Rare Earth Doped Materials

3.1. Praseodymium and Neodymium Doped Materials

The temperature sensing performance of two close lying emission bands of Pr^{3+} ions has been studied using two thermally coupled $^3P_1$ and $^3P_0$ levels to low lying states $^1G_4$ and $^3H_5$ [10, 12, 25, 28, 50]. The energy level structure of Pr^{3+} ion (shown in Figure 3) indicates two thermally coupled levels and their transitions. The increase in emission intensity corresponding to the $^3P_1 \rightarrow ^3H_5$ transition and the decrease in emission intensity for the $^3P_0 \rightarrow ^3H_5$ transition with increase in the samples temperature has been reported by the different researchers. These variations in the emission intensities with temperature are basically due to the thermalization involved between the two close lying $^3P_1$ and $^3P_0$ levels. Maurice *et al.* [50] have reported the temperature sensing behaviour of Pr^{3+} doped ZBLAN glass using FIR technique. They varied the temperature of the sample from -45 ^{0}C to 255 ^{0}C and measured the variation in FIR of the two transitions coming from two close lying $^3P_1$ and $^3P_0$ levels to $^1G_4$ level. The sensitivity of the sensor was observed to be ~0.48%/ ^{0}C at 20 ^{0}C which was the highest value reported at that time.

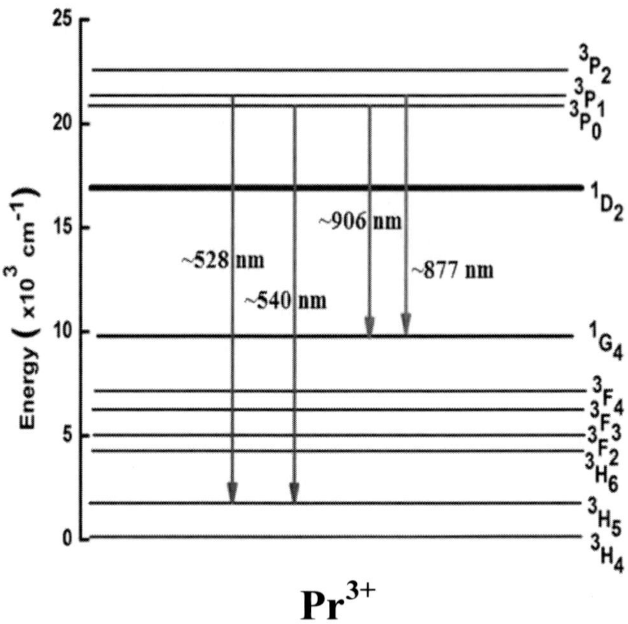

Figure 3. Energy level scheme of Pr^{3+} ion indicating two thermally coupled levels in green and near infrared (NIR) regions.

The FIR technique has also been used to perform the temperature investigation through $^3P_1 \rightarrow ^3H_5$ and $^3P_0 \rightarrow ^3H_5$ transitions of Pr^{3+} ion [28]. They observed a variation in intensity corresponding to the two transitions and used them to estimate the temperature of the sample by using FIR technique. The decay time variation with temperature corresponding to these transitions have also been employed and found that the sensitivity of the FIR method is several orders of magnitude better than that of the FL method [25].

Neodymium ions (Nd^{3+}) may have several possible transitions from two thermally coupled levels $^4F_{3/2}$ and $^4F_{5/2}$ to low lying levels $^4I_{9/2}$, $^4I_{11/2}$ and $^4I_{13/2}$ upon suitable excitation. Several observations on temperature sensing behaviour of Nd^{3+} doped materials have been reported by different researchers [10, 12, 31, 55, 56]. The $^4F_{5/2} \rightarrow ^4I_{9/2}$ and $^4F_{3/2} \rightarrow ^4I_{9/2}$ transitions of Nd^{3+} ions have been used for the temperature sensing purpose. Xu *et al.* [55] have observed the temperature sensing performance using the NIR emission from Nd^{3+}/Yb^{3+}codoped oxyfluoride glass ceramics in 303-623 K temperature range. The FIR technique for the transitions [viz. $^4F_{7/2}/^4S_{3/2} \rightarrow ^4I_{9/2}$ and $^4F_{3/2} \rightarrow ^4I_{9/2}$ (I_{750}/I_{863}); $^4F_{7/2}/^4S_{3/2} \rightarrow ^4I_{9/2}$ and $^4F_{5/2}/^2H_{9/2} \rightarrow ^4I_{9/2}$ (I_{750}/I_{803}); $^4F_{5/2}/^2H_{9/2} \rightarrow ^4I_{9/2}$ and $^4F_{3/2} \rightarrow ^4I_{9/2}$ (I_{803}/I_{863})] coming from two close lying levels to the lower lying levels have been applied. They observed the reasonable agreement between the fitting of experimental and theoretical values. The sensitivity for different thermally coupled transitions in the Nd^{3+}/Yb^{3+}codoped oxyfluoride glass ceramics have been calculated and observed the highest sensitivity for the $^4F_{7/2}/^4S_{3/2}\rightarrow^4I_{9/2}$ and $^4F_{3/2}\rightarrow^4I_{9/2}$ transitions of Nd^{3+} ions (Table-1). The emission bands arising from the upper thermally coupled levels were observed to be superimposed by the emission bands originating from the lower energy levels terminating at the next lower energy level. In order to avoid such overlapping problem appropriate excitation wavelengths are utilized to get the better emission efficiency.

Table 1. Comparision between the sensitivity of Nd^{3+}/Yb^{3+}codoped oxyfluoride glass ceramics for different thermally coupled transitions

Transitions	Sensitivity (K^{-1})	Temperature range (K)
$^4F_{7/2}/^4S_{3/2}\rightarrow^4I_{9/2}$ and $^4F_{3/2}\rightarrow^4I_{9/2}$	$3010.1/T^2$	303-623
$^4F_{7/2}/^4S_{3/2}\rightarrow^4I_{9/2}$ and $^4F_{5/2}/^2H_{9/2}\rightarrow^4I_{9/2}$	$1763.2/T^2$	303-623
$^4F_{5/2}/^2H_{9/2}\rightarrow^4I_{9/2}$ and $^4F_{3/2}\rightarrow^4I_{9/2}$	$1884.6/T^2$	303-623

3.2. Samarium and Europium Doped Materials

The limited reports are available for Sm^{3+} doped materials to be used as temperature sensor. Two thermally coupled levels ($^4F_{3/2}$ and $^4G_{5/2}$) of Sm^{3+} ions are mostly used for optical thermometry. The optical thermometry in Sm^{3+} doped/codoped systems using the transitions from these two close lying levels to ground ($^6H_{5/2}$) level have been studied by several researchers [10, 23, 29, 36, 57]. Rai *et al.* [23, 29] have investigated the temperature sensing performance of triply ionized samarium doped in calibo and oxyfluoroborate glasses. The FL technique has been applied to Sm^{3+} doped oxyfluoroborate glass [23]. On the basis of available experimentally observed data it is concluded that the FL temperature sensing technique can be used to monitor the temperature upto 500 K within the error limit of 15% with sensitivity of about 1.26×10^{-3}%/K at room temperature [23]. The effect of variation of temperature on the fluorescence intensity observed from the $^4F_{3/2}\rightarrow^6H_{5/2}$ and $^4G_{5/2}\rightarrow^6H_{5/2}$ transitions of Sm^{3+} ions has been performed by using the FIR technique and appreciable sensitivity around 1.97%/K has been found [29].

The temperature sensing performance in Eu^{3+} doped solid host material by using the FIR technique has been reported by Kusama *et al.* [58]. Optical temperature sensing behaviour in Eu^{3+} doped materials by using two thermally coupled levels $^5D_1$ and $^5D_0$ of Eu^{3+} ion have been studied extensively [10, 12, 18, 27, 58, 59]. Rai *et al.* [18, 27] have investigated the FIR and FL based temperature sensing performance of Eu^{3+} doped tellurite and calibo glasses by considering the $^5D_1\rightarrow^7F_1$ and $^5D_0\rightarrow^7F_1$ transitions of Eu^{3+} ion. They observed an increment in intensity ratio of the two transitions upto ~500 K in case of tellurite glass and ~550 K in case of calibo glass. They monitored a very small difference between the observed and calculated values of temperature [27]. Also the rate of change of FIR for a change in temperature called sensitivity of the sensor was calculated to be ~ ±2.75%/K at room temperature. They concluded that for a pair of energy levels with larger energy separations the sensitivity of the fluorescence intensity ratio is also larger.

The lifetime of $^5D_0$ level is noted to decrease with increasing the temperature. The lifetime ratio which is independent of source intensity was observed to exhibit the acceptable deviations in calculated and observed values (<10%) upto ~400 K temperature with sensitivity and quantum efficiency of about ~ 0.0145%/K and ~25% respectively [18]. The comparison between two methods provided the suitability of FL technique for lower temperature region only.

3.3. Dysprosium and Holmium Doped Materials

The energy levels ($^4F_{9/2}$ and $^4I_{15/2}$) responsible for blue emission from Dy^{3+} ion are considered as thermally coupled levels having energy separation of about ~1000 cm^{-1}. Very few studies have been done on the Dy^{3+} doped materials for thermometric purpose [10, 33]. Wade *et al.* [10] studied the FIR versus temperature characteristics using Dy^{3+} doped silica fiber in the 22 to 250 ^0C temperature range. The temperature sensing performance of dysprosium doped Al_2O_3 phosphor has been investigated by Rakov *et al.* [33]. They observed a variation in intensity of blue emissions peaking at ~450 and ~471 nm corresponding to the $^4I_{15/2}\rightarrow{}^6H_{13/2}$ and $^4F_{9/2}\rightarrow{}^6H_{13/2}$ transitions respectively upto ~723 K. The intensity ratio of $^4I_{15/2}\rightarrow{}^6H_{13/2}$ and $^4F_{9/2}\rightarrow{}^6H_{13/2}$ transitions (I_{450}/I_{471}) is observed to increase with increasing the temperature. So they applied the FIR technique to see the sensing behaviour of the doped material. The maximum sensor sensitivity around 0.0014 K^{-1} has been found.

The energy levels of interest to study the temperature sensing behaviour of the Ho^{3+} ion are mainly the $^5F_4$ and $^5S_2$ levels of green emission as well as $^3K_8$ and $^5F_3$ levels of blue emission (Figure 4).

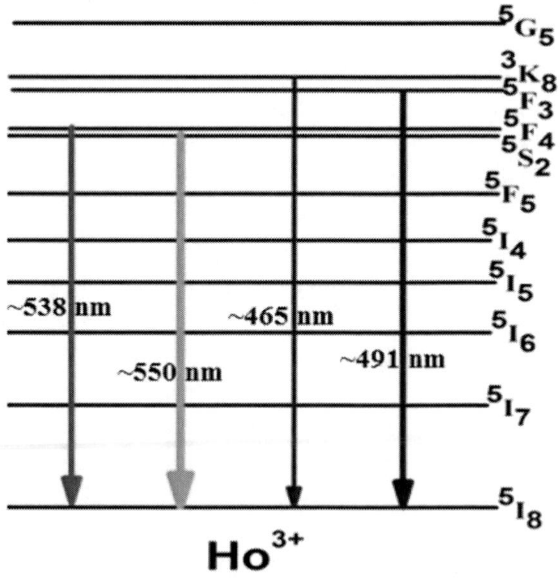

Figure 4. Energy level diagram of Ho^{3+} ion indicating two thermally coupled levels in blue and green regions.

A very few studies on temperature sensing have been done on Ho^{3+} doped glasses, phosphors and ceramics by several researchers [32, 37, 47, 60-62]. Singh *et al.* [32, 61] have investigated the Ho^{3+} doped tellurite glass as temperature sensing probe by taking the variation in FIR of two emission bands arising due to the transitions from the $^5F_4$ and $^5S_2$ levels to a common ground $^5I_8$ level upto ~400K. The temperature sensing behaviour in Ho^{3+}-Yb^{3+} codoped glass ceramic has been detected by Xu *et al.* [37]. The thermometric performance of Ho^{3+}-Yb^{3+}-Zn^{2+} codoped Y_2O_3 phosphor by considering the blue upconversion emissions peaking at ~465 and ~491 nm through the $^3K_8 \rightarrow ^5I_8$ and $^5F_3 \rightarrow ^5I_8$ transitions of Ho^{3+} ions have been investigated by Pandey *et al.* [60]. The calculated energy gap of two thermally coupled levels was observed to be matched very well with its actual value and maximum value of the sensor sensitivity was found arround $30.199 \times 10^{-4}\ K^{-1}$ at 673 K temperature, thereby showing the utility of Ho^{3+}-Yb^{3+}-Zn^{2+} codoped Y_2O_3 phosphor in making a good temperature sensor with better sensitivity.

3.4. Erbium and Thulium Doped Materials

Erbium is the most promising and attractive lanthanide element in the lanthanide series used as a dopant into a variety of hosts for thermometric applications. The two excited levels $^2H_{11/2}$ and $^4S_{3/2}$ are considered to be thermally coupled and variation in intensity with temperature obtained from transitions of these two levels to a common ground $^4I_{15/2}$ state is taken for sensing the temperature. Not only these two thermally coupled levels but the stark sublevels of $^4S_{3/2}$ and $^4I_{13/2}$ levels have also been taken into account for optical thermometry and observed to exhibit the temperature sensing behaviour. A lot of studies on temperature sensing have been performed on erbium doped/codoped materials by several researchers [10, 12, 13, 21, 26, 30, 35, 39-46, 48, 49, 51, 63, 64]. Most of the studies reported are based on FIR temperature sensing technique by utilizing the green emission bands observed through the $^2H_{11/2} \rightarrow ^4I_{15/2}$ and $^4S_{3/2} \rightarrow ^4I_{15/2}$ transitions. On the basis of results obtained by the various researchers, they predicted about the suitability of materials and such transitions for temperature sensing applications. The literature survey shows that not only the two thermally coupled levels but also the stark sublevels of a level with smaller energy gap (>200 cm^{-1}) can be used to monitor the temperature. Thus the erbium doped materials are most superior depending upon the requirements and conditions for developing a temperature sensor upto very high temperature with very good sensitivity.

Optical temperature sensors based on thulium doped materials are very rarely seen [52]. Xu *et al.* [52] have investigated the FIR based temperature sensing behaviour in the Tm^{3+}-Yb^{3+} codoped oxyfluoride glass ceramics. They have considered the NIR emissions of thulium ion around ~700 and ~800 nm observed through the $^3F_{2,3} \rightarrow ^3H_6$ and $^3H_4 \rightarrow ^3H_6$ transitions (shown in Figure 5) and monitored the NIR emissions intensity variation of these two transitions with increasing the temperature. The calculated energy gap (~1952 cm^{-1}) by using the FIR was found to be very close to the experimental energy gap between two levels (~1850 cm^{-1}). Based on the experimental observations they proposed that the $^3F_{2,3}$ and $^3H_4$ levels are thermally coupled. The calculated sensitivity of the doped glass ceramic was noted to increase on increasing the temperature and found maximum at ~703 K.

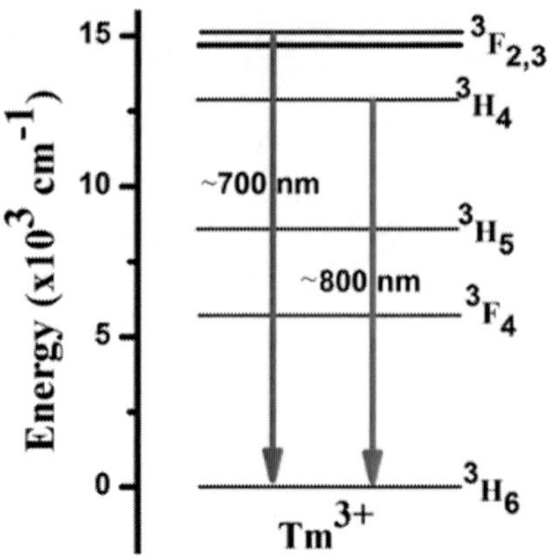

Figure 5. Energy level diagram of Tm^{3+} ion indicating NIR emission through the $^3F_{2,3}\rightarrow{}^3H_6$ and $^3H_4\rightarrow{}^3H_6$ transitions.

3.5. Ytterbium Doped Materials

The triply ionized ytterbium ion (Yb^{3+}) has only two levels, one ground state ($^2F_{7/2}$) and other excited state ($^2F_{5/2}$) separated by ~10,000 cm^{-1}. So the possibility of developing optical temperature sensor may arise by taking stark components of the mentioned levels. Not much research work has been carried out on Yb^{3+} doped materials for such purpose [10, 53]. Wade *et al.* [10] have reported the FIR based temperature sensing performance of Yb^{3+} doped silica fiber using stark sublevels of $^2F_{5/2}$ state of Yb^{3+} ion. Demirkhanyan *et al.* [53] have investigated the possibility of using the Yb^{3+} doped crystals {$Y_3Al_5O_{12}$, NaBi $(WO_4)_2$, $LiNbO_3$} as optical temperature sensor (OTS) based on FIR and absorption coefficient ratio (ACR). They presented the energy level structure of stark states of Yb^{3+} ion in different crystals and concluded that the YAG: Yb^{3+} crystals are more suitable material for OTS in low temperature region (40 - 130 K) with its sufficiently high mean sensitivity (3.4 % K^{-1}).

4. Temperature Sensing Using Two Close Lying Levels of Different Rare Earth Ions

Not only the two close lying thermally coupled levels of a single rare earth ion but also the two close lying levels of different rare earth ions could be used for the temperature sensing purpose. Insufficient study has been done by using the two close lying levels of different rare earth ions whose emission intensity vary significantly with temperature [24, 36, 60]. Optical thermometry in the Ho^{3+}- Tm^{3+}-Yb^{3+} codoped Y_2O_3 phosphor have been observed by taking FIR of two transitions $^1G_4\rightarrow{}^3H_6$ (Tm^{3+}) and $^5F_3\rightarrow{}^5I_8$ (Ho^{3+}) ions having energy difference ~473 cm^{-1} [24]. The estimated error in the measured temperature was found to be ±5.91 K. The maximum calculated sensor sensitivity in this case is deserved to be 3.38×10^{-3} K^{-1} at room temperature. Tripathi *et al.* [36] have also reported the variation in FIR of the $^4G_{5/2}\rightarrow{}^6H_{5/2}$ (Sm^{3+}) and $^5D_0\rightarrow{}^7F_2$ (Eu^{3+}) transitions by using the FIR technique and compared the calculated corresponding temperatures. These investigations established the possibility of obtaining a good temperature sensor probe for higher temperature range with better sensitivity using two different rare earth ions.

5. Optical Heating Generated by the Excitation Source

Table 2. Comparison of the sensitivity and temperature range for different rare earth ions with involved transitions along with sensing technique used

Rare earths	Sensing technique	Transitions	Maximum sensitivity/relative sensitivity	Temperature range [Reference]
Pr^{3+}	FIR FL FIR	$^3P_1\rightarrow{}^3H_5$ and $^3P_0\rightarrow{}^3H_5$ $^3P_1\rightarrow{}^3H_5$ and $^3P_0\rightarrow{}^3H_5$ $^3P_1\rightarrow{}^1G_4$ and $^3P_0\rightarrow{}^1G_4$	0.0097 K^{-1} -- 0.0048 K^{-1}	293-473 K [25] 273-423 K [28] 228-528 K [50]
Nd^{3+}	FIR FIR	$^4F_{5/2}/^2H_{9/2}\rightarrow{}^4I_{9/2}$ and $^4F_{3/2}\rightarrow{}^4I_{9/2}$ $^4F_{7/2}/^4S_{3/2}\rightarrow{}^4I_{9/2}$ and $^4F_{3/2}\rightarrow{}^4I_{9/2}$	0.00454 K^{-1} 3010.1/T^2 K^{-1}	296-670 K [31] 303-623 K [55]
Sm^{3+}	FL FIR	$^4F_{3/2}\rightarrow{}^6H_{5/2}$ and $^4G_{5/2}\rightarrow{}^6H_{5/2}$ $^4F_{3/2}\rightarrow{}^6H_{5/2}$ and $^4G_{5/2}\rightarrow{}^6H_{5/2}$	0.0000126 K^{-1} 0.0197 K^{-1}	300-500 K [23] 373-560 K [29]
Eu^{3+}	FL FIR	$^5D_1\rightarrow{}^7F_1$ and $^5D_0\rightarrow{}^7F_1$ $^5D_1\rightarrow{}^7F_1$ and $^5D_0\rightarrow{}^7F_1$	0.000145 K^{-1} ±.0275 K^{-1}	300-400 K [18] 300-550 K [27]
Dy^{3+}	FIR	$^4I_{15/2}\rightarrow{}^6H_{13/2}$ and $^4F_{9/2}\rightarrow{}^6H_{13/2}$	0.0014 K^{-1}	298-723 K [33]
Ho^{3+}	FIR FIR FIR	$^5F_4\rightarrow{}^5I_8$ and $^5S_2\rightarrow{}^5I_8$ $^5F_{2,3}/^3K_8\rightarrow{}^5I_8$ and $^5F_1/^5G_6\rightarrow{}^5I_8$ $^3K_8\rightarrow{}^5I_8$ and $^5F_3\rightarrow{}^5I_8$	 0.00102 K^{-1} 0.00302 K^{-1}	265-383 K [32] 303-643 K [37] 299-673 K [60]
Er^{3+}	FIR FIR FIR FIR FIR FIR	$^2H_{11/2}\rightarrow{}^4I_{15/2}$ and $^4S_{3/2}\rightarrow{}^4I_{15/2}$ $^2H_{11/2}\rightarrow{}^4I_{15/2}$ and $^4S_{3/2}\rightarrow{}^4I_{15/2}$ $^2H_{11/2}\rightarrow{}^4I_{15/2}$ and $^4S_{3/2}\rightarrow{}^4I_{15/2}$ $^2H_{11/2}\rightarrow{}^4I_{15/2}$ and $^4S_{3/2}\rightarrow{}^4I_{15/2}$ $^2H_{11/2}\rightarrow{}^4I_{15/2}$ and $^4S_{3/2}\rightarrow{}^4I_{15/2}$ $^4S_{3/2}\rightarrow{}^4I_{15/2}$ (stark sublevels)	0.0039 K^{-1} 0.0092 K^{-1} 0.0103 K^{-1} 0.0051 K^{-1} 0.0074 K^{-1} 0.0049 K^{-1}	300-900 K [21] 294-923 K [26] 303-593 K [35] 295-973 K [42] 290-610 K [43] 300-550 K [51]
Tm^{3+}	FIR	$^3F_{2,3}\rightarrow{}^3H_6$ and $^3H_4\rightarrow{}^3H_6$	2829.5/T^2 K^{-1}	293-703 K [52]
Yb^{3+}	FIR	$^2F_{5/2}\rightarrow{}^2F_{7/2}$ (stark sublevels)	0.034 K^{-1}	40 -130 K [53]
Sm^{3+} & Eu^{3+}	FIR	$^4G_{5/2}\rightarrow{}^6H_{5/2}$ (Sm^{3+}) and $^5D_0\rightarrow{}^7F_2$ (Eu^{3+})	--	272-379 K [36]
Ho^{3+} & Tm^{3+}	FIR	$^1G_4\rightarrow{}^3H_6$ (Tm^{3+}) and $^5F_3\rightarrow{}^5I_8$ (Ho^{3+})	0.00338 K^{-1}	303-703 [24]

It is marked that the FIR of two emission transitions changes not only with temperature but also on increasing the pump excitation power [38, 41, 48, 49, 62, 65]. This increase in the FIR due to increase in pump excitation power leads to the concept of optical heating. This internal heating of the material induced by the laser excitation helped in the development of optical nano heater and useful in local hyperthermia based cancer treatment. Hayakawa *et al.* [41] have estimated the laser heating inside the Er^{3+} doped TeO_2-ZnO-Nb_2O_5 glass using femtosecond laser. The laser induced optical heating in the Er^{3+}-Yb^{3+} codoped fluoride nanoparticles by monitoring the $^2H_{11/2} \rightarrow {}^4I_{15/2}$ and $^4S_{3/2} \rightarrow {}^4I_{15/2}$ transitions of Er^{3+} ions have been studied by Tikhomirov *et al.* [49]. Laser induced optical heating in the Er^{3+}-Yb^{3+} codoped Gd_2O_3 phosphor has also been studied by Singh *et al.* [48] and the temperature rise of 504 K is obtained in nanoparticles at a laser power of 290 mW. The Ho^{3+}-Yb^{3+} codoped $Ca_{12}Al_{14}O_{33}$ phosphor has been used to monitor the FIR of green and blue emissions coming from the Ho^{3+} ion [38] and concluded an optical heating induced by laser excitation. Thus from the reported work one can conclude that optical heating generated upto higher temperature around ~500 K from laser excitation could find applications in biomedicine and creation of holes in soft materials.

A comparison between different rare earth ions doped materials is given in Table-2. From the table one can conclude that the triply ionized rare earths doped materials can be used in making the temperature sensor probes with desired temperature range and sensitivity. Also it is marked that triply ionized erbium is the best among all the lanthanides for thermometric applications upto very high temperature range with very good sensitivity.

ACKNOWLEDGEMENT

Authors are grateful to the University Grants Commission, New Delhi, India, for providing the financial assistance [F. No. 39-534/2010(SR)].

REFERENCES

[1] V. K. Rai, L. S. Menezes, Cid B. de Araujo, *J. Appl. Phys.*, 103 (2008) 53514.
[2] A. Patra, C. S. Friend, R. Kapoor, P. N. Prasad, *Chem. Mater.*, 15 (2003) 3650.
[3] V. Singh, V. K. Rai, I. Ledoux-Rak, S. Watanabe, T. K. Gundu Rao, J. F. D. Chubaci, L. Badie, F. Pelle, S. Ivanova, *J. Phys. D*, 42 (2009) 065104.
[4] G. Blasse, B. C. Grabmaier, *Luminescent Materials*, Springer Verlag, Berlin, 1994.
[5] Y. Dwivedi, S. B. Rai, *J. Am. Ceram. Soc.,* 93 (2010)727.
[6] S. K. Singh, K. Kumar, S. B. Rai, *Appl. Phys. B,* 94 (2009) 165.
[7] G. Yi, H. Lu, S. Zhao, Y. Ge, W. Yang, D. Chen, L. H. Guo, Nano Lett., 4 (2004) 2191.
[8] M. Wang, G. Abbineni, A. Clevenger, C. Mao, S. Xu, Nanomed. *Nanotechnol.*, 7 (2011) 710.
[9] H. S. Mader, P. Kele, S. M. Saleh, O. S. Wolfbeis, *Curr. Opin. Chem. Biol.*, 14 (2010) 582.
[10] S. A. Wade, S. F. Collins, G. W. Baxter, *J. Appl. Phys.*, 94 (2003) 4743.

[11] E. D. Rosa, P. Salas, H. Desirena, C. Angeles, R. A. Rodriguez, *Appl. Phys. Lett.*, 87 (2005) 241912.

[12] V. K. Rai, *Appl. Phys. B*, 88 (2007) 297.

[13] B. R. Reddy, I. Kamma, P. Kommidi, *Appl. Opt.*, 52 (2013) B33.

[14] G. A. Hornbeck, *Appl. Opt.*, 5 (1966) 179.

[15] M. Lapp, D. L. Hartley, *Combust. Sci. Technol.*, 13 (1976) 199.

[16] C. J. Yeager, S. S. Courts, *IEEE Sensors J.*, 1 (2001) 352.

[17] Y. Zhan, H. Wu, Q. Yang, S. Xiang, H. He, *Opt. Lasers Eng.*, 46 (2008) 349.

[18] V. K. Rai, C. B. de Araujo, *Spectrochim. Acta A*, 71 (2008) 116.

[19] P. Haro-Gonzalez, S. F. Leon-Luis, S. Gonzalez-Perz, I. R. Martin, *Mater. Res. Bull.*, 46 (2011) 1051.

[20] N. Ishiwada, T. Ueda, T. Yokomori, *J. Bio. Chem.*, 26 (2011) 381.

[21] S. K. Singh, K. Kumar, S. B. Rai, *Sens. Actuators A*, 149 (2009) 16.

[22] S. F. Collins, G. W. Baxter, S. A. Wade, *J. Appl. Phys.*, 84 (1998) 4649.

[23] V. K. Rai, *IEEE Sensors J.*, 7 (2007) 297.

[24] A. Pandey, V. K. Rai, *Appl. Phys. B,* Published online, 2013 (DOI-10.1007/s00340-013-5460-z).

[25] V. K. Rai, S. B. Rai, *Appl. Phys. B*, 87 (2007) 323.

[26] W. Xu, X. Gao, L. Zheng, P. Wang, Z. Zhang, W. Cao, *Appl. Phys. Exp.*, 5 (2012) 072201.

[27] V. K. Rai, A. Rai, *Appl. Phys. B*, 86 (2007) 333.

[28] V. K. Rai, D. K. Rai, S. B. Rai, *Sens. Actuators A*, 128 (2006) 14.

[29] V. K. Rai, C. B. de Araujo, *Spectrochim. Acta A*, 69 (2008) 509.

[30] B. Dong, T. Yang, M. K. Lei, *Sens. Actuators B*, 123 (2007) 667.

[31] C. Li, S. Li, B. Dong, Z. Liu, C. Song, Q. Yu, *Sens. Actuators B,* 134 (2008) 313.

[32] A. K. Singh, *Sens. Actuators A,* 136 (2007) 173.

[33] N. Rakov, L. R. A. Bispo, G. S. Maciel, *Opt. Commun.*, 285 (2012) 1882.

[34] P. H. Gonzalez, L. M. Maestro, M. Trevisani, S. Polizzi, D. Jaque, J. G. Sole, M. Bettinelli, *J. Appl. Phys.*, 112 (2012) 054702.

[35] V. K. Rai, A. Pandey, R. Dey, *J. Appl. Phys.*, 113 (2013) 083104.

[36] G. Tripathi, V. K. Rai, S. B. Rai, *Appl. Phys. B*, 84 (2006) 459.

[37] W. Xu, X. Gao, L. Zheng, Z. Zhang, W. Cao, *Opt. express*, 20 (2012) 18127.

[38] R. K. Verma, S. B. Rai, J. Quant. *Spectrosc. Radiative trans.*, 113 (2012) 1594.

[39] Y. Shen, X. Wang, H. He, Y. Lin, C. W. *Nan, Compos. Sci. Technol.*, 72 (2012) 1008.

[40] S. F. Leon-Luis, U. R. Rodriguez-Mendoza, E. Lalla, V. Lavin, *Sens. Actuators B*, 158 (2011) 208.

[41] T. Hayakawa, M. Hayakawa, M. Nogami, *J. Alloys Compd.*, 451 (2008) 77.

[42] B. Dong, D. P. Liu, X. J. Wang, T. Yang, S. M. Miao, C. R. Li, *Appl. Phys. Lett.*, 90 (2007) 181117.

[43] B. S. Cao, Y. Y. He, Z. Q. Feng, Y. S. Li, B. Dong, *Sens. Actuators B*, 159 (2011) 8.

[44] A. Sayoud, J. P. Jouart, N. Trannoy, M. Diaf, T. Duvaut, *J. Lumin.*, 132 (2012) 566.

[45] N. Rakov, G. S. Maciel, *Sens. Actuators B*, 164 (2012) 96.

[46] C. Li, B. Dong, S. Li, C. Song, *Chem. Phys. Lett.*, 443 (2007) 426.

[47] V. Lojpur, M. Nikolic, L. Mancic, O. Milosevic, M. D. Dramicanin, *Ceram. International*, 39 (2013) 1129.

[48] S. K. Singh, K. Kumar, S. B. Rai, *Appl. Phys. B*, 100 (2010) 443.

[49] V. K. Tikhomirov, K. Driesen, V. D. Rodriguez, P. Gredin, M. Mortier, V. V. Moshchalkov, *Opt. express*, 17 (2009) 11794.
[50] E. Maurice, G. Monnom, G. W. Baxter, S. A. Wade, B. P. Petreski, S. F. Collins, *Optical Review*, 4 (1997) 89.
[51] V. K. Rai, S. B. Rai, Spectrochim. *Acta Part A*, 68 (2007) 1406.
[52] W. Xu, X. Gao, L. Zheng, Z. Zhang, W. Cao, *Sens. Actuators B*, 173 (2012) 250.
[53] H. Demirkhanyan, G. Demirkhanyan, E. Kokanyan, R. Kostanyan, M. Aillerie, *Proc. of SPIE*, 8414 (2012) 84140Q-1(doi: 10.1117/12.923453).
[54] X. Wang, X. Yan, C. Kan, *J. Lumin.*, 131 (2011) 2325.
[55] W. Xu, H. Zhao, Z. Zhang, W. Cao, *Sens. Actuators B*, 178 (2013) 520.
[56] S. A. Wade, J. C. Muscat, S. F. Collins, G. W. Baxter, Review *of Scientific Instruments*, 70 (1999) 4279–4282.
[57] J. P. Feist, A. L. Heyes, *Meas. Sci. Technol.*, 11 (2002) 942.
[58] H. Kusama, O. J. Sovers, T. Yashioka, Japan. *J. Appl. Phys.*, 15 (1976) 2349.
[59] Y. Ohishi, S. Takahashi, *Appl. Opt.*, 25 (1986) 720-723.
[60] A. Pandey, V. K. Rai, *Dalton Transactions*, 42 (2013) 11005.
[61] A. K. Singh, S. B. Rai, *Appl. Phys. B*, 86 (2007) 661.
[62] A. Pandey, V. K. Rai, *Appl. Phys. B*, 109 (2012) 611.
[63] A. K. Singh, S. B. Rai, D. K. Rai, V. B. *Singh*, *Appl. Phys. B*, 82 (2006) 289.
[64] G. Z. Sui, X. P. Li, L. H. Cheng, J. S. Zhang, J. S. Sun, H. Y. Zhong, Y. Tian, S. B. Fu, B. J. Chen, *Appl. Phys. B*, 110 (2013) 471.
[65] A. Pandey, V. K. Rai, R. Dey, K. Kumar, *Mater. Chem. Phys.*, 139 (2013) 483.

In: Spectroscopic Techniques for Security, Forensic ... ISBN: 978-1-63117-404-9
Editors: Y. Dwivedi, S. B. Rai and J. P. Singh

Chapter 12

ADVANCES IN INFRARED (IR) SENSORS FOR SECURITY, FORENSICS AND THE ENVIRONMENT

R. N. Singh[1*], R. K. Sharma[2] and S. S. Negi[3†]

[1]Ex associate Director, IRDE, Dehradun, India
[2]Scientist 'G', SSPL, New Delhi, India
[3]Outstanding Scientist, IRDE, Dehradun, India

ABSTRACT

Modern infrared sensor technology has enormous potential in various fields viz. military, industry, security, forensics, environmental pollution monitoring, etc. Advancements in infrared sensor technology has led to improved detection & surveillance capabilities, effective localization of suspects, investigation of crime scenes, conduct search & rescue operations, checking samples of explosives, during all times and in all weather conditions. Advancements in multispectral and hyperspectral imaging techniques along with non-imaging FTIR spectroscopy enable crime-scene investigation and evidence collection in forensic areas. Ground based and airborne infrared sensor systems both prove to be quite useful for the type of applications envisaged in this chapter.

The present chapter highlights the principles of Infrared imaging, growth of sub system technologies and emerging trends in IR sensor technology. A typical case study for national security using infrared sensor technology has been illustrated. Application areas in forensics & environmental pollution monitoring have been touched upon using Hyperspectral imaging and non-imaging FTIR spectroscopy.

1. INTRODUCTION

There are three distinct areas of interest viz. security, forensics and environment pollution detection to be covered in this chapter. Over the past couple of decades, there has been a growing interest in the use of spectral techniques at visible through infrared (IR) wavelength to enhance the performance of remote sensors. By combining spectroscopy and imaging, a new type of camera has come up having multi and hyper-spectral capabilities which allow users to identify/ differentiate spectral signatures of various objects and materials. They "see the unseen" by splitting the radiation coming out of the object, either reflected or self-emitted, into many spectral bands which they capture and process separately. This way for each pixel in the scene, a spectral signature is recorded. Spectral imaging techniques organize their information into three dimensional spectral data cubes, one belonging to spectral data and the other two to spatial dimensions and can be interpreted as a stack of images, one for each wavelength. Figure 1 illustrates the concept of the spectral imaging. The potential of spectral imaging is enormous and can be easily applied to security, forensics and environmental pollution studies.

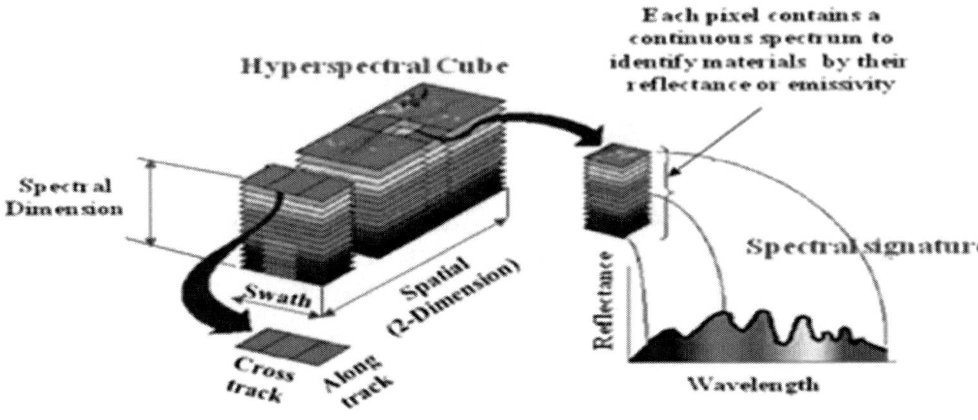

Figure 1. Concept of Hyperspectral Imaging process.

National security is a very wide term. It concerns the vitality of a nation's economy, its ability to generate new ideas, the consensus in the society to carry burdens for national purposes and conviction in the legitimacy of its potential foundation and its institutions. A nation's inability to reach a durable consensus on plans that match the national goals with its financial resources is the greatest national security risk.

However, in the present context, national security refers to homeland security to detect, manage and control security systems, to pre-empting criminals and terrorist's assaults on key assets and critical infrastructure. In a broader perspective may include military/paramilitary capabilities with reference to international border security. Most commonly used systems like daylight camera, night vision & near IR imaging systems have a very limited foul weather viewing capability, poor night time performance & very limited range of operational capabilities. Further, they are likely to yield larger false alarms, giving fatigue to operators and also lack ability to pin-point the place of occurrence on intrusions.

Therefore, we would try to concentrate on the emerging infrared sensors, their current status and challenges that must be overcome to provide security of a nation in the current century. We will discuss both long term breaks through sensing challenges as well as nearer term cost, size and weight improvements to enable new mission capabilities. Components of a complete security and surveillance system are highlighted in table 1 and will be elaborated in a case study subsequently [1].

Table 1. Components of a total security system [1]

Technology	IR Imaging (NIR, SWIR, MWIR, LWIR)	Auto video analytics	Immersive Visual Assessment	Command & Control of wide area
Function	Sensor level detection	Software level detection	Situational awareness	Process control & management
Benefits	• Extends vision beyond visible • Can operate both in day & night and in total darkness • Can operate in poor/ bad weather conditions • Have higher detection ranges • Can reveal details which the human eye and low light imagers cannot. • Cannot be blinded by search light & normal camouflaging is not effective.	• Perform advanced & complex image analysis • Filter sensor data, identify behaviours, categorize targets and potential threats. • Detect multiple alarms caused by a perimeter breach, loitering or an abandoned object.	• 3D immersive video surveillance technology merges live views from a number of cameras with a model of environment • Observer can look back & forth in time and space controlling various cameras and immersing himself in the scene. This allows the user to channel his focus on assessing the situation instead of dillydallying between multiple cameras.	• Access live video associated with alerts received from the access control system. • Thwart potential threats using video analytics to quickly identify and intervene in suspicious activities • Quickly mobilize and coordinate crisis intervention teams • Push incident information to consult individuals. • Engage & share critical information with different units for a more integrated response.
Problems	• Thermal picture not as good as day light images • Are costly • Are voluminous heavier & technically complex adding to cost.			

The second area of interest is forensics wherein we make use of scientific investigations to establish facts in cases of criminal or civil courts of law. This is possible by collecting scientific facts to identify, classify and analyze physical evidences related to criminal investigation, perform tests on weapons or substances such as fiber, hair and tissues to determine significance to investigation, etc.

IR imagers can handle some of the typical cases like checking samples of explosives and other chemicals whether at long distance or close-ups, catching terrorist who detonates bombs by testing the container that hide the bomb rather than the actual explosive, age determination from blood collected at the scene of crime, recovering fingerprints on fabrics which could turn clothes into silent witnesses, detect weapons concealed beneath the clothing of a person or under his luggage. It is critical for law enforcement officers to see in the dark, through smoke, fog, dust, storm and other situations where visibility is poor and to see clearly enough to determine whether someone is holding a bat or a rifle. Infrared imagers and spectral signature techniques do play a highly effective security solution using new generation technologies.

Environmental pollution includes smoke and emissions of green house gases into the atmosphere. Global emissions have been growing at an average of 3% each year since 2000. We are on course to emit several billion tons of carbon dioxide just from burning fossil fuels and producing cement and another a few billion tons or so, of emission from deforestation and other changes in land use. As on today's situation, temperature of environment could raise more than $4^{o}C$ well before the year 2100 and more than $8^{o}C$ before 2150. As arctic ice melts due to increased temperature many parts of the world bear the brunt of extra storms, resulting in steep rise in sea levels. This impact will be more pronounced in a narrow sea coast. Low lying coastal areas will gradually disappear. Other coasts will be eroded faster; storm surges will be higher and will reach further inland. Infrared techniques including hyperspectral imager and Fourier Transform Infrared (FTIR) spectrometry coupled with automatic software processors will help in the detection and identification of polluting clouds in the atmosphere and will help in issuing warnings to areas likely to be most affected.

2. IR Radiations and Imagers

Infrared techniques play very important role from spectroscopic analysis for molecular structural studies to many qualitative and quantitative analysis of explosives, air pollutants, nuclear, biological and chemical analysis of war weapons to playing critical role in modern battlefield. IR radiations extend from red end of visible radiation (0.7μm) to about 1000 μm. Spectroscopically this entire range is divided into near IR (0.7μm to 2μm), middle IR (2μm to 15μm), far IR (15μm to 50μm) and extreme IR (50μm to 1000μm). However, the most popular nomenclature of IR is often based on military applications from infrared imaging consideration as depicted in Figure 2. Near Infrared (NIR) - 0.7μm to 1.1μm, Short Wave Infrared (SWIR) - 0.9μm to 2.5μm, Middle Wave Infrared (MWIR) -3μm to 5μm, Long Wave Infrared (LWIR) - 8μm to 14μm, Very Long Wave Infrared (VLWIR) -14μm to 30μm and extreme Infrared (EIR) - 30μm to 1000μm.

The infrared region touching the microwave region is most popularly known as sub mm region where the most useful windows from military point of view are near 35 GHz and 94 GHz. Infrared radiations are invisible to the human eye, have very small energies but are omnipresent. From imaging considerations NIR & SWIR regions usually employ reflected radiations only. The remaining IR bands use self emitted radiations and are called thermal IR bands. Infrared (thermal) imagers span a wide range of electromagnetic spectrum and provide day & night viewing capability by detecting IR radiations, emitted or reflected from the scene.

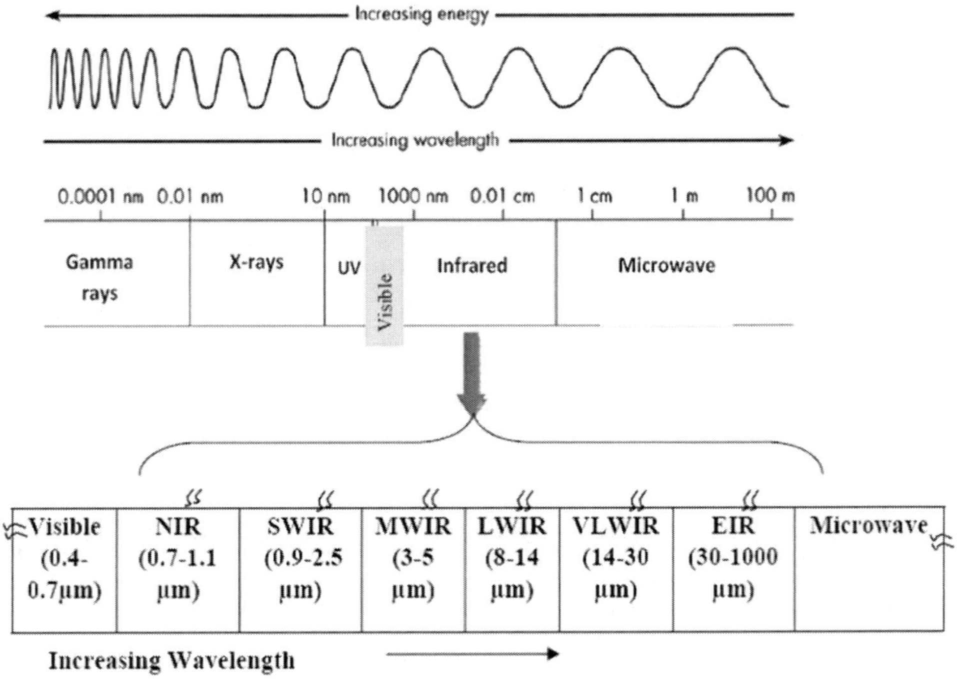

Figure 2. The infrared spectrum showing its sub-bands.

The NIR and SWIR images are identical to day light camera images but are monochrome because many colour pigments are transparent to these radiations. The NIR/SWIR cameras are used in low light security applications using short focal length lenses and many times a number of NIR light emitting diodes (LEDs) for illumination at short distances. They serve as night vision goggle (NVG), NIR beacon to show helicopter landing, to evaluate the health of crops and forests, to inspect textile weaving defects, to examine tampered documents and to sort out bruised vegetables and fruits.

Most common detectors for NIR & SWIR ranges are either monolithic Complementary Metal Oxide Semiconductor (CMOS) & Charge Couple Devices (CCDs) or hybrids of Indium Gallium Arsenide (InGaAs) & Si PIN Focal Plane Arrays (FPAs). Earlier NIR sensors were image converter tubes having S-1 photocathode (sensitivity up to 60mA/lm) and silicon based sensors. In SWIR imaging, silicon imaging sensors are unable to stimulate the detectors due to too low photon energy. Therefore SWIR sensor materials are more exotic materials and include InGaAs, Indium Antimonide (InSb) and Mercury Cadmium Telluride (MCT). InGaAs materials have high quantum efficiency (QE) and low dark current at room temperature. MCT based SWIR detectors for long range applications need thermoelectric cooling.

SWIR radiations can penetrate through thin paints, haze, air pollutant and other thin solids including ordinary glass. Multispectral SWIR cameras can identify materials simply by looking at them in a multitude of bands just by comparing their spectral signature with standard spectra of known materials. For a given aperture, SWIR has two to three times better spatial resolution than the MWIR band, alongside better atmospheric penetration than the visible band. Compared to thermal imagers, SWIR cameras also have a better dynamic range. However, SWIR camera images of human face look quite different due to liquid water absorption of SWIR radiation [water is available right under the surface of human skin]. For imaging through thick obscurants, range gated imaging could be resorted to and can also detect the absence of heat.

The MWIR & LWIR imaging cameras see heat from scenes. MWIR radiation sensors are InSb, MCT, Strained Layer Superlattices (SLS), Quantum Well Infrared Photodetector (QWIP), and Quantum Dot (QDs), etc. All these detectors require cooling to cryogenic temperatures. MWIR imagers see better through high humidity air at ranges of several kilometers or more compared to LWIR cameras. Special MWIR cameras with suitable optical filters can image gas plumes that are completely invisible to human eye. This is crucial for safety inspection of power plants & refineries.

Recently, Prof. Wang Qijie of Nayang Technical University Singapore has added graphene photosensor to the list of SWIR detectors. Pure graphene is 1000 times more light sensitive than current sensors and is able to detect wavelengths from visible to MWIR making it suitable for all types of camera up to middle IR region. It is ultra light sensitive and it also uses 10 times less energy and operates at lower voltages.

In the LWIR band, there is more radiation emitted from terrestrial object as compared to MWIR, but the amount of radiation varies less with temperature. There is less contrast in the image in the LWIR band. The LWIR sensor are MCT, SLS, QWIP, Quantum dot infrared photodetector (QDIP), and uncooled microbolometers like Vanadium Oxide (VOx) or Amorphous Si (A-Si) etc. The uncooled sensors are revolutionizing the IR camera industry as they are miniaturized in size and are cost wise cheaper. They are widely used in security & surveillance applications in fixed installations, on vehicles and on unmanned aerial vehicles (UAVs). Fire-fighters use them to see through smoke and identify hot-spots that can indicate a fire burning inside a wall or behind a door.

Table 2. Bandpass attributes [3]

S.No.	Phenomenology	SWIR	MWIR	LWIR
1	Night vision in total darkness	Better than visible	Better than SWIR	Better than MWIR
2	High humidity	Better than visible	Better than SWIR & LWIR	Moderate
3	Vision through smoke	Better than visible	Better than SWIR	Better than MWIR
4	Vision through fog	Better than visible	Better than SWIR	Better than MWIR
5	Resolution	Best	Better	Poor
6	Cold weather night vision	N/A	Moderate	Best
7	Image 1.06μm and 1.5μm eye safe lasers	Yes	N/A	N/A
8	Red painted SYMBOLS	Good	Poor	Poor

LWIR cameras can see through thick smoke, in sub-zero temperature and imaging scene that contain the sun or a reflection of sun. No other band can entirely meet the imaging objective under such conditions [2].

Each band has some individual strengths and interesting phenomenology to exploit them. Table 2 gives band pass attributes for SWIR, MWIR, & LWIR. The best system depends on the application and system constraints. At this point, it should be clearly understood that SWIR cameras offer a better image during dusk & dawn compared to night time which is self explanatory.

2.1. Principles of IR Imaging

Since every object above absolute zero temperature radiates energy, therefore the difference in the amount and type of radiation emitted by warm and cool objects in a scene form the foundation of this emerging infrared technology. Estimates of this differential energy can be made using the famous Plank's formula for a black body, which has been subsequently modified by Kirchhoff to include emissivity factor for gray bodies. Plank's Law describing the spectral distribution of black body [Figure 3] is given as

$$W_\lambda = 2\pi hc^2/\lambda^5.(1/e^{hc/\lambda kT} - 1) \quad (1)$$

and for a gray body it is modified as

$$W_\lambda' = \varepsilon(\lambda).W_\lambda = \varepsilon(\lambda).2\pi hc^2/\lambda^5.(1/e^{hc/\lambda kT} - 1) \quad (2)$$

Where, W_λ = spectral radiant emittance for a black body; W_λ' = spectral radiant emittance for a gray body; λ = wavelength of light (μm); h= Plank's constant; c= velocity of light (m/s); k= Boltzmann constant; T= absolute temperature in Kelvin; $\varepsilon(\lambda)$ = emissivity of an object at wavelength λ

While integrating Plank's equation over wavelengths from λ = zero to λ = infinity gives the total energy radiated into a hemisphere by a black body of 1cm^2 area & at an absolute temperature T and differentiating Plank's law and solving for maximum gives the Wein's displacement law to find out the wavelength (λ_m) of maximum spectral radiant emittance. Thus the Stefan-Boltzmann law gives total radiation.

$$W = \sigma T^4 \quad (3)$$

Where, σ = Stefan - Boltzmann constant $=2\pi^5k^4/15c^2h^3=5.6697\text{X}10^{-12}$ Wcm^{-2}k^4

and Wein's displacement law gives $\lambda_m.T = 2897.6 \pm 0.4$ μK (4)

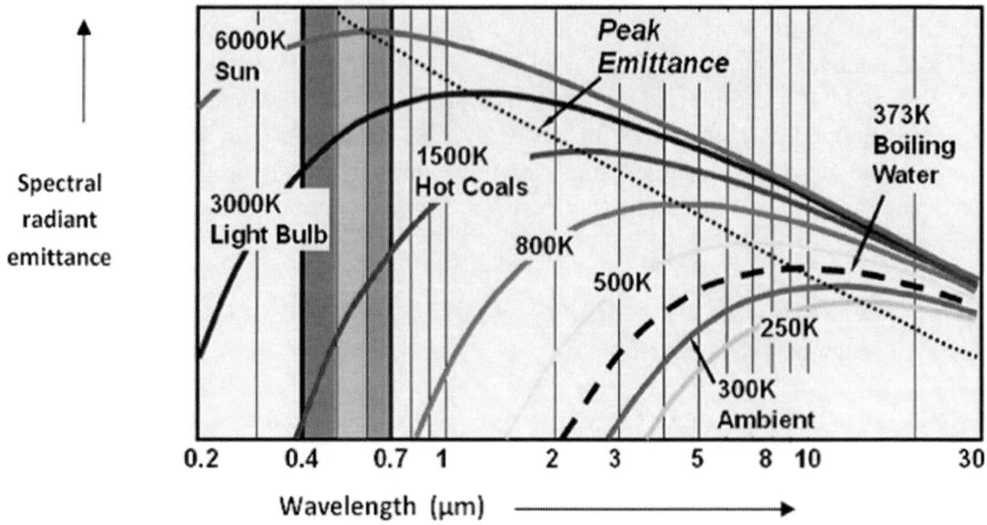

Figure 3. Planck's black body radiation curve.

2.2. Reflectance Spectra

To understand fundamentals of reflectance spectra, we need to define few terms like radiance (L), irradiance (H), reflectance (ρ), and spectral radiance ($L_s(\lambda)$). Radiant flux emitted per unit solid angle per unit area is called radiance of the source and is measured in Watt.cm^{-2}Sr^{-1} where as irradiance is the radiant flux incident per unit area of a sensor and its unit is Watt/cm^2. Reflectance characterizes the fraction of incident light reflected by the surface and can have any value between 0 and 1. To a first order the spectral radiance on our eye or any other sensor is the product of the impinging scene radiance ($L_i(\lambda)$) and the material reflectance spectrum ($\rho(\lambda)$) and is given as [4]

$$L_S(\lambda) = \rho(\lambda) . L_i(\lambda) \tag{5}$$

Where, $L_S(\lambda)$ is spectral radiance

It may be noted that both ρ and L depend on wavelength λ. From above equation it is clear that if the illumination of spectrum is known, then the material reflectance spectrum can easily be recovered from the observed spectral radiance over those regions of the spectrum in which the illumination is non-zero.

3. Infrared (Thermal) Imagers

There are two distinct categories of infrared imagers: reflected radiation based (NIR and SWIR) and self emitted radiation based (MWIR, LWIR, VLWIR). The increased requirement for multi and hyper spectral infrared imaging systems significantly complicates the system and also its design.

4. Basic Elements of IR Imaging

There are four basic parts of an infrared camera: the IR source of radiation (target and background or external illuminating source), the intervening atmosphere, the sensor comprising optics, IR detector & cooler and primary electronics, and a signal processor, image processor and display device. Sometimes auxiliary controls like field of view (FOV) changing mechanism, reticles control and camera focusing etc. are also essential elements and may vary from one application to others [Figure 4].

IR radiation from the scene is captured by the optics module and focused onto an IR detector placed at the focal plane of the optics module. The detector type and its configuration, the number of elements it contains, pixel size, their Dee Star (D^*) and Modulation Transfer Function (MTF), dewar cold shield and filters are all important from the optical point of view. The signal processor, image processor and display devices produce a real time image in which the electrical signal generated represents the scene intensity differences due to various objects in the FOV. The visual image is presented to the observer in the desired display mechanism.

4.1. The IR Source of Radiation

IR radiation from a target comprises two parts: (i) the emitted radiation from target & background, and (ii) the reflected/ scattered radiation from target due to external source (sun, earth, sky and other sources).

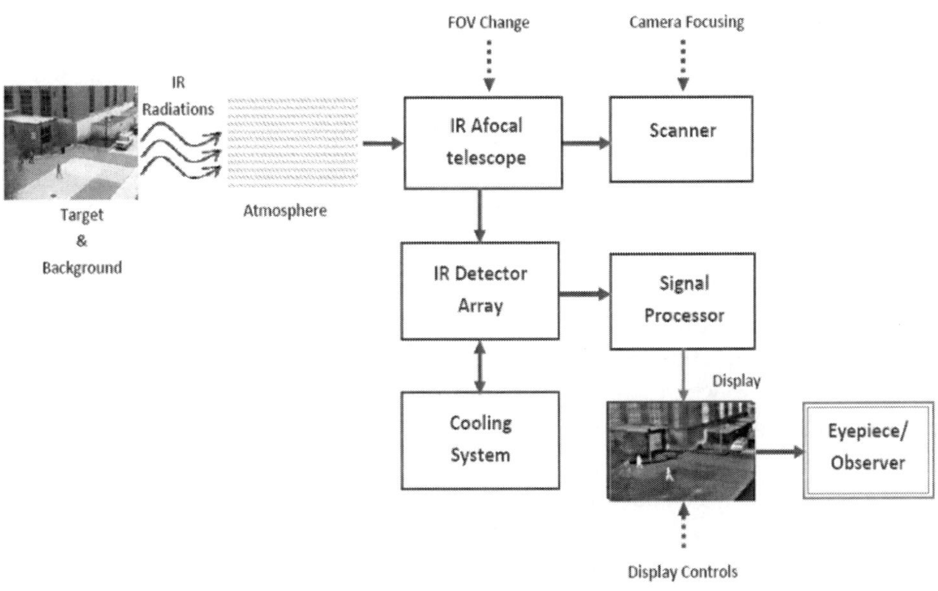

Figure 4. Schematic arrangement of elements of thermal imagers (TI).

In general, an infrared source of radiation could be each & every object starting from backgrounds like terrain features, industrial & technical houses, clouds, atmosphere to tactical object of interest like human being, vehicles, bridges, aircrafts, ships, etc. Emitted radiation is a function of the target's temperature and its emissivity and is identical during day and night at a given temperature and can be estimated from Planck's curve (Figure. 3). The area under the curve is a measure of the total emittance. These radiations clearly lie in the MWIR and LWIR region. IR targets can be considered to be either point target or extended target. Ideally, point targets do not exist however, any target at infinity can be said to be a point target.

The radiance of atmospheric path may serve as a clutter to the long range thermal imagers being used by armed forces and degrades their performance. The NIR & SWIR images produced by photons reflected from the scene are basically the extension of visible radiations extending into near IR & SWIR and their origin lies in the sunlight, moonlight, starlight and other night sky illuminations.

4.2. Atmospheric Attenuation

The irradiance $H(\theta,\Phi)$ on a sensor received from a target at a distance R is a function of source radiant intensity $J(\theta,\Phi)$, the distance R between source & sensor and atmospheric attenuation τ_A, and is given as

$$H(\theta,\Phi) = [J(\theta,\Phi).\ \tau_A]/\ R^2 \tag{6}$$

It is obvious from above equation that all infrared systems are susceptible to atmospheric conditions. Infrared radiations are likely to get attenuated in passage through terrestrial atmosphere by the process of absorption and scattering by gas molecules, aerosols, rain, snow, smoke, fog, haze and smog. Main absorber of these radiations is water vapour (2.7 μm, 3.2 μm and 6.3 μm), carbon dioxide (2.7 μm, 4.3 μm and 15 μm), ozone (4.8 μm, 9.6 μm and 14.2 μm), nitrous oxide (4.7 μm & 7.8 μm), carbon mono-oxide (4.8μm) and methane (3.2 and 8.7μm). In the lower atmosphere, absorption by carbon mono-oxide and nitrous oxide is negligible; water absorption band at 6.3 μm and CO_2 absorption band at 2.7 μm & 15 μm, limit the IR radiation transmission in the MWIR & LWIR windows. Scattering by suspended molecules, aerosols, fogs, hazes and clouds, called Mie Scattering, plays significant role in the IR transmittance thereby affecting the range of IR cameras.

Atmospheric turbulence is due to the time varying temperature inhomogeneities which causes refractive index fluctuations and may result in blurring and distortion of the images. It is especially noticeable when viewing smaller objects at longer distances. Theoretically, atmospheric transmission of IR radiations and the path radiance could be computed using softwares LOWTRAN or MODTRAN.

One of the important considerations in system design is that the system should have a high enough target signal to clutter ratio to provide high probability of detection and low probability of false alarm. False alarm may be caused by natural occurring events that may resemble the target observable signal.

4.3. The Sensor Module

The sensor module is made up of three sub modules viz. IR optics & scanner, IR detector-cooler assembly and primary electronics. Infrared optics and scanner module collects the scene radiated / reflected energy and focuses it onto IR detector. IR detector converts it into electrical signal and passes it to primary electronics.

4.3.1. IR Optics

Optical design of an IR system involves the selection of a suitable lens/mirror configuration to meet the basic system specification. IR optics may be a refractive, a reflective, or a catadioptric (a combination of both, lenses and mirrors) types. Many a time a combination of different materials are used to make an aberration free optics and to provide a diffraction limited performance using spherical, aspherical and diffractive elements.

Infrared optics special features are: use of special IR materials, development of broadband efficient anti reflection DLC type of coatings, athermalisation, and control of narcissus and other cosmetic effects. Optical designers frequently refer to such terms as optical aperture, F/no., FOV, instantaneous field of view (IFOV), optical transmission, effective focal length, back focal lengths and front focal lengths. These are standard optical terms and may be found in any book on applied optics/thermal imaging.[5] Aperture (entrance pupil) defines the limiting resolution of optics, more the aperture, better is the angular resolution of the optics. F/no. controls the speed of optics and is given by the ratio of equivalent focal length to the diameter of the entrance pupil.

$$F/no = F/D_o \tag{7}$$

where F is the focal length, D_o is the optical aperture.

IFOV is a measure of the spatial resolution of the detector and is given by:

$$IFOV = \text{detector element size/ effective focal length} \tag{8}$$

and,

$$FOV = IFOV \times \text{total no. of elements in the detector FPA} \tag{9}$$

Depending on the application, optics module may be having a single or multiple discrete FOV or a continuous Zoom capability. Time for change from one field of view to the other should be quite small so that the user does not lose the target during this change over. Zoom lenses are described by the ratio of their longest to shortest focal length popularly known as zoom ratio.

Sometimes one may require using scanners and efficient cold stops which are not common in the visible, NIR and SWIR systems. These additional components make optical design and optical assemblies quite complex. Special optical materials may include germanium (Ge), silicon (Si), zinc sulphide (ZnS), zinc selenide (ZnSe) and a number of chalcogenide glasses made from Ge, As and selenium in different ratios. In LWIR optical systems, a combination of germanium-ZnS or ZnSe is used, while a combination of Si, Ge, ZnS and ZnSe could be used in MWIR. For NIR and SWIR optics, normal glasses are used.

A typical layout for SWIR, MWIR and LWIR optical systems are shown in Figure 5 (a), 5(b) and 5(c) respectively.

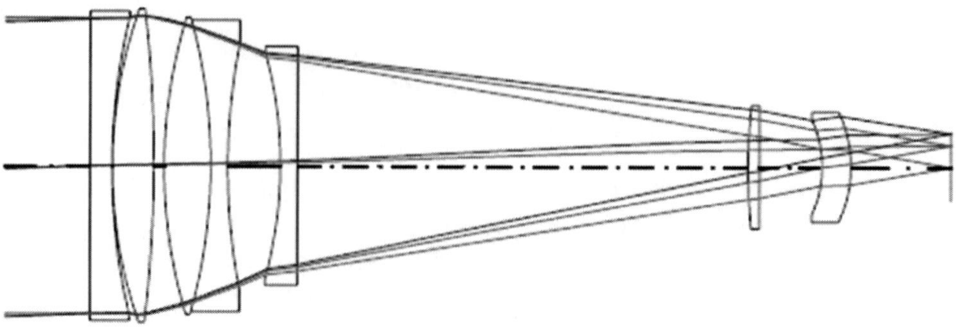

(a)

SWIR all glass optics

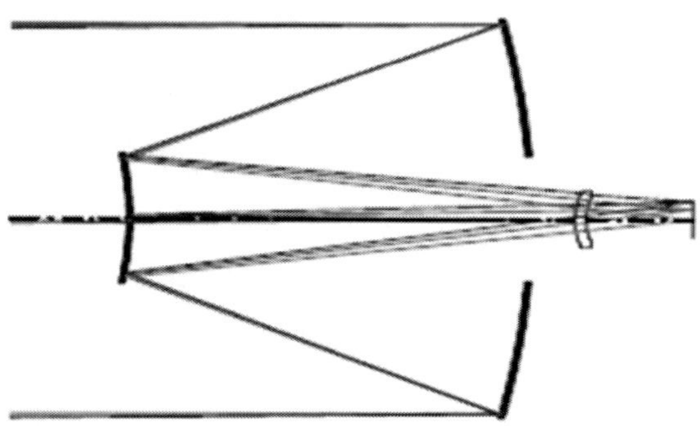

(b) MWIR catadioptric system

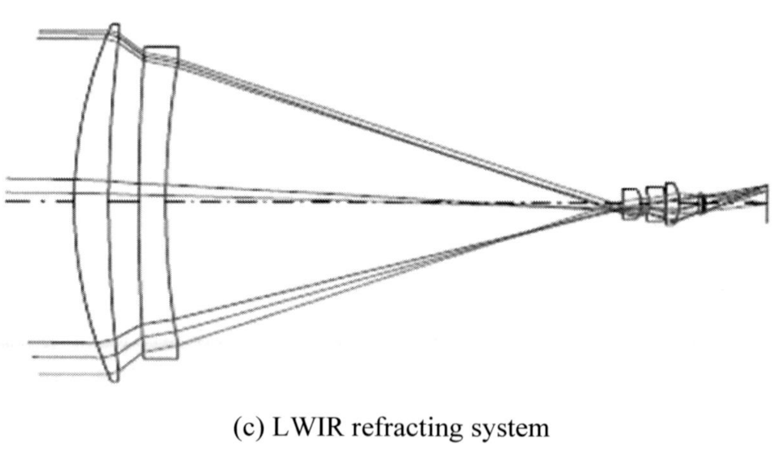

(c) LWIR refracting system

Figure 5(a). Glass refracting system having EFL: 300mm, aperture: 75mm, FOV: 1.8°x1.2°, wavelength band = 0.9 - 1.7 μm (b) Catadioptric system having EFL: 300mm, aperture: 75mm, FOV: 1.8°x1.2°, wavelength band = 3 - 5 μm (c) refracting system having EFL: 300mm, aperture: 100mm, FOV: 1.8°x1.2°, wavelength band = 8 - 12 μm.

The refractive indices of IR materials are generally quite high and therefore require multilayered antireflection coatings to increase their effective transmission in the preferred spectral band. If the system is likely to be used in an adverse open field, then antireflection coating on front optical surface must have a diamond like carbon (DLC) coating to provide resistance to abrasion, salts, acids, alkalis and oil and protection from high velocity airborne particles, sea water, engine fuels, high humidity and last but not the least, of improper handling.

Use of scanners in infrared imagers is a thing of the past where in small arrays of IR detectors were used to cover reasonable FOV, both in azimuth & elevation directions. However, with the availability of staring FPAs of reasonably large formats (320 × 256/640 × 512/1280 × 1024, etc.) there is no need of using scanners. Generally, staring array images are aliased due to the inactive space between consecutive detectors in both directions. Microscanning is implemented to remove the aliasing effect and to get over-sampled smooth images, thus enhancing the spatial resolution (Figure 6). In the process of micro-scanning images are displaced by a fraction of a pixel in both x & y directions and are then interlaced before displaying on to the monitor. Now that the pixel sizes are going down day by day, micro-scanning also may be undesirable

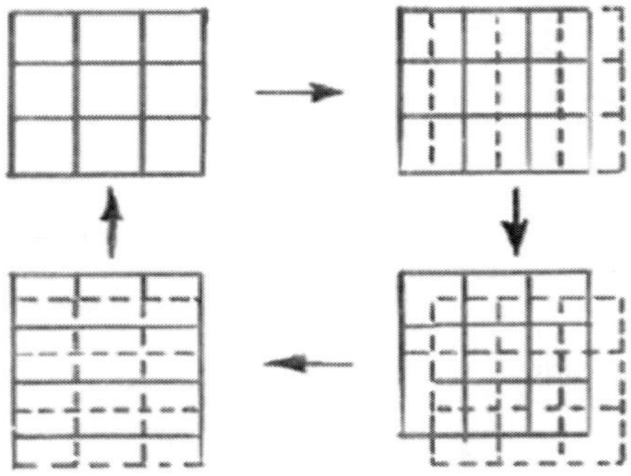

Figure 6. Micro-scanning process.

4.3.2. IR Detectors

An infrared detector is simply a transducer of radiant energy. It converts radiant energy into some other measurable form: this can be an electrical current, a change in some physical property of the detector, or the blackening of a photographic plate. IR detectors are classified as photon and thermal detectors.

In photon detectors, the radiation is absorbed within the material by interaction with electrons. The observed electrical output signal results from the changed electronic energy distribution. The photon detectors show a selective wavelength dependence of the response per unit incident radiation power. They exhibit both perfect signal to noise ratio (S/N) performance and a very fast response (of course to achieve this they require cryogenic cooling). Depending on the nature of interaction, photon detectors are called as intrinsic, extrinsic, photoemissive, quantum well, quantum dot and superlattices.

The second class of IR detectors is thermal detectors where radiations falling on the detector are absorbed by the material and raises its temperature. This heating effect produces a change in one of the physical properties of the material. The most vulnerable physical properties of the materials are electrical resistance (bolometric), spontaneous polarizability (pyroelectric) or expansion of gases leading to three types of thermal detectors: a) thermistor bolometer, b) pyroelectric, and c) Golay cells. In contrast to photon detectors, thermal detectors typically operate at room temperature, have modest sensitivity and slow response but they are lighter, consume less power, have instant ON functionality, and are cheap & easy to use. Thermal detectors are generally wavelength independent (Figure 7); the signal depends upon the radiant power (or its rate of change) but not upon its spectral content.

The geometries of detectors can vary from single element and linear arrays to two-dimensional staring arrays as shown in Figure 8.

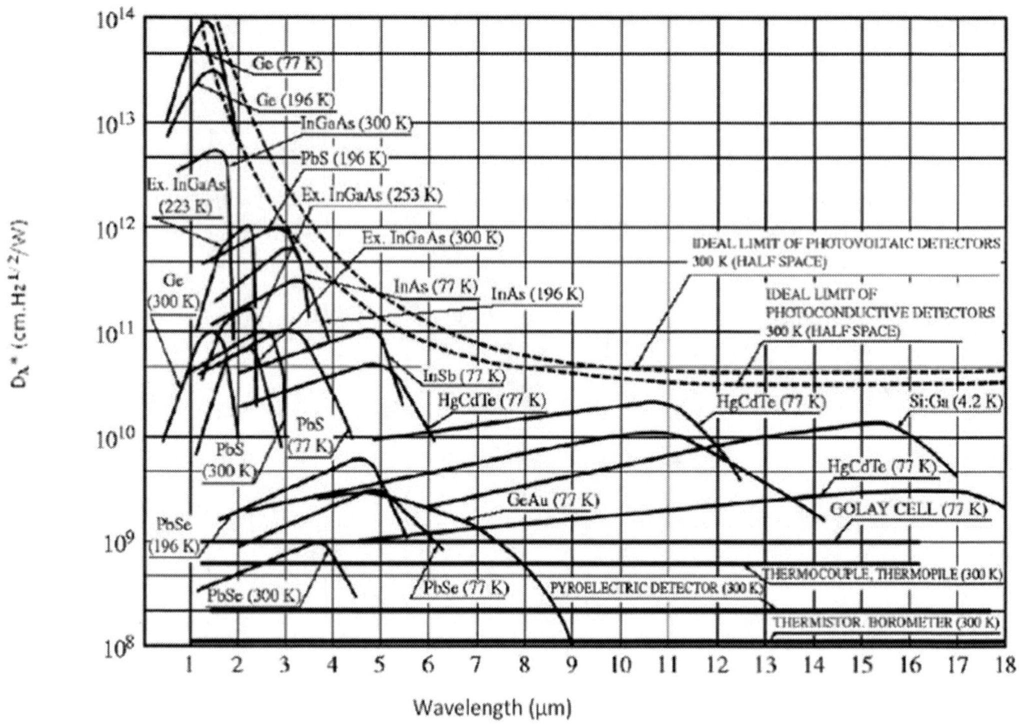

Figure 7. Spectral response of selected IR materials.

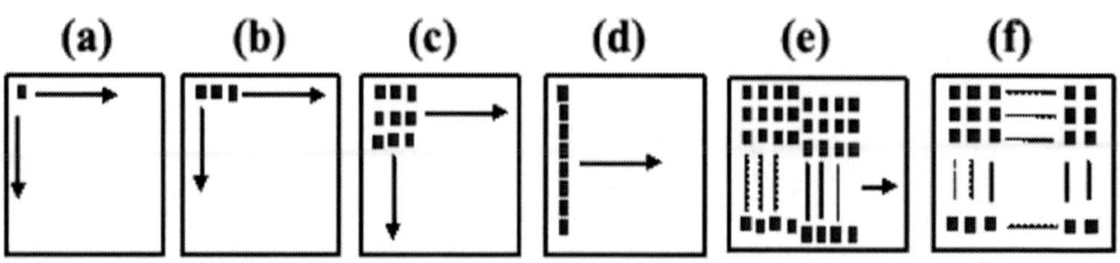

Figure 8. Infrared detector configuration types (a) Single Detector (b) Serial Detector (c) Serial-Parallel Detector (d) Parallel Detector (e) Linear FPA (f) Staring FPA.

Linear arrays containing 15 × 2, 60, 100, 120, 180 and 240 × 1 elements of cooled MCT, InSb were the common module detectors and are phasing out. Figure 8(a) to 8(c) are approaching obsolescence. In their place linear FPAs containing 288 × 4, 480 × 6, 576 × 7 cooled MCT array (for LWIR) and similar linear FPAs, and 256 × 256, 320 × 256 and larger format 640 × 512 FPA 2D matrix of InSb and MCT (for MWIR) are being used. These detectors are generally referred to as 2nd generation detectors. According to common understanding third generation IR detector system will provide enhanced capabilities such as larger number of pixels, multi-color or multiband capabilities, higher frame rates and better thermal resolution and other on-chip signal processing functions. They may be room temperature operated like a microbolometer array (VO_x & a-Si) or cooled ones like MCTs, QWIPs, InSb or PtSi FPAs.

IR detectors using InSb, MCT, QWIPs, and SLS materials need to be operated at cryogenic temperatures employing specially designed Dewars and coolers (Figure 9). Some of the important infrared detector coolers are: Liquid transfer system, JT coolers, Stirling engine, thermoelectric coolers, and radiative coolers. From logistics, weight and operational considerations Stirling coolers are universal choice. Thermoelectric coolers are used to stabilize uncooled detectors at room temperature. Flexure bearing and moving magnet technology based coolers have MTBF up to 50000 hrs, which is much more than Stirling engines (MTBF 5000-10000 h).

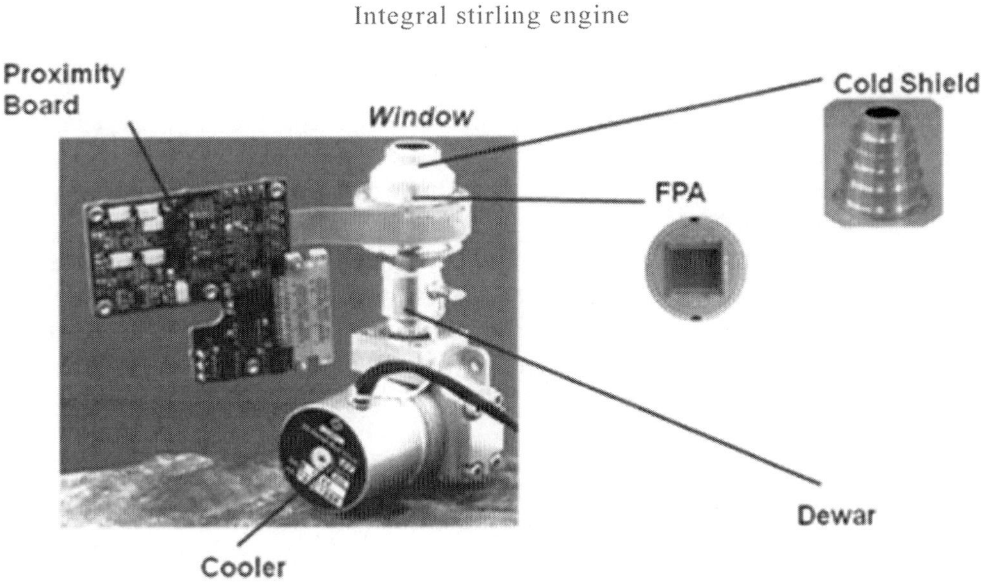

Figure 9. Integrated Detector Dewar Cooler Assembly (IDDCA) [4].

4.3.3. Primary Electronics

Infrared radiations from a scene are converted into representative analog signals by the detector. Proximity electronics constitute the bias voltages for the detector & cooler electronics, pre-amplifier, timing & control circuits, etc. In case of FPAs provision of buffer circuits and correlated double sampling and BITE circuits may also be added to primary electronics.

4.4. Signal Processor and Display

A signal processor takes input from its primary/proximity electronics of IR detector and processes it to form a real time image. Signal processor includes amplification, multiplexing, digitization of video from proximity electronics, fixed pattern noise compensation, field averaging, stagger correction and digital scan conversion. Many a time image processing containing image restoration, image enhancement, edge enhancement, texture analysis, etc. may be incorporated depending on the application.

Thermal images captured by multi-element IR detectors could be presented to the observer on any one of the popular devices like cathode ray tubes (CRTs), liquid crystal display (LCD), electroluminescent displays and plasma based displays. Selection of particular type of display depends on the resolution, size, weight and power requirements for specific applications. CRT displays are being phased out due to their high power requirement and are being replaced by LCD/Organic light emitting diode (OLED). These displays have high resolution, high brightness, low power consumption and are lightweight.

5. Performance Parameters of Thermal Imager

The Performance of a thermal imager (TI) is characterized by two parameters: thermal sensitivity (or radiometric resolution) & spatial resolution. Sensitivity is the smallest signal that can be detected, whereas resolution is the smallest detail that can be discerned by a thermal imager. Thermal sensitivity is defined in terms of 'Noise Equivalent Temperature Difference' (NETD) and spatial resolution is defined in terms of 'Modulation Transfer Function' (MTF). Both these parameters put together give 'Minimum Resolvable Temperature Difference' (MRTD) which correlates well with the field measurement of recognition capability of thermal imagers. MRTD is the measure of human visual acuity through a thermal imager and is related to sensitivity (NETD) and resolution (MTF) as under

$$MRTD \propto \frac{NETD}{MTF} \tag{10}$$

5.1. NETD

The NETD is the temperature difference required to achieve a signal to noise ratio of unity in video output of the detector. From practical point of view, the NETD is the minimum signal that can be detected by the system. Sensitivity is affected by detector responsivity, optical transmission, f-number, detector area and integration time.

S*ensitivity depends on D*, optics diameter, optical transmission, and detector size as given below:*

$$NETD \propto \frac{\sqrt{(Detector\,area \times Equivalent\ noise\ bandwidth)}}{Optical\,transmission \times Optical\,aperture \times System\,resolution \times Detector\,D^* \times Thermal\,contrast} \quad (11)$$

where, D* is the specific detectivity of a detector and is defined as the signal to noise ratio of a detector when one watt of radiation is incident on a detector sensitive area of 1 cm^2 and noise is measured in an electrical bandwidth of 1Hz.

5.2. MTF

MTF is the ability of the system to reproduce various input frequencies faithfully. MTF is defined as the transmission of modulation (contrast) from object space to the image space. It decreases with increasing frequencies. Sensor is modelled as a Linear Space Invariant System and each component MTF can be multiplied with each other to create the system MTF. The system MTF can be calculated as below:

$$MTF_{sys} = \prod_{i=1}^{n} \mathrm{MTF_i} \quad (12)$$

Where MTF_i represents MTFs due to optics, detector, signal processing, vibration blur (LOS stabilization), display and eye. The frequency where MTF approaches to about 5% of its maximum value is considered to be the highest resolvable angular resolution.

5.3. MRTD

For thermal imaging system, the most useful performance parameter is MRTD which gives the subjective resolving capability of the imager-observer system and is defined as the minimum target to background temperature difference in NATO standard 4-bar pattern of 7:1 aspect ratio which can just be resolved by an observer by viewing the 4-bar through the imager. MRTD is a single measurable quantity that includes all the major imaging characteristics of the system including NETD, MTF, eye integration time, frame rate, target spatial frequency, IFOV and SNR. A typical MRTD curve is shown in Figure 10.

6. Emerging Trends

Future thermal imagers design would be aimed at achieving a more compact, low power, and low cost with improved observation capabilities like detection & identification in difficult conditions (low contrast, camouflage, bad weather, etc.). These requirements are posing new challenges to thermal imaging technology. As on today, only relying on radiation intensity is not enough, there is a need to exploit additional attributes of incoming radiation like spectral characteristics, polarization, etc.

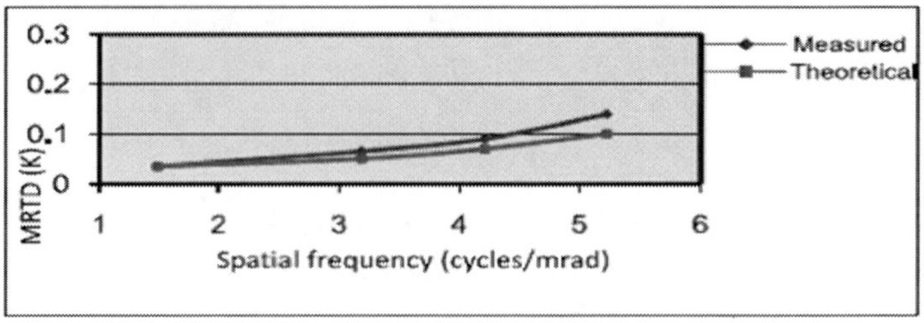

Figure 10. MRTD curve for a typical thermal imager.

6.1. Multispectral & Hyper-Spectral Imaging

The primary physical quantities associated with an optical field are the *intensity*, *wavelength*, *coherence*, and *polarization*. Conventional panchromatic cameras measure the intensity of optical radiation over some wave band of interest. Spectral imagers measure the intensity in a number of wave bands, which can range from one or two (three is common for a color camera) through multispectral systems having more than 10 spectral channels to hyperspectral systems that may measure 300 spectral channels or more. Spectral sensors tend to give us information about the distribution of material components in a scene.

6.2. Multi-sensor Fused Imagery

Multi-sensor image fusion tends to combine information from different images and provides much more information than can be derived from a single sensor. The fusion of information from multiple sensors with different physical characteristics enhances the understanding of scene and provides the basis for better decision-making. It improves the overall performance in image based application. IR and visible images from TI & day cameras can be fused for improved detection & recognition performances, as shown in Figure 11. Fused image shows a better picture than either of the CCD or TI image.

6.3. Polarimetric Imaging

Polarimetric imaging measures the polarization states of light coming from all points in a scene. This information, particularly in the infrared region, can help classify materials and identify objects of interest for remote sensing and military applications. While the spectral information tells us about materials, polarization information tells us about surface features, shape, shading, and roughness. It has the capability to show enhanced contrast even when there is little contrast in the intensity of imagery. Application of polarization imagery ranges from remote sensing to microscopy to industrial monitoring. Capturing and storing infrared polarization images is done in a similar way that intensity information is gathered, i.e., by using a pixelated focal plane array (FPA). Multiple images of a static scene are taken with a combination of oriented polarizers and waveplates in the optical path. Differences between image pairs quantify the distribution, for example, of right circularly polarized light in the scene.

CCD image

Infrared image

Fused image

Figure 11. An example of image fusion.

6.4. High Frame Rate Thermal Imagers

With the advancement in FPA Readout Integrated Circuits (ROIC), and signal processing electronics, now high frame rate infrared images can be developed. Such cameras can be very useful in capturing fast moving targets such as a missile or a bullet in flight, without introducing any motion blur.

6.5. High Operating Temperature (HOT) Detectors

To increase the efficiency of the integrated detector dewar cooler assembly (IDDCA) trends are to go for IR FPA that operates at elevated temperatures around 100 K. They are high spatial resolution infrared focal plane arrays that operate at 160 to 250K. High operating temperature FPAs serve to reduce cost, increase reliability, and provide a sensor package where multiple arrays can be distributed on the system platform.

7. Typical Case Study for National Security

Protecting a country's borders is vital to its national security. No matter what hour of the day, the job is an enormous undertaking that becomes even more challenging under the cover of darkness. It is a key homeland security problem and is one of the most difficult surveillance tasks. The surveillance area to be covered will often be very vast, typically quite remote and can experience very harsh weather conditions. The border lines may go through many isolated and uninhabited area. In order to monitor intrusions crossing of such a vast border by people, an automated system is necessary in addition to the regular patrolling. Therefore, a network of surveillance systems comprising high resolution electro-optical sensors and radars for all weather, day & night operation with wireless connectivity and intelligent and covert monitoring of porous borders to assist military/ paramilitary personnel in their mission of zero tolerance to safety and security is envisaged. Standard CCD & SWIR cameras may be used for day time and dusk & dawn time operation. To achieve a more effective day & night detection solution thermal imagers may be used.

7.1. Features of Border Security System

Challenges involved in protecting the international borders are numerous. Remote locations require reliable and sustainable power and communications system. Detecting threats from the air, on the ground or under the ground, each has its own set of challenges. The systems deployed in the border areas must be capable of surviving extreme hot or cold, rain, humid and arid environments.

7.2. IR Sensor technology (for 24 × 7 Surveillance)

Advancement in IR sensor technology over last more than 3 decades provides a wide range of IR cameras for round the clock surveillance of border areas. During dark nights no other surveillance technology performs so well as thermal imagers when detecting intrusion and security breaches. The heat sensing capabilities of thermal imaging creates valuable scene details in which people and vehicles standout in contrast to the background because of the temperature differences across the object in the scene. For automated system, a surveillance device must contain in addition to TI other technologies like video analytics to make IR sensor more effective and less prone to false alarm during 24 × 7 surveillance.

7.3. System Configuration

Total security surveillance system consists of four major units: (a) observation tower having EO payloads mounted on a servo platform (b) control station for payload control & auto video analytics to detect and create an initial alert (c) master control station for verification of initial alert received from video analytics and provide ability to track a suspected movement, and d) command and control for dissemination of information about initial alert and to coordinate rapid local response, as shown in Figure 12.

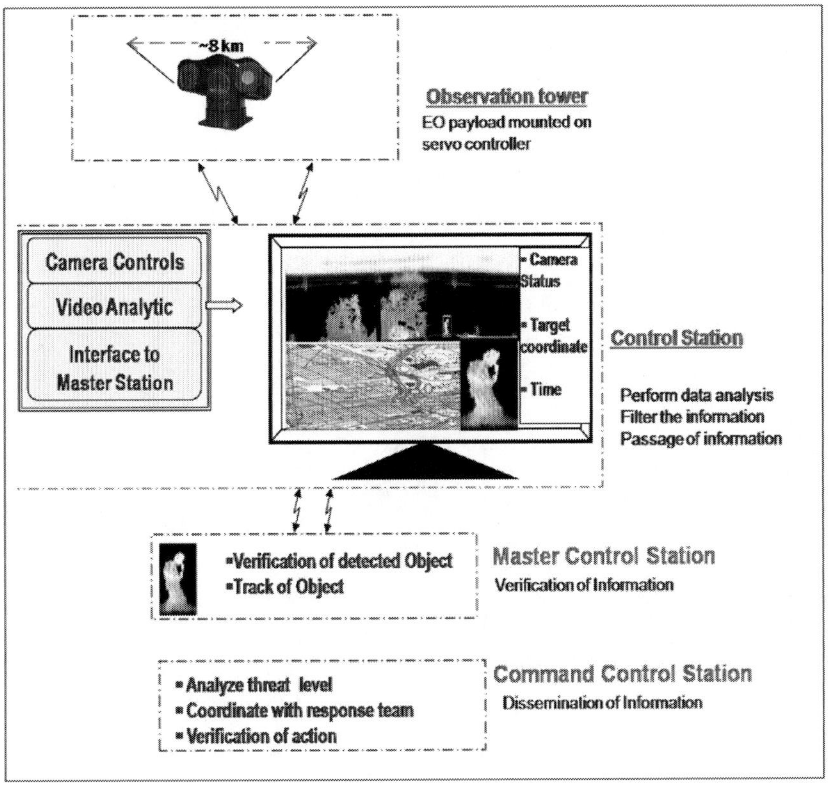

Figure 12. System configuration of a typical border surveillance & security system.

Sensors CCD/ SWIR and MWIR/ LWIR would acquire images of the border areas in panning mode and transmit image data to control station in a real time at a distance of a few kilometers (5-10kms) from the observation tower. The video analytic system placed at control station will acquire the data. It may generate the panoramic view of the complete pan area free from any disturbance. Automatic alert function will be performed and subsequently the detected change will be displayed on zoomed window as well as on the panoramic view with symbology overlay. Thus, it creates an initial alert without relying on personnel to monitor a number of screens and create ambiguous situation.

The detected object will be tracked in image sequences to generate the target profile. The snap of the detected object, its range and geo-location will be generated for possible wireless transmission to master control room. Master control station may be located at a far off distance apart from the control station, which will verify the alarm raised by control station and subsequently pass the valid information to command & control station. Command & control station will analyze the threat level and accordingly coordinate with the response team. It will also verify the action taken by response team.

Depending on situation & location of border, following two configurations can be exploited:

7.4. For Manned Observation Posts

Figure 13 shows the configuration of border surveillance system for manned observation posts.

- Observation tower may be located in close proximity of monitoring station (~50-100m apart),
- Operator control can be placed at monitoring station through wired link,
- Existing power sources at observation post can be used to power-up the surveillance equipment,
- Live video feed to control station can be provided through fiber-optic/ wireless link,
- 24 × 7, seven days video recording is possible to provide at the monitoring station.

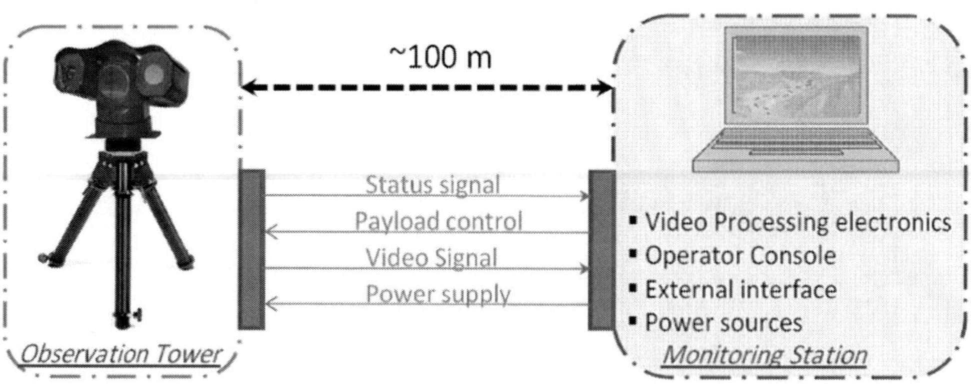

Figure 13. Manned observation post for surveillance.

7.5. For Unmanned Remote Observation Posts

Figure 14 shows the configuration of border surveillance system for unmanned observation posts.

- Observation tower may be located in remote/unattended site with 5-10 km distance apart from the monitoring post.
- Pay load consists of CCD/ SWIR, thermal camera and other sensors may be mounted on an enclosed rugged servo/pan-tilt platform at a height of approx 10-12 meters.
- Live video feed to monitoring station could be provided through a fiber-optic/ wireless link.
- Operator control can be placed at monitoring station through wireless link and the video could be processed for automatic intrusion detection and panoramic view generation.
- Suitable power sources at observation post can be provided.
- 24 × 7, seven days video recording will be provided at observation tower for future analysis,
- An auxiliary wide area surveillance camera could be provided at the observation tower for security of surveillance equipment.

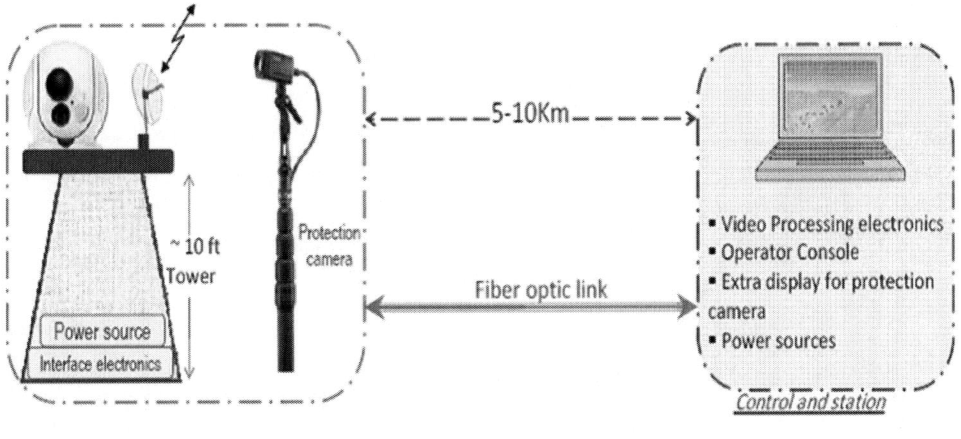

Figure 14. Configuration for unmanned/remote observation posts.

In short we can conclude that:

- Thermal imaging is a valuable tool for detecting intruders during dark nights at long ranges due to its inherent heat sensing capabilities.
- Auto video analytics creates an initial alert without relying on personnel judgement.
- Master control information based on video analytics data to track a suspect movement is more authentic.
- Command & control centre is useful at larger sites involving a number of local response centres.

Table 3. Typical specifications for an observation post

Various units	Description
Sensors	▪ CCD/SWIR Camera ▪ Thermal Camera (MWIR/ LWIR) IR FPA : 640 × 512 InSb cooled to 77K IR Band: 3-5μm Pixel Size: 15 μm NETD ≈ 20mK FOV: 1.2° × 0.8°, an f/4 optics ▪ GPS and DMC ▪ Pan-tilt facility, ±170°, 60°/sec speed
Video processing features	▪ Image enhancement ▪ Automatic intrusion detection ▪ Panoramic view panned area ▪ Man-machine interface (MMI) control & external interface ▪ 24 × 7, 7 day recording system
Modes of operation	▪ Manual slew mode ▪ Automatic sector scanning pattern
Video interface	▪ Analog video (625 line , 50HZ) to feed the console ▪ Two analog video sources each for fiber optics and wireless (UHF)
Power source	▪ Renewable energy source

8. Application in Forensics: Detection of Trace Materials by Infrared Imaging

Applications of infrared imaging in forensics are very wide and varied. Starting from collection of information and evidences related to criminal investigation, to analyze the evidences and thereby identify the criminal is a very tough job. Some of the typical applications may include catching a terrorist, detection of explosives such as anthrax & TNT, age estimation from blood sample, recovering finger prints from place of crime & on fabrics which could turn clothes into silent witnesses, to determine sex of skeletal remains from foot bones, etc.

Scientific techniques have reached to such a level of sophistication that the forensic scientist can definitively link a suspect to a crime scene. There is no "getting away with murder" when in tiniest traces of body fluids, hair and clothing particulates or even soil or vegetation clues are left behind during a traumatic event.

In any forensic investigation, the first step is to analyze the crime scene, looking for traces of fluids or solids that could provide clues. Urine, sweat, saliva, vaginal fluids and semen are naturally fluorescing materials and can be easily detected using UV light. As compared to above fluids, blood stains are entirely different in behaviour. Criminal's first attempt is to clean the trace of body fluids. Blood absorbs UV light; therefore they can be viewed as a dark stain against a brighter background. Diluted body fluids will render UV useless for such analysis.

However, they cannot go undetected using such techniques as Raman spectroscopy & IR spectroscopy. Alternatively FTIR can detect non-destructively and in a non-contact fashion the strong absorption of hemoglobin at 1650 (6.06μm) and 1540 cm^{-1} (6.49 μm) against the non-absorbing background of common surfaces and textiles.

Using a blackbody source (Nernst glower, a globar or space heater) and a conventional thermal camera, one can reveal the visual contrast between stain and no-stain regions by optimizing the detector response by choosing a select chemical filter that maximizes the discrimination between blood and unstained surfaces.

Hyper-spectral sensors can identify material and compounds left in the ridges of a finger print based on the inherent chemical composition of the sample. Using a hyper -spectral imager at SWIR band in the range 900 - 2500 nm spectral band, hyper-spectral data cubes for a finger print are obtained using a "push-broom" technique that requires movement to capture the line scan images.

Sensor could be mounted on a suitable platform with a very stable source of illumination and a translation stage for moving the sample. The finger print is scanned in a short time and a 3D cube displays the two spatial positions as well as the spectral content at each point in the scene. This spectral data can be compared to the spectral library of known material of interest.

Using infrared photography it is possible to find blood traces behind many layers of black paint, white oil based paint, white spray paint as well as hiding under layers of purple, orange, blue, yellow and green paints.

It may be summarized that by passive remote sensing one can

- reliably identify industrial released hazardous gaseous compounds
- observe stack of gas plumes
- do remote sensing of aircraft emissions
- detect warfare agents under battle field conditions
- use for civil defence and fire-brigade operation.

Thus the passive remote sensing is a powerful tool for the observation of released hazardous chemical agents and also for chemical process control.

9. Environmental Pollution Detection

Emissions from vehicle exhaust, particularly from land transports contain a substantial portion of harmful gases such as carbon monoxide (CO), oxides of nitrogen (NO_x), volatile organic compounds and particulate matter. Air pollution can cause acute respiratory problems to children and chronic bronchitis related problems to adults, premature mortality and reduced life expectancy.

Long term exposures to severe air pollution can cause adverse health effects to people, it may even damage vegetation and buildings in a heavily polluted area. The six Green house gases namely (CO_2, CH_4, N_2O, HFCs, PHCs and SF_6) emanating from motor vehicles, air conditioners and refrigerators, etc. are a prime contributor to the current warming up of the globe. Green house gases remain in the atmosphere for a long time and contribute to global warming.

There are many ways to measure and monitor these gases remotely even online. Out of six green house gases the N_2O, SO_2 and NO have characteristic absorption bands in ultraviolet (UV), but CO_2 has a characteristically strong band in the MWIR at 4.2 to 4.5μm with its high peaks at 4.23 & 4.28 μm. CO_2 also shows a weaker band at 2.7μm. Passive remote sensing is the only detection method which allows mobile, fast, and standoff (wide range) detection of hazardous chemical agent.

Contaminated surfaces and drinking water are other more serious environmental problems. Leaking sewage collection lines, storm water drain discharges and illegal connection of storm water drainage system can be identified by their thermal infrared signature. As these sources of pollution leak, seep or empty into creeks, streams, rivers, and lakes, their thermal signature vary from their surroundings. Typical fluids flowing into a stream or lake appear warmer than the surfaces of the larger body of water due to the relative warmth of the ground a short distance below the surface. These temperature differences are visible in thermal IR spectrum.

The emergence of infrared Hyperspectral imagery (IR-HSI) in recent years enables their use in remote environmental monitoring of gaseous plumes. IR-HIS combines the unique advantages of traditional remote sensing methods such as multispectral imaging and non imaging Fourier transform infrared (FTIR) spectroscopy, while eliminating their drawbacks. The most significant improvement introduced by HSI technology is the capability of standoff detection and discrimination of effluent gaseous plumes without the need for a clear reference background or any other temporal information.

ACKNOWLEDGMENTS

Authors would like to acknowledge the confidence shown by the organizers of Indo-US Workshop on "Spectroscopic Techniques in National Security, Forensics and Environment" held at the department of Physics, Banaras Hindu University, Varanasi during 18th to 20th January 2013 in asking us to write a chapter on "Advances in Infrared Sensors for Security, Forensics and Environmental Pollution".

During the course of writing this chapter authors received maximum help from Shri Anurag Kumar Srivastava, Scientist 'D', IRDE Dehradun in literature survey, preparation of the manuscript and its finalization. We also received help from Shri Manvendra Singh, Scientist 'E' IRDE in working out the sample case for security and Shri Vivek Singh, Scientist 'D' for preliminary discussions. Authors would also like to acknowledge help received from the chief librarian and his team at IRDE in compilation of available literatures.

REFERENCES

[1] Glen Francisco and J. Tillman, "Integrated homeland security system with passive thermal imagers and advanced video analytic", *Infrared Technology & Applications,* volume XXXIII, proc. SPIE 6542, 65423E (2007).

[2] Austin Richards, "IR imaging optics meet varied needs", *Photonics Spectra,* 62, (August 2012).

[3] John Lester Miller, "Future sensor systems needs for staring arrays", *Infrared Physics & Technology,* 54, 164-169, (2011).

[4] Gary A. Shaw and K. Burke, "Spectral Imaging for Remote Sensing", *Lincoln Lab Journal,* 14, No. 1, (2003).

[5] R. N. Singh, "Thermal Imaging Technology Design & Applications", *University Press,* Hyderabad, India (2009).

In: Spectroscopic Techniques for Security, Forensic ... ISBN: 978-1-63117-404-9
Editors: Y. Dwivedi, S. B. Rai and J. P. Singh

Chapter 13

DISSOCIATION OF A CO MOLECULE INDUCED BY 10 KEV ELECTRONS: KINETIC ENERGY RELEASE DISTRIBUTIONS

***Raj Singh**, *Pragya Bhatt, Namita Yadav and R. Shanker*†**
Atomic Physics Laboratory, Department of Physics,
Banaras Hindu University, Varanasi, India

Dissociative ionization of the CO^{q+} (q=2-4) molecular ions produced in collisions of CO with 10 keV electrons is studied using time-of-flight mass spectrometer and position sensitive detector with multi-hit ability. The kinetic energy release distributions for different channels produced from the dissociation of CO^{q+} (q=2-4) are obtained. We found that a pure Coulomb explosion model is insufficient to explain the observed kinetic energy release distributions for the Coulomb explosion channels.

1. INTRODUCTION

The electron impact-induced multiple ionization and fragmentation of molecules is of fundamental importance in various areas of science and technology [1-3]. Experiments in which all fragment ions produced from a particular collision event are detected in coincidence provide valuable information about the charge state, geometry, dissociation pathways and kinetic energy release (KER) of the intermediate multiply-charged molecular ion [4, 5]. Particularly, the observed kinetic energy distributions (KERDs) play a very important role in determining the possible potential energy surfaces for a given fragmentation pathway. The fragment ions formed from a multiply-charged molecular ion have sufficient kinetic energy to create significant chemical and physical changes in nearby media. The understanding of such processes is very important at many palaces of science, for example, in astrophysics, atmospheric sciences, surface studies, plasma physics and in radiological study of biological systems [6-8]. Most work so far has concentrated on the study of the partial ionization cross-

* Email: rajsinghraup@gmail.com.
† Corresponding author: shankerorama@gmail.com.

sections (PICSs) [9-10]. Few studies have been done on the fragmentation dynamics of the water molecule with the impact of keV ions. In this paper, we have studied the KERDs of different channels arising from the dissociation of CO molecule under impact of 10 keV electrons using a time-of-flight mass spectrometer coupled with a position sensitive detector in multihit mode.

2. Experimental Setup and Data Analysis

The details of experimental setup are given in our recent publications [11, 12]. In brief, a monoenergetic continuous beam of electrons was extracted from a commercial electron gun and was made to collide with an effusive beam of CO molecules. The pure CO gas (99.99%) was introduced into the vacuum chamber through a hypodermic needle with inner diameter of 0.01 cm and length of 1.2 cm. The base pressure of the vacuum chamber was maintained at better than 7×10^{-7} Torr which increased to 8.5×10^{-6} Torr with CO gas load. A mechanical fine-needle-valve was used to control the number density of the CO molecules. A single stage linear time-of-flight (TOF) mass spectrometer was used to detect and analyze the fragment ions. The electron beam, the gasjet and the axis of TOF spectrometer were aligned mutually perpendicular to each other. The collision products (electrons and positive ions) were extracted from the collision zone by application of a homogeneous electric field; electrons were detected by a channeltron mounted opposite to the time- and position sensitive multi-hit dual microchannel plate (MCP) ion detector. The electron signal is used as the timing reference for ion arrivals to a dual MCP. The ion detector has a multi-hit capability of detecting up to four concomitant ion hits. The momentum vectors of fragment ions detected in coincidence were calculated from their respective TOFs (t) and positions (x, y) on the MCP. Using the momentum vectors, the kinetic energy of each individual ion was determined and the total KER was obtained from the sum of the kinetic energies of individual fragments involved in the considered dissociation channel.

3. Results and Discussion

The background subtracted TOF spectrum of ions produced from direct and/or indirect ionization of CO molecule under impact of 10 keV electrons is shown in Figure 1.

We observe sharp peaks of CO^+ and CO^{2+} ions having thermal energies. These ionic peaks arise from direct ionization of the CO molecule. Additionally, five ion fragments (O^+, C^+, O^{2+}, C^{2+} and C^{3+}) are also formed which arise from dissociative ionization of the multiply charged CO^{q+} ($q \leq 4$) ions. The CO^+ peak is the most intense peak in the TOF mass spectrum. The two peaks adjacent to CO^+ peak on the right side are attributed to $^{13}C^{16}O$ and $^{12}C^{18}O$ isotopes whose natural abundances are found to be 1.1% and 0.25%, respectively. The CO^{2+} is a quasi-stable molecular ion which is found to have a life time of 1.19 μs in our experiment. Many workers have reported its mean life time ranging from 1 to 30 μs in TOF measurements [13 and references therein]. However, the long lived (>3.8 s) CO^{2+} ion has been observed in the ion storage rings; it has also been shown theoretically that the life time of this ion depends on the involved metastable electronic states [14]. From the ion-ion coincidence map, five dissociation channels ($C^+ + O^+$, $C^{2+} + O^+$, $O^{2+} + C^+$,

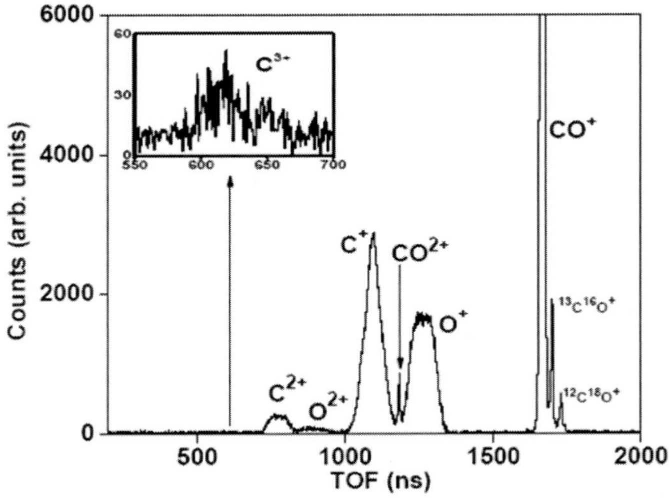

Figure 1. TOF spectrum (background subtracted) of the produced ions following direct and dissociative ionization of a CO molecule by impact of 10 keV electrons. The C^{3+} ion peak is shown in the inset.

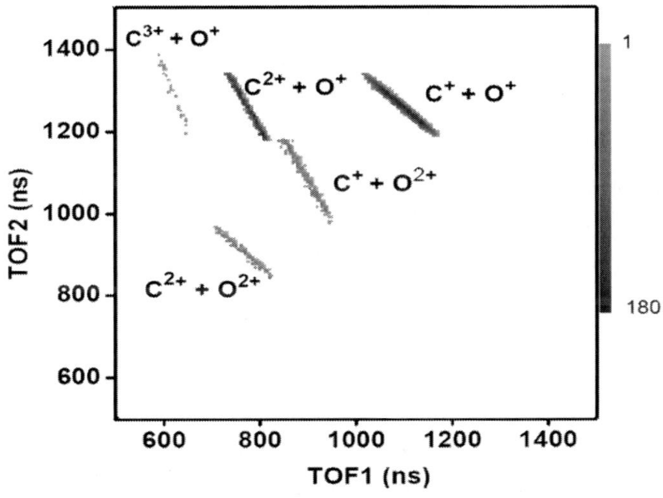

Figure 2. Ion-pairs spectrum resulting from dissociative ionization of CO^{q+} (q=2-4) produced in 10 keV electron collisions with CO molecule.

$C^{2+} + O^{2+}$ and $C^{3+} + O^+$) originating from the complete Coulomb dissociation of CO^{q+} (q=2-4) are observed (see, Figure 2). The KER values for these channels are calculated by using the Coulomb explosion (CE) model at equilibrium internuclear distance of 1.128 Å of CO molecules [15]. The calculated KER values and the relative intensities for these channels are also given in Table 1. The KERDs are discussed in detail in the following sections:

3.1. Dissociation of CO^{2+}

A typical ion-ion coincidence spectrum resulting from the collision of 10 keV electrons with CO is displayed in Figure 2; the resulting fragmentation channels are labeled in the Figure. The $C^+ + O^+$ channel arising from the dissociation of CO^{2+} is the most dominant dissociation channel among all observed channels (see, Table 1). The KERD for this channel is shown in Figure 3. The peak of the KERD is found to be at 2.8±0.3 eV which is much smaller than the KER value predicted by the CE model (see, Table 1).

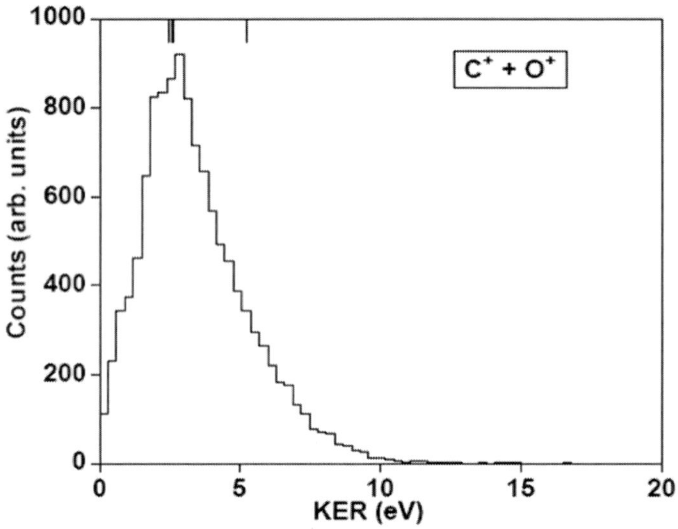

Figure 3. KER distribution for the channel: $C^+ + O^+$ observed in the dissociation of CO^{2+} in 10 keV electron impact with CO.Theoretical values derived from the relevant states involved in this dissociation channel (see, Table 2) are shown in Table 4 by the vertical lines at the top of the graph.

To account for the KER distribution for this channel, we consider only three low-lying states of CO^{2+} ion from [16]; the molecular states, their associated KER values, vertical energy for different CO^{2+} states and the dissociation limit for these states are given in Table 2.

The values of $C^+(^2P) + O^+(^2D)$ and $C^+(^2P) + O^+(^4S)$ dissociation limits are 39.30 eV and 35.98 eV, respectively, relative to ground state of CO molecule at the equilibrium inter-nuclear distance [17]. Table 2 indicates that the $^1\Sigma^+$ and $^1\Pi$ states significantly contribute to the KERD while the contribution of $^3\Pi$ state is probably smaller being far-off from the KERD peak. The metastable states $^1\Sigma^+$ and $^1\Pi$ most probably dissociate into $C^+(^2P) + O^+$ (^2D) by tunneling through the potential barrier because their coupling with $^3\Sigma^-$ repulsive state yield KER value above 5.0 eV [18, 16]. However, the $^3\Pi$ state predissociates via $^3\Sigma^-$ repulsive state into $C^+(^2P) + O^+(^4S)$ [16]. The KERD for this channel has been reported by many workers with the impact of ions, low energy electrons, intense laser fields and photons (see, Table 3).

Most of them have reported the peak value of KERD above 5.0 eV. However, from the ion impact, Ben-Itzhak *et al* [19] and Tarisien *et al* [15] have obtained the KER component below 5eV; they have concluded that this KER component arises due to the dissociation of metastable CO^{2+} molecular ion via tunneling process. It is noted that the KERD has a long tail which may arise due to the participating higher states of CO^{2+} ion. From the above discussions, it may be stated that the low value of KER corresponds to the involvement of those dicationic states that are metastable. In view of the variation of KER values obtained by different workers with their excitation sources and energies, some complex dynamics is envisaged between decay via tunneling through the potential barrier on one hand and the interaction time on the other hand.

Table 1. The KER values calculated from the CE model (KER_{CEM}) and relative intensities for different dissociation channels produced from dissociation of CO^{q+} (q=2-4) in 10 keV electron impact with CO

Channels	KER_{CEM} (eV)	Relative intensity (%)
$C^+ + O^+$	12.8	87.9
$C^{2+} + O^+$	25.5	8.7
$C^+ + O^{2+}$	25.5	2.2
$C^{2+} + O^{2+}$	51.0	1.0
$C^{3+} + O^+$	38.3	0.2

Table 2. The possible molecular states of CO^{2+} dissociating into $C^+ + O^+$ along with the theoretically calculated values of KER

Molecular states	Vibrational levels	KER (eV)	Vertical energy(eV) *(ref. 16)*	Dissociation limit *(ref. 17)*
$^3\Pi$	0	5.31	41.29*ref. 25*	$C^+(^2P) + O^+(^4S)$
$^1\Sigma^+$	0	2.39	41.69	$C^+(^2P) + O^+(^2D)$
	1	2.64	41.94	
$^1\Pi$	0	2.52	41.82	$C^+(^2P) + O^+(^2D)$
	1	2.70	42.00	

Table 3. Comparison of kinetic energy release in different dissociation channels produced by impact of 10 keV electrons on CO with the earlier reported experimental results

Dissociation Channel	KER (eV)							
	Electron impact		Ion impact			Intense Laser field	Photon impact	
	Present	200 eV[a]	11.4 MeV/u O^{7+} ions[b]	1.2 MeV/u Ar^{8+} ions[c]	1.0 MeV/u F^{4+} ions[d]	820 nm (10^{14}W/cm^2)[e]	44 eV[f]	306.4 eV[g]
$C^+ + O^+$	2.8±0.3	7.8	6.5	6.2	12.0	6.74	6.0	9.5
$C^{2+} + O^+$	11.5±0.4	-	-	17.5	27.0	11.87	-	-
$C^+ + O^{2+}$	17.0±0.4	-	-	28.0	39.0	12.17	-	-
$C^{2+} + O^{2+}$	23.5±0.5	-	-	48.0	49.0	22.11	-	-

[a]Ref. [16], [b]Ref. [15] [c]Ref. [20] [d]Ref. [19] [e]Ref. [24] [f]Ref. [21] [g]Ref. [26].

3.2. Dissociation of CO^{3+}

From Figure 2, we observe two fragmentation channels: $C^{2+} + O^{+}$ and $C^{+} + O^{2+}$ which arise from the fragmentation of CO^{3+} molecular ion. It is seen that the relative intensity for the $C^{2+} + O^{+}$ channel is relatively larger (about four times) than that for the $C^{+} + O^{2+}$ channel (see also, Table 1). Similar results have been noted by other workers [19-21]. The difference in the relative intensity between the two channels is attributed to the nature of dissociation dynamics because the lower dissociation limit for $C^{+} + O^{2+}$ lies about 11.0 eV above the lowest dissociation limit for $C^{2+} + O^{+}$channel [22]. The KERDs for both channels are shown in Figure 4. The peak values of KERD for the channels $C^{2+} + O^{+}$ and $C^{+} + O^{2+}$ are found to be at 11.5±0.4 eV and 17.0±0.4 eV, respectively.

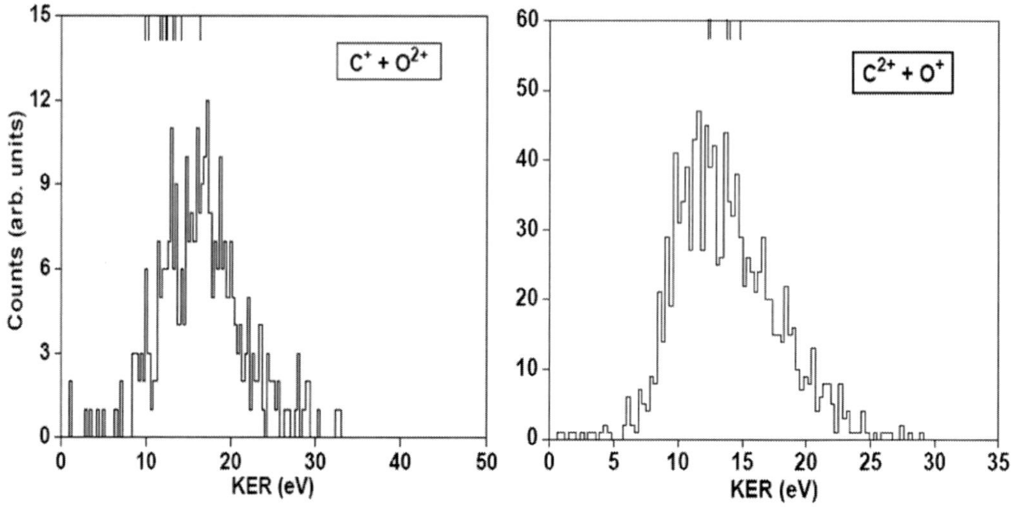

Figure 4. KER distributions for the channels: $C^{+} + O^{2+}$ and $C^{2+} + O^{+}$ observed in the dissociation of CO^{3+} in 10 keV electron impact with CO.Theoretical values [23] derived from the relevant states involved in this dissociation channel are shown in Table 4. These values are marked by the vertical lines at the top of the graph.

These KER values are smaller than the KER values calculated from the CE model. Different electronic states of CO^{3+} molecular ion, the calculated KER values and the dissociation limits for both channels are listed in Table 4 (taken from [23]). For both channels, the theoretically calculated KER values show a reasonably good agreement with our observed KER values. Some workers have also reported the KER values for these channels under the impact of ions as shown in Table 3. It is found that their KER values are larger than the KER values obtained in the present experiment. Further, Wei *et al* [24] have observed the KER value for both channels from the interaction of intense laser fields with the CO molecule which is shown to be about 12.0 eV for both channels. It is seen that our KER value for channel $C^{2+} + O^{+}$ finds a good agreement with that of Wei *et al* [24], while our KER value for $C^{+} + O^{2+}$ channel is larger than that of Wei *et al* [24]. It may be noted here that the larger KER values observed in the experiments with the ion impacts are possibly due to the excitation of higher electronic states of CO^{3+}molecular ion in these experiments.

Table 4. The possible electronic states of CO^{3+} molecular ion dissociating into C^{+}+ O^{2+} and C^{2+}+ O^{+}along with theoretically calculated values of KER (taken from [23])

Molecular states	KER (eV)	Dissociation limit
$^2\Pi$	12.27	$C^{2+}(^3P) + O^{+}(^2D)$
	8.91	$C^{+}(^2P) + O^{2+}(^1S)$
	11.83	$C^{+}(^2P) + O^{2+}(^3P)$
$^4\Pi$	14.07	$C^{2+}(^3P) + O^{+}(^2D)$
	13.62	$C^{+}(^2P) + O^{2+}(^3P)$
$^6\Pi$	16.47	$C^{+}(^4P) + O^{2+}(^3P)$
$^2\Sigma^+$	13.74	$C^{2+}(^3P) + O^{+}(^2D)$
	10.38	$C^{+}(^2P) + O^{2+}(^1S)$
	13.30	$C^{+}(^2P) + O^{2+}(^3P)$
$^2\Sigma^-$	13.04	$C^{2+}(^3P) + O^{+}(^2D)$
	12.59	$C^{+}(^2P) + O^{2+}(^3P)$
$^4\Sigma^-$	12.41	$C^{2+}(^3P) + O^{+}(^2D)$
	11.97	$C^{+}(^2P) + O^{2+}(^3P)$
$^2\Delta$	14.73	$C^{2+}(^3P) + O^{+}(^2D)$
	14.28	$C^{+}(^2P) + O^{2+}(^3P)$
	12.08	$C^{+}(^4P) + O^{2+}(^3P)$

3.3. Dissociation of CO^{4+}

The fragmentation of CO^{4+} molecular ion gives rise to two dissociation channels C^{2+} + O^{2+} and C^{3+} + O^{+} (see, Figure 2). It is clear from the Figure as well as from Table 1 that the symmetric channel C^{2+} + O^{2+} occurs more intense than the asymmetric channel. The KERD for the channel C^{2+} + O^{2+} is shown in Figure 5. The peak value of KER is observed at 23.5 ± 0.4 eV which is also much smaller than the KER value calculated by using the CE model. The KERD for the channel C^{3+} + O^{+} is not shown due to its low statistics; however, it is found that it has a distribution between 7 eV to 35 eV.

The molecular states of CO^{4+} ion, their associated KER and the dissociation limits for different molecular states are listed in Table 5 (The value of KER and dissociation limit are calculated using the values taken from [20, 24]). It is obvious from Table 5 that the calculated KER values explain the KERD; the high energy side of KERD may be arising either from higher states of CO^{4+} molecular ion or from the lower dissociation limits. It is observed that our KER value is much smaller than those of the ion impacts in general, whereas it shows a reasonable agreement with the KER value reported from the interaction of intense laser field (see, Table 3). In passing by, it may be mentioned here that there are no experimental data available in the literature for dissociative ionization of CO^{4+} ions under impact of electrons of considered energy to compare with. Hence, further studies are required to gain detailed knowledge on the fragmentation dynamics of CO^{4+} molecular ion.

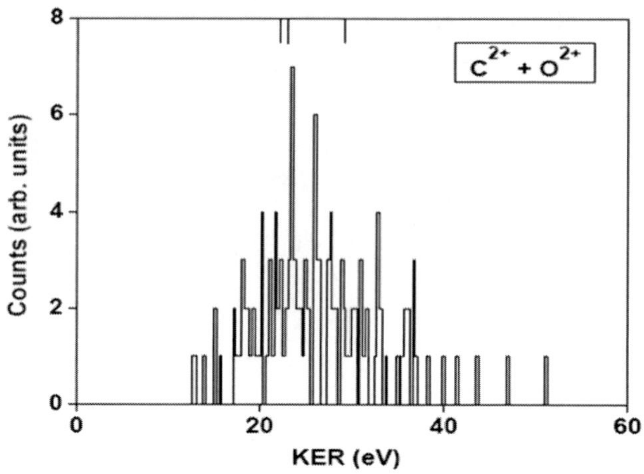

Figure 5. KER distribution for the symmetric channel: $C^{2+} + O^{2+}$ observed in the dissociation of CO^{4+} in 10 keV electron impact with CO.Theoretical values derived from the relevant states involved in this dissociation channel are shown in Table 5. These values are marked by the vertical lines at the top of the graph.

Table 5.The possible molecular states of CO^{4+} dissociating into C^{2+}+ O^{2+} and theoretically calculated values of KER from [20, 24]

Molecular states	KER (eV)	Dissociation limit
$^3\Pi$	22.98	$C^{2+}(^1D) + O^{2+}(^3P)$
$^3\Delta$	21.99	$C^{2+}(^1D) + O^{2+}(^1S)$
$^1\Delta$	29.27	$C^{2+}(^1D) + O^{2+}(^1D)$

Conclusion

In this chapter, we have studied KERDs for different channels originating from the dissociation of CO^{q+} (q=2-4) molecular ions using TOF spectrometer with position sensitive detector. Five dissociation channels arising from the dissociation of CO^{q+} (q=2-4) ions are observed and identified. It is found that the channel $C^{2+} + O^{+}$ occurred more frequently than the channel $C^{+} + O^{2+}$ produced from the dissociation of CO^{3+} molecular ion. The dominance of symmetric channel over asymmetric channel is clearly observed for the dissociation of CO^{4+} molecular ion. The KERDs for different channels produced from the dissociation of CO^{q+}(q=2-4) are obtained and compared with the CE model and with those reported by other workers. For all channels, the observed KERs are found to be smaller than the KER values predicted by CE model. The KERD for the $C^{+} + O^{+}$ dissociation channel mainly arises due to the tunneling process.

Acknowledgment

The work was supported by the department of science and technology (DST), New Delhi under the research project: SR/S2/LOP-09/2006. R. S., N. Y. and P. B. thankfully acknowledge the partial supports by Council of Scientific and Industrial research, New Delhi (CSIR) and Board of research in Fusion Science & Technology, Ahmedabad (BRFST) during the progress of this work

REFERENCES

[1] G. Draganić, Radiat. *Phys. Chem.* 72, 181 (2005).

[2] R. G. Carlson et al., *Science* 283, 2062 (1999).

[3] D. E. Shemansky and D. T. Hall,*J. Geophys. Res-space phys*. 97, 4143 (1992).

[4] F. A. Rajgara, M. Krishnamurthy, D. Mathur, T. Nishide, T. Kitamura, H. Shiromaru, Y. Achiba and N. Kobayashi, *Phys. Rev. A* 58, 032712 (1998).

[5] B. Bapat and V. Sharma, *J. Phys. B* 40, 13 (2007).

[6] U. Werner, K. Beckord, J. Becker, H. O. Folkerts and H. O. Lutz, Nucl. *Instrum.Methods Phys. Res.* B 98, 385 (1995).

[7] B. Boudaiffa, P. Cloutier, D. Hunting, M. A. Huels, and L. Sanche, *Science* 287, 1658 (2000).

[8] S. D. Price, *Phys. Chem. Chem. Phys.* 5, 1717 (2003).

[9] H. C. Straub, P. Renault, B. G. Lindsay, K. A. Smith and R. F. Stebbings, *J. Chem.Phys.*108, 109 (1996).

[10] S. W. J Scully, J. A. Wyer, V. Senthil, M. B. Shah, and E. C. Montenegro, *Phys. Rev.A* 73, 040701(R) (2006).

[11] R. Singh, P. Bhatt, N. Yadav, and R. Shanker, *Meas. Sci. and Tech.* 22, 055901(2011).

[12] P. Bhatt, R. Singh, N. Yadav, R. Shanker, *Phys. Rev. A* 85, 042707 (2012).

[13] M. A. Mangan, B.G. Lindsay, and R. F. Stebbing, *J. Phys. B* 33, 895 (2000).

[14] L. H. Andersen, J. H. Posthumus, O. Vahtras, H. Ågren, N. Elander, A. Nunez, A. Scrinzi, M. Natiello, and M. Larsson,*Phys. Rev. Lett.*71, 1812 (1993).

[15] M Tarisien et al., *J. Phys. B* 33, L11 (2000).

[16] M. Lundqvist, P. Baltzer, D. Edvardsson, L. Karlsson, and B. Wannberg, *Phys. Rev. Lett.*75, 1058 (1995).

[17] T. Masuoka, *J. Chem. Phys.* 101, 322 (1994).

[18] V. Krishnamurthi, K. Nagesha, V. R. Marathe, and D. Mathur, *Phys. Rev. A* 44, 5460 (1991).

[19] I. Ben-Itzhak, S. G. Ginther, V. Krishnamurthi, and K. D. Carnes, *Phys. Rev. A* 51, 391 (1995).

[20] J. Rajput and C. P. Safvan, *Phys. Rev. A* 75, 062709 (2007).

[21] P. Lablanquie *et al.*, *Phys. Rev.* A 40, 5673 (1989).

[22] G. Handke, F. Tarantelli, and L. S. Cederbaum, *Phys. Rev. Lett.*76, 896 (1996).

[23] D. Mathur, E. Krishnakumar, K. Nagesha, V. R. Marathe, V.Krishnamurthit, F. A. Rajgara, and U. T. Raheja,*J. Phys. B* 26, L141 (1993).

[24] G. Wei, Z. Jing-Yi, W. Bing-Xing, W. Yan-Qiu and W. Li, *Chin. Phys. Lett.* 26, 013201 (2009).

[25] G. Dawber, A. G. McConkey, L. Avaldi, M. A. MacDonald, G. C. King, and R. I. Hall, *J. Phys. B* 27, 2191 (1994).

[26] Th. Weber*et al.*, J. Phys. B 34, 3669 (2001).

About the Editors

Prof. S. B. Rai

Shyam Bahadur Rai received his M.Sc. and Ph. D. degrees in 1970 and 1974 from Banaras Hindu University, India. Currently, he is professor and Head of Department of Physics in Banaras Hindu University. Prof. Rai is recipient of several National & International awards and fellowships including prestigious Alexander Von Humboldt and INSA – DFG Exchange Fellowships. He has served as president of Indian Science Congress (Physical Science) and Vice President of Laser and Spectroscopy Society of India. He has vast teaching/research experience of 35 years and had completed 15 research- projects of Govt. of India. He has supervised 16 Ph.D. students and more than 50 projects of Master students. He has published ~270 articles in field atomic/molecular physics and material science.

Dr. Y. Dwivedi

Yashashchandra Dwivedi received the Ph.D. degree from Banaras Hindu University (BHU) India in the subject of Lanthanide doped materials. He was a postdoctoral fellow at University of Sao Paulo, Brasil. At present, he is working as Assistant Professor at Physics Department, National Institute of Technology, Hariyana, India. He has been awarded prestigious ISCA young scientist award along with four other young scientist/best poster presentation awards. He was member of advisory committee of several national and international conferences. He has published 48 international articles and 7 book chapters. He also edited a book 'Synthesis, Characterization and Application of Multifunctional Materials'. His research is mainly focus on linear and nonlinear spectroscopic investigations of lanthanide doped materials for the development of white LEDs, temperature/UV/NIR sensors, upconversion lasers etc.

Prof. J. P. Singh

Jagdish Prasad Singh is a Research Professor, Emeritus at Mississippi State University and President and CEO, JPS Advanced Technology LLC, Starkville, Mississippi, USA. He received MS and Ph.D. degree in Physics from Banaras Hindu University, India in 1974 and 1980. He was a Post Doctoral Research Associate at McMaster University, Canada and Kansas State University. He has more than 170 referred publication, 175 conference presentation and 5 book chapters. He is an author of a book title "Laser Induced Breakdown Spectroscopy". Dr. Singh is Fellow of Optical Society of America (OSA) and LASSI. His research interest includes Laser spectroscopy, Optical Fiber Sensors, Laser Diagnostics for Combustion, and Hazardous Waste Management.

EDITORS' CONTACT INFORMATION

Prof. Shyam Bahadur Rai,
Head and Professor,
Physics Department,
Banaras Hindu University, India

Dr. Yashashchandra Dwivedi,
Instituto de Fisica de Sao Carlos,
University of Sao Paulo, Brazil

Prof. Jagdish Prasad Singh,
Institute for Clean Energy Technology,
Mississippi State University, US

INDEX

A

B

C

D

E

I

J

K

L

M

N

O

P

Q

R

S

T

U

V

W

X

Y

Z